Cultural Transmission and Evolution

MONOGRAPHS IN POPULATION BIOLOGY

EDITED BY ROBERT M. MAY

1. The Theory of Island Biogeography, by Robert H. MacArthur and Edward O. Wilson
2. Evolution in Changing Environments: Some Theoretical Explorations, by Richard Levins
3. Adaptive Geometry of Trees, by Henry S. Horn
4. Theoretical Aspects of Population Genetics, by Motoo Kimura and Tomoko Ohta
5. Populations in a Seasonal Environment, by Stephen D. Fretwell
6. Stability and Complexity in Model Ecosystems, by Robert M. May
7. Competition and the Structure of Bird Communities, by Martin Cody
8. Sex and Evolution, by George C. Williams
9. Group Selection in Predator-Prey Communities, by Michael E. Gilpin
10. Geographic Variation, Speciation, and Clines, by John A. Endler
11. Food Webs and Niche Space, by Joel E. Cohen
12. Caste and Ecology in the Social Insects, by George F. Oster and Edward O. Wilson
13. The Dynamics of Arthropod Predator-Prey Systems, by Michael P. Hassell
14. Some Adaptations of Marsh-nesting Blackbirds, by Gordon H. Orians
15. Evolutionary Biology of Parasites, by Peter W. Price
16. Cultural Transmission and Evolution: A Quantitative Approach, by L. L. Cavalli-Sforza and M. W. Feldman

Cultural Transmission and Evolution: A Quantitative Approach

L. L. CAVALLI-SFORZA

AND M. W. FELDMAN

PRINCETON, NEW JERSEY

PRINCETON UNIVERSITY PRESS

1981

Published by Princeton University Press, Princeton, New Jersey
In the United Kingdom: Princeton University Press, Guildford, Surrey

Library of Congress Cataloging in Publication Data will be found on the last printed page of this book
This book has been composed in VIP Baskerville
Clothbound editions of Princeton University Press books are printed on acid-free paper, and binding materials are chosen for strength and durability.
Printed in the United States of America by Princeton University Press, Princeton, New Jersey

Preface

This book arose from the conviction that theory is the backbone for development of any scientific discipline. There is today a need for a theory of cultural change. A number of scholars have discovered that concepts like mutation, selection, and random drift, which emerged from the quantification of the theory of biological evolution, may also be useful in elucidating evolutionary phenomena in many disciplines. Accumulated experience with the study of biological evolution has taught us that central to a satisfactory theory of evolution is the sound knowledge of the laws of biological transmission. Similarly, knowledge of cultural transmission should be important in understanding cultural change. Although cultural transmission has received little attention, it obviously differs greatly from biological transmission. Its study may provide a theoretical framework for future investigation in quantitative anthropology and social science. What emerges from the theoretical analysis is the idea that the same frame of thought can be used for generating explanations of such diverse phenomena as linguistics, epidemics, social values and customs, and the diffusion of innovations. With all of these we suffer from inadequate knowledge of the mechanisms of human behavior. A consequence of this ignorance is the confounding that occurs between imposition and independent choice, between genetic and cultural transmission, or between cultural transmission and cultural selection. The development of a quantitative theory forces us to be explicit about the distinctions between these concepts, even if they raise questions which cannot be resolved empirically.

We have chosen to develop a mathematical theory, and we are well aware of the serious disadvantages that result from this decision. The necessary over-simplification is usually so great, especially in the applications to human behavior, that there is often a danger of distortion. Our position, however, is that a mathematical theory is always more precise than a verbal one, in that it must spell out precisely the variables and parameters involved, and the relations between them. Theories couched in nonmathematical language may confound interactions and gloss over subtle differences in meaning. They avoid the charge of oversimplification at the expense of ambiguity. Another reason for favoring a mathematical treatment is our belief that the theory of biological evolution owes much of its present strength to its mathematical background, primarily in population genetics. Quantitative predictions can provide the potential to test the validity of the quantitative theory.

A disadvantage, however, of choosing a mathematical presentation is that of potentially limiting the audience. We have tried to make the presentation as elementary as possible, for example, only elementary calculus is used, and most of the analysis involves just simple algebra. We have left a more detailed treatment of the problem to published scientific papers: here we have concentrated on simple numerical and graphical examples. It would have been our choice to use many more real examples, but there are few for which adequate data are available and from which the relevance of our transmission models might be evaluated.

The size of this manuscript grew continuously during the writing. We therefore thought the work would benefit from a partition into more than one book. The major portion of the first volume is dedicated to a mathematical treatment of cultural change, unaffected by individual differences. The first chapter contains definitions and a predominantly verbal exposition of the major concepts we have used. Chapters 2 to 5 deal with our theory of the transmission and evolution of cultural traits. Chapters 2 to 4 discuss discrete traits, and Chapter 5 examines continuous traits. A short epilogue summarizes some of the major qualitative ramifications of our theoretical treatment.

Another volume will take account of individual, inherited differences in learning ability. The introduction of individual differences, for instance in capacity to learn, requires a quantification of some classical genetic concepts, such as "norm of reaction," and allows us to make predictions about that elusive entity, genotype-environment covariance. Cultural and genetic evolution can be directly compared and their extremely complex interactions studied. Inevitably this involves reference to some controversial issues, such as the determination of IQ, and the recently expanding field of sociobiology.

When we commenced our work some ten years ago the topic of cultural transmission was clearly far from the mainstream. This is no longer the case: scholarly work on quantitative theories of cultural inheritance, transmission, and evolution are now increasingly common. There remains a need for accurate empirical observation, without which theories may prove to be frustrating exercises. But at least in the discipline of linguistics and in certain aspects of sociology and anthropology it is possible to make quantitative observations that can be used to test theory.

Our hope is that quantitative studies of modes of cultural transmission and their long-term consequences will stimulate discussion on the theoretical interpretation of cultural phenomena and present a conceptual framework for their understanding. Mathematical models may eventually be expanded and refined and in turn lead to new empirical studies of cultural phenomena.

ACKNOWLEDGMENTS

We are grateful to our friends and colleagues whose critical encouragement contributed to improving the manuscript at various stages. We thank especially K.-H. Chen, F. B. Christiansen, S. Dornbusch, S. Feldman, R. Holm, R. C. Lewontin, L. Mann, R. Pulliam, B. Singer, M. Uyenoyama, and W. Wang for their detailed comments. Henri Fingold was skillful and very patient in incorporating our innumerable changes into the typescript. Excellent computational assistance was provided by Juliana Hwang and Barbara Andersen, and art work by Rashida Basrai and Barbara Hyams. The generosity of the J. S. Guggenheim foundation (in supporting M.W.F.) greatly facilitated the writing of the manuscript.

Stanford
June 1980

List of Symbols

a: correlation coefficient for assortative mating (continuous traits)

b: regression coefficients for assortative mating (continuous traits)

b_0, b_1, b_2, b_3: probability that trait H is present in children of parental pairs $h \times h$ (father $\times$ mother) (b_0), $h \times H$ (b_1), $H \times h$ (b_2), $H \times H$ (b_3).

B: $b_0 + b_3 - b_1 - b_2$ (deviation from additivity of parental contributions in vertical transmission (discrete traits))

C: $b_1 + b_2 - 2b_0$ (sum of parental contributions (discrete traits))

D: drift coefficient

E: expectation

f: coefficient of oblique or horizontal transmission for discrete traits

g: group-effect (coefficient of oblique or horizontal transmission for continuous trait).

H, h: presence and absence of cultural trait under study.

m (not subscripted): assortative mating for discrete traits

m_i, m_{ij} (always subscripted): migration coefficients

N: population size in the study of drift; sample size in the study of correlations

p,q: gene frequencies ($q = 1-p$)

r: generic symbol for linear correlation coefficient. For discrete traits, computed from $\chi N^{-1/2}$

r_{PO}: correlation between parent and offspring

r_{FO}: correlation between father and offspring

r_{MO}: correlation between mother and offspring

$r_{\bar{P}O}$: correlation between midparent (mean parent) and offspring

r_{SS}: correlation between sibs (intraclass)

r_{HS}: correlation between halfsibs (intraclass)

r_{FHS}, r_{MHS}: correlation between paternal and maternal half sibs

s: selection coefficient ($1 + s$ is Darwinian fitness of H versus fitness of h taken equal to 1)

t: time in generations

u, v: frequency of individuals of type H,h

V_t: variance of quantitative trait of generation t

α_p, α_m: paternal and maternal transmission coefficients in additive scheme (discrete traits)

ϵ: normal variate with mean 0 and variance σ^2

μ: mutation rate

μ_t: mean (expected) of quantitative trait at generation t

σ^2: mutation variance for quantitative traits

$*$: to mark frequencies and other variables after selection as opposed to before selection

$\hat{}$: to mark equilibrium values of variables

Contents

Cultural Transmission and Evolution

CHAPTER ONE

Introduction

1.1 MAN AS A CULTURAL ANIMAL

Culture, derived from a Latin word meaning "cultivation, care," has developed many different meanings. The one we use here is closest to that in Webster's dictionary: "the total pattern of human behavior and its products embodied in thought, speech, action and artifacts, and dependent upon man's capacity for *learning* and *transmitting* knowledge to succeeding generations." (It seems to us redundant to add the last part of this definition: "through the use of tools, language, and systems of abstract thought.") The word "behavior" may well require clarification, as it can be interpreted very broadly to encompass all activities of an organism, whether innately determined, or learned in response to environmental conditions or stimuli.

Ethological research suggests that the limitation to humans in this definition is questionable. In recent decades the uniqueness to man of speech, toolmaking, symbolizing, and some other "higher" activities has been challenged. There is not much doubt that speech, toolmaking, symbolizing, and aesthetics are most highly developed in man. But analogues are being found more and more often, even if in primitive forms, in animals not phylogenetically close to man. The same can be said of such cultural activities as the capacity to learn and transmit knowledge to succeeding generations. Examples include the cultural diffusion of innovations e.g., toolmaking and potato washing among monkeys and apes, and the learning of songs in birds, and perhaps even the technological activities (e.g., "gardening") of termites. What may be unique to man is the capacity to transmit knowledge to other individuals remote in space and time by means of such devices as writing, mainly, the transference of abstract instructions and explanations in ways that do not require face-to-face observation and direct imitation. Thus, in the following, we ask the reader to bear in mind that, although the processes of cultural diffusion, change, innovation and adaptation are most highly developed in man, the possibility of their existence in other species should not be ignored.

1.2 THE ADAPTIVENESS OF BEHAVIOR

In the context of evolution, a physical or behavioral trait is said to be adaptive if the possession of the trait by an individual confers an increase in that individual's chance of survival or reproductive success. Even if this is regarded only in an average sense, over all the situations that confront the individual, evolutionary adaptations are apparently so successful that human observers have often attributed their origin to an act of creation by an infinitely intelligent being.

Not being "infinitely intelligent," natural selection cannot "imagine" situations that never occur. Thus we should expect to observe only the most pronounced adaptations to the most common environmental challenges. But the success with which organisms cope with their environment appears to be so great that evolutionists often resort to teleological reasoning and search for some deep adaptive reason for every observed behavior, no matter how apparently trivial. There is no idiosyncrasy of a living organism to which an adaptive significance cannot be ascribed by evolutionists. Indeed there *might* be some adaptive meaning behind specific behaviors, but this need not always be the case, and moreover, the adaptive meaning of a morphological, physiological or behavioral trait is often far from obvious. The demonstration that a given trait is adaptive requires critical and careful experimentation. In general it is extremely difficult to quantify the advantage conferred by the presence of a given behavior over its absence.

It seems possible to assign to many of the enormous variety of behaviors in man and animals a precise adaptive meaning; for others it may be difficult or even impossible to discover one. Indeed, the meaning of a specific behavior may vary in different species, or even within a species. For example, shaking the head means "yes" in much of India and "no" in Western societies. Tail wagging has a different meaning in dogs and cats. Yet this does not contradict the notion that behaviors like tail wagging are acquired through natural selection, if it is assumed that the *common* adaptive feature is cases like this is, for instance, the communication of emotions.

The expression of emotions to, and recognition by, members of the same species has repercussions for so many biological processes that some scholars have been tempted to regard specific manifestations of communication as entirely innate. However, social interaction provides the opportunity to learn (even if only by continuous reinforcement or conditioning) the meaning and appropriate display of a great number of behaviors that communicate emotions. Whereas it is possible that for some mechanisms of communication, such as smiling in babies, the behavior is innate (blind babies smile), it is also likely that the behavior is elicited, augmented, and, in the long run, maintained by interaction with others. Thus in considering adaptiveness of behavior it is essential to distinguish between the general capacity to learn and any specific manifestation that might be the result of an interaction of innate tendency and environmental stimuli.

The dynamics of the changes within a population of the relative frequencies of the forms of a cultural trait under defined cultural interactions is the subject of this book. Laws of transmission of these forms among individuals of the same or different generations will be assumed, and the resulting changes in frequencies followed. It is these frequency dynamics and kinetics that constitute our major interest in this study.

1.3 LEVELS OF LEARNING

In all living beings, including man, there are some innate behaviors. Reflex actions, for example, belong in this category. We cough if stimulated by foreign objects in most parts of the respiratory tract, and sneeze in response to nasal irritations. We respond with characteristic jerks of our legs and arms if hit in the right place by a doctor's hammer. Such reactions can be regarded as completely innate. But most of human behavior is not entirely preprogrammed, and is at least in part learned. Indeed, the class of behaviors that have both innate and learned components is probably large. Animals may be so innately predisposed to learn some specific behavior that they almost never fail to learn it. Recognition of the mother goose by goslings occurs by about 24 hours after hatching, and any moving thing sensed by the young may then be accepted as the mother goose—anything from scientist Lorenz to a toy train engine. This highly specific and irreversible learning, which is possible only during a very narrowly circumscribed developmental period or critical period, is called "imprinting." Despite the absence of direct evidence (from genetic crosses, for example), it is not difficult to accept the hypothesis that in birds the capacity to imprint is genetically determined.

With imprinting, the organism is a passive participant in the learning process. Classical conditioning and instrumental conditioning, which involve more active participation by the learner, have been extensively studied by psychologists. Thus rats can be trained by the use of rewards to run through a maze. Generally, these experiments are conducted in such a way that one subject does not observe another perform the task. There is, however, another type of learning in animals: observational or imitative. Experiments reviewed for example, by J. Michael Davis (1973), B. G. Galef (1976), and D. Mainardi (1980) suggest that observation by one animal of a task performed by another may affect the rate of acquisition of successful task performance by the latter, and even its response in a novel situation (see also the recent book by Bonner, 1980). Learning by a process of direct teaching or instruction has achieved its greatest expression in man. But Washoe, the chimpanzee, has learnt sign language from human caretakers (Gardner and Gardner, 1969), although the interpretation of the observation is controversial. The presence of a formally designated teacher is not a prerequisite for transmission in the models we shall study. Informal contacts with other members of the social group may be sufficient for learning to take place.

We will use the term "cultural" to apply to traits that are learned by any process of nongenetic transmission, whether by imprinting, conditioning, observation, imitation, or as a result of direct teaching.

1.4 INNATE AND LEARNED TRAITS

The evolution of traits that are cultural depends ultimately on the way in which such traits are transmitted among individuals within a generation, and between generations. The modeling of this transmission process is therefore important for the development of an evolutionary theory. In evolutionary biology, "transmission" generally connotes Mendelian genetic transmission, and studies of evolution under this mode of transmission constitute a large part of population genetics. Much less has been written about the evolution of traits whose transmission is not genetic, and this problem is one of the sources of inspiration for the present volume. However, the distinction between biological transmission and transmission by learning has been the subject of research and speculation in the fields of ethology, psychology, and education. The most general conclusion to emerge from these studies is that, for animals that can learn, and especially for humans, it is difficult to partition the process of transmission into purely genetic and purely cultural components.

Consider the learning of language, in which genetics might seem *a priori* to be unimportant. We know there are some genetic differences between Chinese and English, yet in the absence of prior influences Chinese and English children learn Chinese and English equally well. That Chinese and English people speak Chinese and English, respectively, is clearly not a matter of genetic differences in linguistic propensities. Yet it would be a mistake to completely disregard innate tendencies in the learning of language. Apparently no animal can achieve the capacity that man has for using spoken language for communication. It is unavoidable to conclude that this difference between man and animals has a genetic basis despite the present impossibility of proof. There have recently been claims of success in teaching higher primates the meaning of several hundred words or their symbolic equivalents. These animals appear to be able to use words purposefully and to combine them in new ways, thus indicating they are capable of some level of abstraction. Thus today we are less sanguine in making sharp distinctions between man and other animals with respect to language, a capacity with which earlier anthropologists glorified only our species. It is possible that the obstacles to the development and use of speech among our closest phylogenetic neighbors are in some part due to the different anatomy of mouth and throat, which limits their phonetic repertoire. It is especially where there are large structural differences between species that the assumption of genetic differences for those behaviors that are limited by structure is justified. However, in general, the observations that (1) two groups differ in a specific capacity, and (2) there are genetic differences (for some other traits) between the groups, does not imply that the differences in the capacity are genetically determined. This may seem to be an elementary logical caution, but some ethologists, sociobiologists and behavioral scientists have inferred from observations (1) and (2) that differences in capacities between subgroups of the same species are genetically determined.

It is worth examining this last point in detail. It is reasonable to assume that many behavioral differences between man and primates are genetic, even if there is no direct supporting evidence. Millions of years of reproductive separation can generate considerable genetic differences, some of which can be noted on a very cursory examination. Also, when contemporary human groups are considered, some genetic differences can be observed among them—for instance for skin color, body shape and size, and, in some cases, for proteins. But there are also differences in many behavioral traits, and, while the former are evidently biological the latter confront the observer with a problem: are behavioral differences determined culturally or genetically? Even now, the fact that there are biological as well as cultural differences between human groups is often used to make the unwarranted inference that a given behavioral trait is genetically determined. This opinion is no longer as prevalent as it has been, but among both scientists and lay people there still exists a substantial proportion who have not recognized this error of logic. The basis of the error is, of course, the equation of causation with correlation. If two groups differ in skin color, which is known to be genetic, and in their level of wealth or technological development, why should the latter not also be genetic? This kind of simplistic thinking is hard to discourage because there are very few direct tests of the hypothesis. As we shall see, cultural transmission can simulate genetic transmission, making it difficult to separate them in a careful analysis.

1.5 CULTURE AS THE OBJECT OF EVOLUTION

We accept as culture those aspects of "thought, speech, action [meaning behavior], and artifacts" which can be learned and transmitted. This encompasses a large and heterogeneous collection indeed, and at first glance it may seem futile to look for the common properties of such a diverse group. But it should be remembered how difficult it was (and for some still is) to accept that viruses, bacteria, plants, animals, and man should have in common the set of laws composing the theory of biological evolution.

The feature common to all the above "cultural entities" is that they are capable of being transmitted culturally from one individual to another. Transmission may imply copying (or imitation); copying carries with it the chance of error. Thus we have in cultural transmission the analogs to reproduction and mutation in biological entities. Ideas, languages, values, behavior, and technologies, when transmitted, undergo "reproduction," and when there is a difference between the subsequently transmitted version of the original entity, and the original entity itself, "mutation" has occurred. Whether this change is a result of random copying error or has been intentionally made does not determine its subsequent fate, since the altered cultural entity, rather than its progenitor, is now the model for other individuals who will transmit it.

Reproduction and mutation ensure that evolutionary change will take place. However, if these were the only effective factors, biological evolution would proceed randomly without adaptive meaning. Natural selection is the mechanism that generates biological adaptation. In cultural evolution, however, there is in addition a second mode of selection, which is the result of capacity for decision making. We amplify these comments in the next sections.

1.6 THE MEASUREMENT OF SELECTION IN BIOLOGY

Is the degree to which a variant cultural trait is accepted by a substantial fraction of people and survives over generations a measure of its adaptiveness? In biology we use the word "fitness," or "Darwinian fitness," to describe the adaptiveness of a biological trait. The Darwinian fitness of a specific biological trait in a specific environment *relative* to the absence of the trait (or another form of the same trait) is measured by the number of offspring that survive to maturity produced by those individuals who carry the trait relative to those who do not.

Organisms that are better than others at leaving progeny that survive to reproduce will be represented to a greater extent in the next generation. The offspring will receive many parental characteristics by the process of biological inheritance, and those characters which made the parents better adapted on average will be passed to children and tend to make them also better adapted on average than their contemporaries. Increased adaptation in biology is the same as greater capacity to leave progeny, all other things being equal, and is therefore measured in terms of a capacity to survive and reproduce. This effectively means the capacity to transform nonliving matter into living matter similar to oneself. Unless it is realized that adaptation is really measured in terms of expectation of progeny, it might appear that there is a circularity in the definition of "better adapted."

Consider the following example of estimation of Darwinian fitness of a biological trait a genetic disease called cystic fibrosis (abbreviated CF). Patients affected by this disease nowadays have about a 20% chance of reaching the age of 18 years, versus a corresponding chance of 96% for the healthy population. (An exact computation of these probabilities is made difficult by the fact that they are continuously changing; we take approximate values and assume for simplicity that when 18 years is reached, the chance of survival of a patient is equal to that of a normal individual.) This difference in mortality makes the CF patient's chance of reaching adulthood, relative to that of a healthy person, equal to 0.20/0.96 or 21%. Thus, as adulthood is a necessary prerequisite for reproduction, if CF patients were as fertile as normal individuals, the fitness of CF relative to "normal" individuals would be on average 21%. It so happens that CF patients do not have normal fertility, male patients being practically sterile. If females were also sterile, then the fitness of CF individuals would be zero. If females have about normal fertility, considering that males and females are about equal in numbers at reproductive age, the fitness of CF would be half of 21%, or 10.5%. This reduction of 89.5% below normal is called the "selection coefficient" against CF and can be regarded as the complement (with respect to 100%) of fitness.

This calculation is of necessity somewhat tedious despite its considerable oversimplification in terms of demography. R. A. Fisher (1930) has supplied a classical formula, borrowed from Lotka's theory of demographic equilibrium, that produces an estimate of Darwinian fitness on the basis of age specific birth and death rates. (See L. Cavalli-Sforza and Bodmer, 1971, for an elementary exposition.) Darwinian fitness (or selection coefficients) of organisms in their natural habitat is not always easy to measure in practice. Even under carefully controlled experimental conditions, fitness differences smaller than a few percent are very difficult to evaluate.

In spite of these difficulties of measurement, evolutionary theorists use Darwinian fitness as a measure of "adaptiveness" of genetic types in a given environment. If a similar acceptable index for measuring the adaptiveness of cultural alternatives could be given, empirical observations might be more easily reconciled with theory.

A technique that has been used with success in the measurement of moderately large differences in fitnesses employs simple mathematical models for the change in frequencies of genes over time. The predicted trajectories of these frequencies through time are then compared to the observed data to produce estimates of the fitness parameters (see, e.g., Wallace, 1968). This procedure is not powerful enough to distinguish small fitness differentials between different genotypes despite a great deal of painstaking data collection (Yamazaki, 1971). Recently, more sophisticated statistical techniques that break down selective differences into components specific to each stage of the life cycle have been used to advantage in experimental situations (Prout, 1965, Christiansen and Frydenberg, 1973), but again these are most useful when selection is at least moderate. When selection is weak it is extremely difficult to infer from data whether or not the trait is under natural selection.

Despite the difficulty of controlled laboratory experiments on cultural change, the trajectory approach should in principle be applicable to cultural evolution in the same way that observed dynamics of epidemics are used to estimate parameters in predictive models. Thus it might be possible to identify a new cultural type in the process of substituting for an old one, or of being accepted in lieu of nothing, if there was no true "old" type, but just the absence of some specific trait. Such observations might be most easily made in the evolution of technological changes. As in biology, the difficulty of using data to demonstrate that a specific aspect of culture is under selection is profound. Indeed, what constitutes a "neutral" cultural process is not well understood.

We have, thus, seen two ways of measuring Darwinian fitness of a biological trait outside the context of culture. One is by evaluating the probability of survival of carriers of the trait jointly with their fertility by comparing the average number of their children who survive to sexual maturity. The other way of estimating fitness is to follow, if possible, the relative frequency of the trait in actual evolution over several generations. In both cases what is evaluated is the fitness of a trait relative to an alternative: in practice, both are subject to strong experimental and statistical limitations. Naturally, observation over several generations is very difficult in man, because almost the only durable residue of people who died a long time ago are bones that turn out to be relatively depauperate in simply inherited traits. For most culturally transmitted traits, the absence of a natural time unit like that of a generation in biological transmission makes the use of an equivalent of the first method difficult to devise. The second method (by measuring rates of acceptance over time) remains, of course a possibility. That we need to view such estimates with caution will become clear as we discuss in greater depth the process of cultural acceptance.

1.7 TWO LEVELS OF SELECTION AND TWO ORDERS OF ORGANISMS

A culturally acquired behavior becomes part of the overall *phenotype,* that is, the trait or sum of traits observed in an individual. During the acquisition of a mode of behavior, different degrees of individual choice (probabilities of acceptance/rejection) may exist. Among the many thousands of things we are usually taught are, for example, to avoid offensive words, to go to school, to clean our teeth, to select reasonably nutritious food, to look left and right before crossing roads, and so on. Much of the teaching has taken place by an age and under conditions which make it questionable whether there was actually a chance to make a real choice. It is probable that, in many instances, behavior is imposed rather than taught, and that there is no real possibility of deciding for or against acceptance.

Not all of us are taught these things, however, and not all of us who are taught them accept them. Failure to receive or accept some of them—for example skill in crossing roads or in recognizing a poisonous berry or in knowing what to do if bitten by a poisonous snake—may be fatal. Survival skills like these, and even rules of hygiene, all have an effect on survival probability and thus on Darwinian fitness. The pressure to go to school may affect the chance of landing a good job, getting married, having children, and earning the money to have them properly cared for. Thus, going to school may eventually affect the fertility component of Darwinian fitness.

Darwinian fitness does not, however, necessarily play a direct role in the adoption of a cultural trait: some other examples may help to illustrate this point. The Japanese style of ancestral tablets gained very wide acceptance, in preference to the earlier Chinese form, during the Japanese occupation of Taiwan, yet the Japanese form of ancestor worship did not. Buddhism was extremely successful in its spread east from India. On the anthropologically more trivial side, Coca-Cola, frisbee, volleyball, and yo-yo's are examples of "innovations" that have spread rapidly through whole countries or continents. It is obvious that in none of these examples does participation appreciably alter the probability of surviving or having children.

Clearly, then, some kind of non-Darwinian selection is operating here. Let us call this selection *cultural,* and define it on the basis of the rate or probability that a given innovation, skill, type, trait, or specific cultural activity or object—all of which we shall call, for brevity, *traits*—will be accepted in a given time unit by an individual representative of the population. Of course, there may be differences among individuals of the population in the probability of their learning or accepting the trait: this probability can change with time and conditions and may be a complex function of many variables. But as a first conceptual approximation we prefer to think of cases in which it is relatively constant across individuals. This probability of acceptance as a measure of *cultural selection* must be clearly differentiated from the *Darwin-*

ian or *natural selection* due to the cultural trait. In practice, one can distinguish between them by noting that cultural selection refers to the acquisition of a cultural trait, while Darwinian selection refers to the actual test by survival and fertility of the advantage of having or not having the trait. Some of the examples of cultural traits given above probably have little if any effect on Darwinian fitness. But this possibility should not be excluded. Indeed, the two types of fitness may operate in opposite directions to one another. Parachute jumping and other dangerous sports must have a positive cultural fitness, even if it is not very high (because at least some people indulge), in spite of the fact that they probably reduce Darwinian fitness. Other customs that conceivably have a positive Darwinian fitness may have very low cultural fitness: this may have been true of jogging before the recent epidemic.

Does natural selection have direct control over culture? Many ethologists would argue that, through its control over "the physical basis of behavior," natural selection dictates cultural activities so that the latter are in fact only superficially culturally determined, but in reality are innate. The alternative view is that, in humans, and to a lesser extent some higher animals, the chain of events connecting most behaviors to physical structures is very long, complex and indirect, so that genetic preprogramming cannot determine all behaviors that demand some kind of choice or decision making. As has been pointed out, it may often be most difficult to decide experimentally where on the continuum between completely preprogrammed and completely learned a cultural trait lies, and which of natural or cultural selection is more important in determining the state of this trait in a population. In some of our models, both cultural transmission and Darwinian fitness enter the evolutionary formulation; in some there is potential conflict between the two, while in others the two types of selection may converge.

Ideally, we would like to define both types of selection operationally, by giving recipes for their measurement. This is why an example of a Darwinian fitness calculation was presented earlier, together with its limitations. These practical difficulties, which are amplified in attempts to estimate cultural fitness, need not be of great concern here, because Darwinian fitness is a well established, clear-cut concept, even if its measurement is usually imprecise. In addition, the following line of reasoning shows how cultural fitness may be reduced to a Darwinian fitness of another "second-order" organism.

A particular cultural trait is chosen for study for the sake of convenience of observation and measurement, or because of its specific interest with respect to a more general context. Usually the trait is abstracted from a larger, more complex unit which can be truly defined as a *cultural object*. Modern technology offers many highly developed examples of cultural objects, which are almost as complex as a living organism: a jet, a car, or a washing machine, for example. These could be considered as "organisms," because they are reproduced by some kind of assembly line, a procedure which puts together a number of pieces specifically molded or machined, so as to look as closely as possible like a prototype. There is thus definitely a reproduction, which does not take place ordinarily by binary fission, but by a complex one-prototype–multiple-copies kind of mechanism ultimately under human control. The production of a violin is closer to a binary fission, if only one violin is made at a time by an artisan. The artisan making a violin, or the engineers and factory workers making a car, are the living organisms or the *first-order organisms* that produce the pieces and assemble them. Without them, no *second-order organisms* like the violin and the car would be produced.

Specific traits of these cultural objects, like the fins or chromium-plated accessories of cars of some years ago, or the lower gas consumption or good acceleration of cars today, may be a basis upon which choice of a specific car model is made. These properties can be regarded in the present terminology as criteria simultaneously for cultural selection by the first-order organism and for natural (Darwinian) selection to which the second-order organism is subjected. Those car models or makes of musical instruments that are selected because of some aesthetic and technical qualities that appeal most to prospective customers will prosper. But for the first-order organism that makes the cultural choice, there are also potential consequences of the cultural choice at the level of natural selection. We measure this by *Darwinian fitness of the first-order organism.* In the case of a car, this will be determined mostly by the car's safety. The effect on Darwinian fitness of the buyer of a violin might take more imagination to visualize, and is likely to be small. The cultural fitness of stylistic traits of the car, or the violin, which determine acceptability by the *first-order organism,* constitute the Darwinian fitness of the *second-order organism.* If the chooser ignores the safety characteristics of a car, his/her own Darwinian fitness will nevertheless be influenced by them. It may be that, in the long run, certain stylistic features, while attractive to the eye, may be recognized as unsafe. Then the lower Darwinian fitness for the first-order organisms (man) may translate into a lower fitness also for the second-order organism, the car, provided the negative weight given to loss of safety overcomes the positive one given to style.

Outside technology, the recognition of cultural objects that have some unity may be more difficult but is frequently possible. Correct use of the knife and fork may be recognized as part of the cultural object called "table manners." Many such traits of etiquette differ greatly among populations and have no obvious adaptive reason. The complex itself, however, has probably originated from considerations of aesthetics, respect, hygiene, and convenience, even if many individual rules seem to make little sense. A single trait may be transmitted as part of the complex, and the complex may have adaptive meaning even if its individual parts are trivial or even counterproductive, as we will consider later in discussing customs.

Even language and its components (words, rules, and sounds) can be regarded as cultural "objects," and the cultural fitness (the appeal to the speaker) of various alternatives for its components, rules, and so on, determines the Darwinian fitness of these components of language in the sense that they can be considered as second-order organisms.

The problem of natural versus cultural selection has been discussed at length here because it is a central issue not usually addressed in the context of a quantitative theory of evolution, although its relevance has been recognized (e.g., Campbell, 1965, 1975; Durham, 1976). By this we do not mean to deny the difficulties inherent in the measurement of cultural selection and in confounding it with other events of evolutionary importance.

1.8 SOME EXAMPLES FROM THE EVOLUTION OF LANGUAGES

With few exceptions, studies of specific cultural traits have been almost entirely descriptive. Our aim here is to explicate the evolutionary role of a few forces that impinge on cultural traits and ultimately to determine whether, even within this restricted context, some general principles are possible. With this in mind, the following brief discusssion will be oriented toward major paradigms rather than a complete survey.

Our first topic is the evolution of language, an issue less fraught with emotional overtones than, say, social interactions and inequality, or altruism. Language also has the advantage of reliable, accurate measurement not only because of the nature of the cultural object under study but also because of the tradition of rigor that has characterized the discipline. Indeed it may be that this tradition has produced the taboo on the study of linguistic evolution—perhaps it has been deemed premature to expect reasonable progress in such studies. Attempts in the present and in the last century have met with at best limited enthusiasm and often outspoken criticism. Only very recently has there been a rebirth of interest in linguistic evolution. It is remarkable that some of the principles that seem to emerge from such studies are perfectly acceptable to evolutionary biologists. Yet, within the field of linguistics, such generalizations are still presented with caution, despite the backing of considerable direct and indirect evidence. It would seem that an independent rediscovery of basic evolutionary principles is in progress in linguistics, but that the mainstream of the field has not yet shown readiness or interest in accepting the principles. For example, Weinreich, Labov, and Herzog, in a 1968 paper, said that "not all variability and heterogeneity in language structure involves change; but all change involves variability and heterogeneity. The generalization of linguistic change throughout linguistic structure is neither uniform nor instantaneous. . . ." This statement recognizes that the unit of variation is not the population, but the individual, and that there exist steady-state equilibria in which the proportions of individuals (or of the various traits) differ from zero or one and are stable over time. These principles are, of course, basic in biological evolutionary theory.

The recent sociolinguistic studies by Labov (1972) and by others represent real attempts to describe individual variation within populations, even if this was not, perhaps, the primary aim of their research. Prior to this the usual approach in linguistics had been typological, with little interest in individual variation. Typically, dialect atlases or studies in linguistic taxonomy are based on a single (or at most two) informants per location. At that stage of the development of science when only average or modal analysis is important, the taxonomical approach is both valid and economical. In biology, for example, it remains the norm in such fields as physiology. But it is clearly incompatible with the study of evolution. Our interest here is in those areas of linguistics in which statistics have been used in the attempt to explicate the evolutionary process.

Linguistic change can be studied at any one of four levels: the grammar can change, new words can enter a language and may replace the old, the meaning of a word can change, pronunciations may change.

Over the past 150 years some emphasis was placed on the regularities detected in phonological change. No doubt the impetus for this area of research was provided by the discovery of sound correspondences across sister languages, from which one can reconstruct the sound systems of the parent languages. The best known of such correspondences, probably because of its early date of discovery, is a set called Grimm's Law (though a correspondence set cannot really be considered a "law" in the usual sense of this word in science). Grimm's discovery has to do with the evolution of obstruent consonants from Proto-Indo-European into the various Germanic languages. Many such correspondences have been worked out now for the major language families of the world.

As an illustration of such correspondences, note that the letters for the vowels A, E, I, O, and U represent very different sounds in English than they do in languages like French or Spanish. This is because, between the time of Chaucer and that of Shakespeare, a large scale change took place in English called the Great Vowel Shift. Before this time the vowel letters were pronounced as in German or Italian now. During the Great Vowel Shift they acquired their present values. We can detect traces of this Shift by noting some internal correspondences in English. The first vowel letter in words like "s*a*ne, sl*ee*p, f*i*ve" is now pronounced with the post-Shift values, differently from "s*a*nity, sl*e*pt, and f*i*fth." The vowels in the latter set of words are closer in quality to Chaucer's pronunciation as well as contemporary German or Italian.

Many other rules of phonological change have been described

although the number of situations and languages to which they apply varies considerably. Thus certain clear trends can be detected but their causes are unknown. It is not even known whether in each case the change has been gradual or abrupt. But Wang (1976) and his associates have collected evidence from a wide range of languages, including Chinese, Dravidian, English, and German, showing that some of these changes may be abrupt—the equivalent of biological mutations—and that the frequency of individuals using the new sound tends to increase regularly over time according to an S-shaped curve. Curves of this type, for example the logistic, may be explained by a process of cultural selection in which there is at first contact between two different dialects, and this is followed by spread, in which conscious or subconscious choice occurs.

The reasons for the preference of one sound over another are obscure. The principle of parsimony might be inferred: those phonological changes which decrease the amount of effort expended by the speaker might be preferred. However, least effort is not the only determinant of the process, and more subtle reasons might include the requirement that similar sounds be clearly distinguished when used in similar words that have different meanings. Abrupt phonological changes can be compared with biological mutations, but what about gradual, continuous change? Here again an analogy can be drawn with continuous polygenic traits, such as height, or other linear body measurements. For such biological traits genetic influences are known, in many species, but most of the variation is probably due to the combined action of many genes, each with small effect, interacting with the environment. The outcome is an effectively continuous distribution of trait values. It has been convenient to describe the biological mutation process in terms of a continuous distribution of probabilities of the specified amount of genetic change measured on the scale of the trait, and the Gaussian distribution has most often been used for the theoretical exposition of such mutation distributions in polygenic variation. A similar concept can be applied whenever cultural changes are gradual and continuous. In the study of language, when sounds can be translated into spectra of sound frequencies, the scale of measurement is naturally continuous. Discontinuity arises not so much at the level of production, but at the level of perception.

The sources of linguistic variation are manifold. The sounds which come before and after a given one influence its pronunciation. The mood of the speaker can be important and so can social context, as shown by Labov (1972). The same word may be pronounced differently when read from a word list, from a prepared text, if uttered in a formal speech, in informal talk, or under conditions of excitement. Thus there is not only considerable variation *among*, but also *within* individuals because of the differences in "styles," (as the above situations are called), and comparisons between individuals must be made for the same style of speech. Even so there are differences between individuals, and even greater differences between groups. Sex and age must also be considered. Figure 1.8.1 shows unpublished observations made by E. Migliazza, who recorded American adult males and females and an equal number of British speakers, all reading the same word ("pot") from a word list. The differences between sexes, between origins, and between individuals are apparent from the graph, in which two measurements made on each word by sonograph are plotted. The lowest and the second-lowest resonance peaks of the vocal tract, known as the first and second vowel formants, are shown in the figure. The formants are known to vary considerably between vowels, and every vowel has a characteristic set of formant frequencies. It is these formant frequencies, calibrated against the quality of the speaker's voice, that are recognized by the human ear as characteristic of the specific vowel produced. Thus formants contribute essentially toward characterizing a vowel utterance and can be employed for quantitative study.

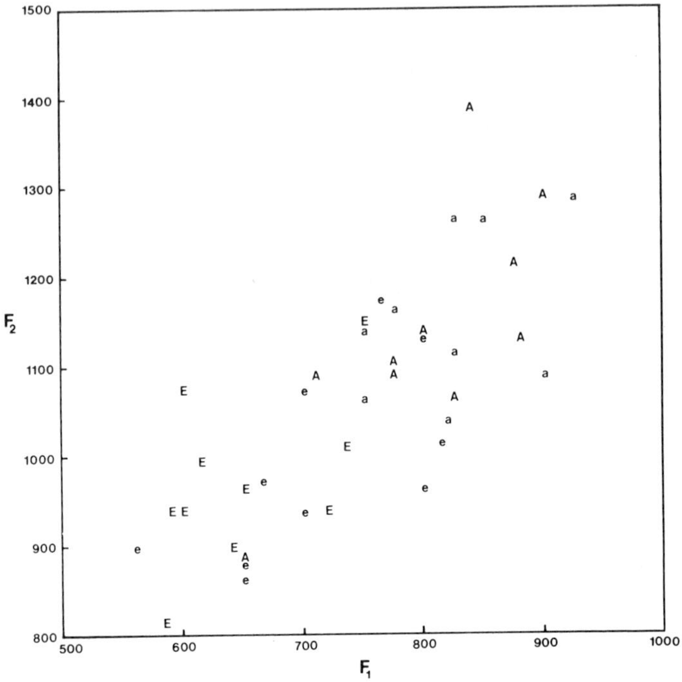

FIGURE 1.8.1. The word "pot" pronounced by 20 English (10 males indicated by *E*, 10 females *e*) and 20 American (*A* male, *a* female) adults. The individual pronunciation of the word is recorded on the basis of the two first formants, or frequency peaks (F_1, F_2), given in cycles per second on the two axes. Each word was read twice from a word list; the average values observed for a given individual are plotted in the graph. From unpublished observations by E. Migliazza.

In light of this clearly continuous nature (at least at the present level of phonological observation), it would indeed be surprising if there were no random phonological variations from generation to generation, due merely to an error in copying by the learner. Even if the child attempts to reproduce the sound he hears as faithfully as he can, the reproduction cannot be perfect and will have some, perhaps very minor, differences from the model. Assuming that transmission is from parent to child (which, of course, is not the only important transmission, especially in language), when the child is grown up and a parent, his version of the vowel, containing small differences from his parents whom he copied, will be the model for another child. Under these conditions changes are bound to be generated and accumulate over time. An utterance that can give rise to a misunderstanding will under normal circumstances be corrected somehow during the learning process so that deviations will remain within bounds. This type of cultural selection which eliminates trait values too far from some fixed mode is called stabilizing and is depicted in Figure 1.8.2A.

Although the existence and/or magnitude of this mechanism of *continuous* change has not been clearly demonstrated, it seems from a theoretical point of view to be such an important possibility that we must take it into account in our understanding of linguistic evolution. But other, more abrupt mechanisms of change are conceivable, which, unlike these gradual changes, are likely to be clearly perceived by the speakers, and be selected for

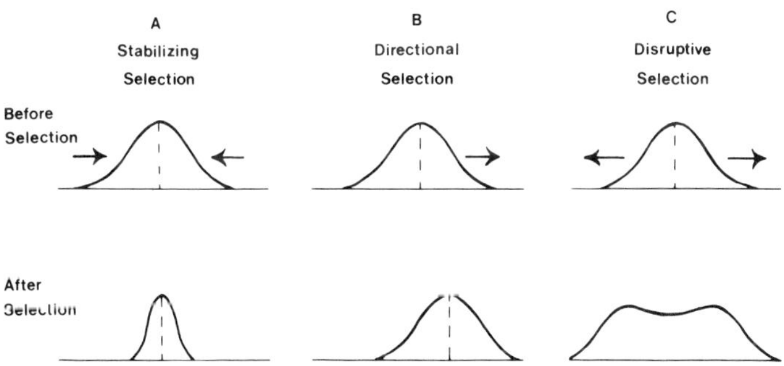

FIGURE 1.8.2. In biological evolution it is customary to distinguish three basic types of natural selection: stabilizing, directional, disruptive. The distributions shown before and after selection refer to a quantitative continuous trait but could be extended to discrete traits. Arrows indicate the direction of selection pressures. These concepts can be carried over to cultural selection unchanged.

or against on the basis of a true process of cultural selection. When such a selection pressure is maintained in the same direction for a substantial period of time, the process is called "directional cultural selection" (Fig. 1.8.2B). The traditional sound correspondences, such as those described in Grimm's Law, as well as the cases of abrupt sound changes presented by Wang and his associates, are probably fair examples. We may not know which is the specific reason that favored *f* versus *p*, etc., in certain groups, just as we are also very often unable to give a functional or adaptive explanation of evolutionary trends in biological evolution. As we have suggested, the variation between individuals in the same subgroup should remain bounded. If different subgroups are subjected to different bounds, or different modes of selection (Fig. 1.8.2C), pronunciation of words begins to change, until after several centuries mutual intelligibility between subgroups can be compromised. At this stage the parallel to the biological taxonomic notion of speciation occurs: groups can no longer understand one another. Of course, during these processes new words may be invented, very strongly modified, or hybridized (analogs of mutation), or may be introduced from other languages (the analog of immigration). Sometimes, especially following important political events, such as conquests, more extensive hybridization of language occurs. But even after millennia the common origin of words usually remains recognizable. This has allowed easy recognition, for instance, of the fact that all Indo-European languages belong to the same group. Even if the majority of words in two closely related languages are still clearly descendent from a common source, in which case they are said to be "cognates," mutual intelligibility of the languages may be compromised as soon as the similarity between two languages, both phonetic and lexical, falls below a certain level.

The study of the percentage of cognates common to two languages has been developed by Swadesh into a technique called "glottochronology," used to establish the time of evolutionary separation between two languages on the basis of the percentage of cognate words. Examples of cognates are the Latin and Greek "pater," the English "father," the German "vater," the French "père," and the Italian "padre," all of which come from a common source. Words like "spleen" in English, and "splen" in Greek and Latin, are cognates for the lymphatic organ, but are not cognate with the word in French ("rate") or in Italian ("milza"). The latter is cognate with a barrowed Germanic word, "Milz" adopted after the invasion of Italy in the sixth century by the Longobards, a Germanic people. The probability that a word is substituted in a language by an unrelated or noncognate word is different for different words. Frequency of use of the word is among the many largely unknown factors influencing this probability, and the frequency of use of specific words changes with culture. Swadesh (1952) has tried to minimize the influence of variation in probability of substitution by choosing a list of words supposedly having the same probability of substitution. The analysis of classical languages supplied a time yardstick based on the probability of cognate substitution in a given time interval. From data obtained at different times, Swadesh concluded that the frequency of cognates between two languages, p, decreased exponentially with time: $p = \exp(-rt)$. From the curve of the observed values, the constant, r, could be evaluated such that, according to Swadesh (1952), in 1,000 years of separation there is a fall from 100% to 66% in the probability of cognates between two languages. Thus $r = .00042$, and the exponential curve could be used to evaluate the time of separation of two languages, based on the percentage of cognates they share.

The method has been the subject of much criticism, and is certainly not as accurate as was originally thought. Under certain conditions some languages may have a slower rate of evolution as measured by the percentage of cognates, while under others the change may be faster. The frequency of contacts with people speaking foreign languages and the mode of contact may be important determinants of the rate of cognate substitution. Moreover, it has been shown conclusively that the probability of substitution varies considerably from word to word. Because of this, the overall curve of change of cognates with time is not a negative exponential (Kruskal, Dyer and Black, 1971); see also

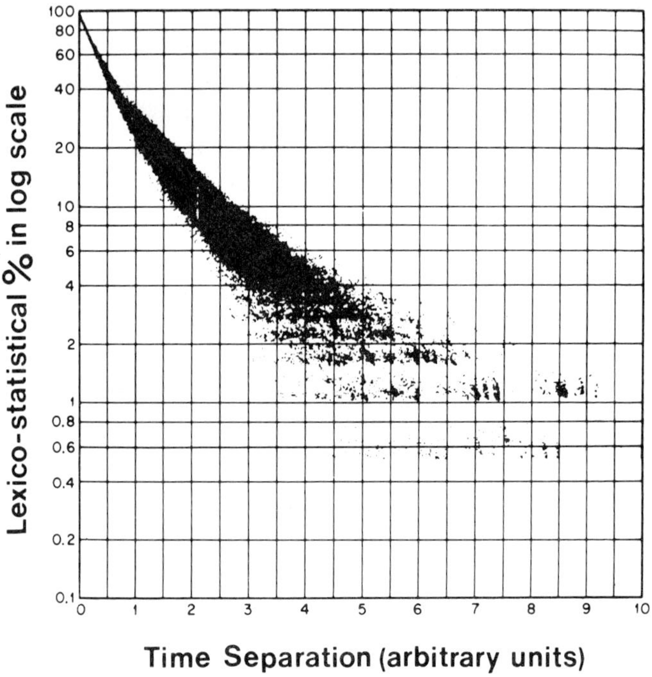

FIGURE 1.8.3. Word substitution in the evolution of 371 Malayo Polynesian languages. The percentages of shared cognates (based on 196 words) between the 68,635 possible pairs among these languages are plotted on a logarithmic ordinate. Separation time (arbitrary units) is on the abcissa. The rate of evolution, r, of each word and the relative time of separation of each pair of languages were estimated by maximum likelihood. On the basis of thirteen language pairs for which some historical evidence of separation time is available, the arbitrary time unit of the abscissa was then equated to 3500 years. Note that the relationship is *not* exponential, and that especially for long separation times there is considerable scatter, making the estimation of separation dates very uncertain. The above plot and analysis are from Kruskal, Dyer, and Black (1971).

A. J. Dobson, (1978) and Fig. 1.8.3. The estimate of separation time remains possible, but is likely to be especially bad for pairs of languages that are least closely related. Thus the higher glottochronological estimates are the most uncertain. And, if they are based on the usual assumption of exponential change, the longest separation times are likely to be grossly underestimated.

The analysis of the shape of the curve can be refined. For instance, it can be shown that the observed distribution of word substitution rates, as estimated by Kruskal et al. (1971), gives information on the shape of the curve of percentage of cognates with time (Sgaramella-Zonta and Cavalli-Sforza, unpublished) The argument is very similar to one used later for the evolution of words, not in time but in space, which we will discuss with some more detail in section 3.6. Under certain conditions there are simple relations between synchronic and diachronic variation of cognate frequencies.

1.9 THE DIFFUSION OF INNOVATIONS

The new word that becomes part of a language, or a modification of an old word caused by changes in pronunciation or deletion of parts of old words, or fusion of words in earlier use, is an innovation and can be considered as an analog of mutation in biology. The usefulness of this analogy and its limits will be investigated later. "Innovation" has been used to describe widely varying situations, not only with regard to technological invention. Rogers and Schoemaker (1971), in their extensive review of the field, include among their examples "a new speech form among oil drillers, a declaration of warfare among nations, a rumor aboard a submarine, and snowmobiles among Lapps."

When the process of diffusion of an innovation is followed for a sufficiently long time the frequency of use of the innovation almost always follows an S-shaped curve. At the beginning the number of acceptances rapidly increases (in fact almost exponentially). There follows an approximately linear increase, and finally the increase slows down and is barely perceptible. Of course, these are characteristics of the *cumulative* number of acceptances; the number of acceptances per unit time (*rate* of acceptance) is the derivative of the cumulative curve, and ordinarily increases to a maximum, then decreases to zero. Figure 1.9.1 illustrates a widely used theoretical version of this functional behavior, the logistic. An actual example is shown in Figure 1.9.2.

The analogy between innovations and biological mutations is

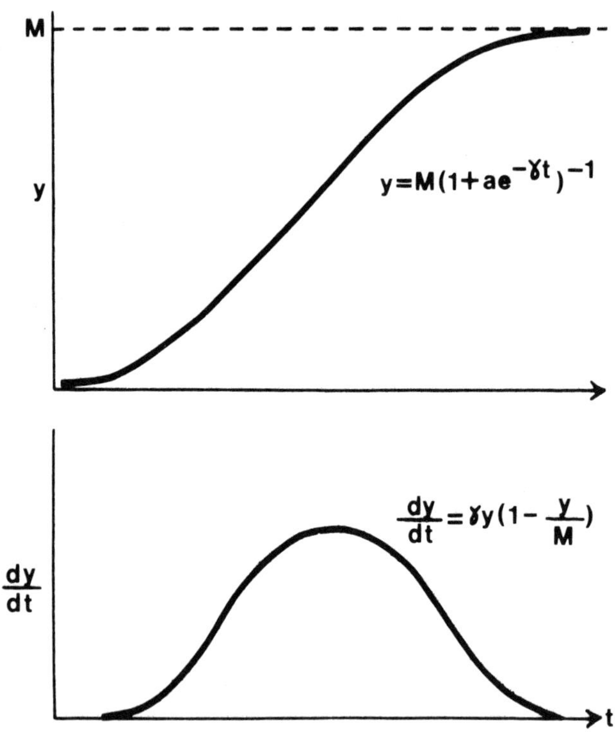

FIGURE 1.9.1. The logistic curve, above, and its derivative, below.

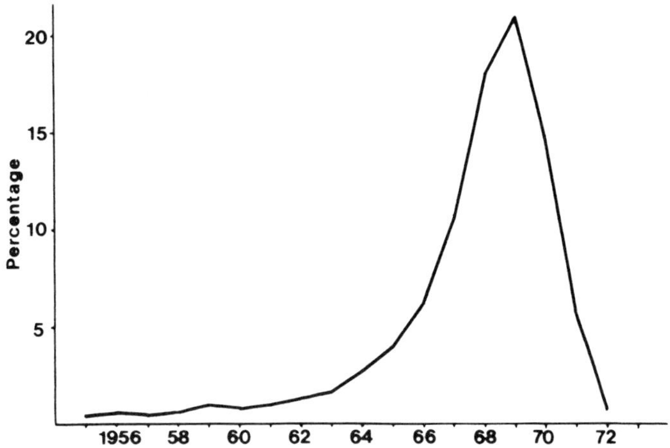

FIGURE 1.9.2. An example of a cultural epidemic: heroin addiction in Washington, D.C. The ordinate is the percentage of all patients entering the Narcotics Treatment Administration in December 1972, totaling about 13,000, plotted by the year in which they began heroin use. In 1969 a major commitment was made by the District of Columbia in an attempt to solve the problem, supplying methadone treatment to addicts and strengthening police action. After R. L. DuPont and M. H. Greene (1973).

inevitably superficial. Yet it remains of interest to note that the quantitative dynamics of the spread of innovation can be formally modeled in the same way as those of an advantageous mutation, which, in the simplest case of logistic change, can be treated as follows. Assume that a mutation has been introduced into a population, and that it is advantageous for the carrier. At generation t there are a_t individuals carrying the mutation and b_t individuals without it. $n_t = a_t + b_t$ is the total number of individuals. Further, assume that the carriers of the mutated gene multiply at the rate k_1, that is, each produces k_1 direct descendents in the next generation, so that the number a_{t+1} carrying the mutation in the next generation is

$$a_{t+1} = k_1 a_t.$$

The carriers of the old gene multiply at the rate k_2 (smaller than k_1), so that

$$b_{t+1} = k_2 b_t.$$

Let a_0, b_0 be the initial numbers (i.e. at time $t = 0$) of the two types. Then

$$a_{t+1} = k_1 a_t = k_1^2 a_{t-1} = \cdots = k_1^{t+1} a_0, \tag{1.9.1}$$

$$b_{t+1} = k_2^{t+1} b_0. \tag{1.9.2}$$

The ratio of mutant to original type is

$$r_{t+1} = \frac{a_{t+1}}{b_{t+1}} = \left(\frac{k_1}{k_2}\right)^{t+1} \frac{a_0}{b_0}, \tag{1.9.3}$$

Thus, if we write $k_1/k_2 = e^\gamma$ we have

$$r_t = r_0 e^{\gamma t}. \tag{1.9.4}$$

The ratio k_1/k_2 is the Darwinian fitness of the mutant relative to the old type. Since we assumed the mutant to be advantageous, $k_1 > k_2$ and $\gamma > 0$. (γ is the logarithm of the Darwinian fitness). Now, instead of the ratio r, consider the proportion of mutants in the whole population, $p_t = a_t/(a_t + b_t)$, at time t. Then

$$p_t = \frac{r_t}{1 + r_t}, \; r_t = \frac{p_t}{1 - p_t}. \tag{1.9.5}$$

Substituting (1.9.5) into (1.9.4) we have

$$p_t = \frac{p_0}{1 - p_0} e^{\gamma t} \left(1 + \frac{p_0}{1 - p_0} e^{\gamma t}\right)^{-1}. \tag{1.9.6}$$

Thus p_t increases monotonically from p_0 (at $t = 0$) until in the limit ($t \longrightarrow \infty$) it approaches $p = 1$. The graph of p_t as a function of t follows the S-shaped curve in Fig. 1.9.1.

Another biological model can be used to produce the same logistic curve and in some senses may offer a more satisfactory interpretation of the diffusion of innovations. The model is that of an epidemic, that is, the spread of an infectious disease within a generation. In its simplest form the model commences with a population of n individuals of whom, at time $t = 0$, a_0 have the disease and are *infectious*. The remaining $n - a_0$ are *susceptible* and contract the disease, upon contact with a diseased individual, with rate γ. The change in the number $a(t)$ of diseased can be written

$$\frac{da}{dt} = \gamma a(n - a). \tag{1.9.7}$$

A more complete description of the model is given in the next section. Integration of this differential equation produces

$$a(t) = \left[1 + \frac{1 - a_0}{a_0} e^{-\gamma t}\right]^{-1}, \tag{1.9.8}$$

which is easily seen to be the same as (1.9.6), with $a_0 = p_0$.

The analogy of an infectious disease to the spread of an innovation within a generation is perhaps closer than that of an innovation with an advantageous mutation. However, the mathematical theories formulated for any of the three variables—number of individuals carrying an *infectious disease*, proportion carrying an *advantageous mutation*, or number of individuals who have *adopted an innovation*—are formally identical when studied at this elementary level.

Still another situation modeled by the logistic is that of the *growth* of a population in an environment where a maximum number N of individuals can be maintained. In such a situation, it is assumed that the rate of growth of the population decreases as the population itself increases. If $n(t)$is the number of individuals at time t, then we write

$$\frac{dn}{dt} = \alpha n\left(1 - \frac{n}{N}\right). \tag{1.9.9}$$

Here there are two parameters: α is the initial growth rate (when the presence of other individuals has little effect on the fertility of any single individual) and N is the maximum population size reached. If $\alpha > 0$ and $n(0) < N$, then $n(t) \rightarrow N$ as $t \rightarrow \infty$. Of course, α is equivalent to the contact rate in the epidemic, or the logarithm of the Darwinian fitness of the mutant in the first model.

Strict validity of the logistic is based on a number of assumptions that are not always met. It is worth reviewing them in detail, indicating useful alternatives when the basic assumptions do not apply, given the wide use of the logistic in the interpretation of observed data.

(i) *An assumption of the simple logistic is that acceptance of an innovation is a one-step process.* In fact it is important to distinguish at least two steps in the process of adoption. First there is *awareness,* or knowledge of the innovation. Then comes the actual *adoption,* which usually involves the making of a decision. As an example (Figure 1.9.3), there was an average delay of between 2 and 3 years between awareness and adoption of 2,4,D, a weed killer. Similarly, in epidemics it is useful to distinguish at least two time points: the time at which the disease is contracted, and the end of the disease process, that is, healing or death. Indeed epidemiologists often use a quantitative two-stage treatment: susceptible → infectious → removed (i.e. dead or healed), which could be transferred almost directly to the two stages of diffusion of innovations, namely awareness and adoption. The next section of this chapter summarizes this simplest treatment of epidemics.

For innovations, the two steps involve substantially different processes. The first process, growth of awareness, depends on communication, which involves both a source—for instance, the makers of a new product, salespeople, or evangelistic users—and transmitters —for example, newspapers, journals, radio, or TV. The second step involves a process of decision, which may be partially motivated by economic, emotional, or other types of

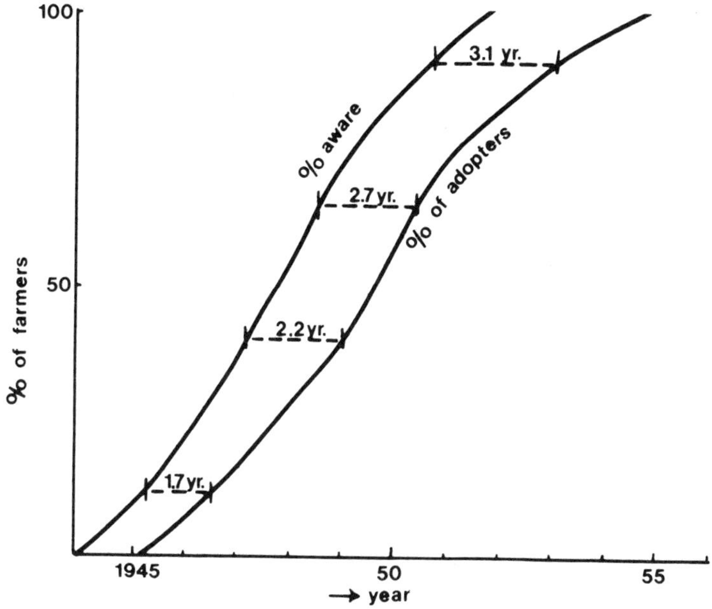

FIGURE 1.9.3. The kinetics of awareness and of adoption of 2-4D weed spray among Iowa farmers. Note that the two curves are not quite parallel, but the interval between them increases somewhat with time. (Modified after Beal and Rogers, 1960, as cited in Rogers and Shoemaker, 1971.)

factors, and is a typical example of what we have called cultural selection. It is often found that an adopter of an innovation has been persuaded by an example set by one or more earlier adopters who are psychologically influential. Person-to-person interaction seems to be an important part of the adoption step. Alternatively, of course, the advice received from one or more experts may be influential. (See, e.g., Rogers and Schoemaker, 1971; Coleman, et al., 1966.)

(ii) *The basic assumption of the logistic model is that the probability of adoption is proportional to the fraction of individuals who have already adopted the innovation, and the residual fraction who have not yet adopted it.* The proportionality constant might be a function, perhaps just of the innovation itself, or of the probability of contact between an individual who has not yet adopted and one who has, or of the probability that the example of the adopter is followed. There are many reasons why these need not be constant. The adoption of a new product on the market by a large fraction of the potential users may cause substantial economic changes, which in turn may put extra pressure on the late converters to adopt the new product. The spread of a dangerous drug may cause a more rigid enforcement of legislation against it, or modification of it to avoid or minimize its unwarranted use.

(iii) *Actual transmission is exclusively due to knowledge gained by observation or contact with people who have already adopted the innovation.* If there is a constant flow of information about the innovation (perhaps due to a sustained advertising campaign), then the adoption curve need not follow the logistic. In fact, in the study of adoption by doctors of a new antibiotic (Coleman et al., 1966) the increase of adoptions among doctors *not* in group practice (or those having little contact with colleagues) seemed to follow a negative exponential curve. This is to be expected when the information source is constant, for then the number of new adopters per unit time is proportional only to those who have not yet adopted, that is,

$$\frac{dy}{dt} = \beta(n - y). \tag{1.9.10}$$

In this case the curve of adoptions is

$$y(t) = n(1 - e^{-\beta t}). \tag{1.9.11}$$

However, for physicians who had considerable contact with colleagues, the curve was different (Fig. 1.9.4). In addition to the constant source of information (β), these doctors may have had access to internal sources of reciprocal information: this resulted

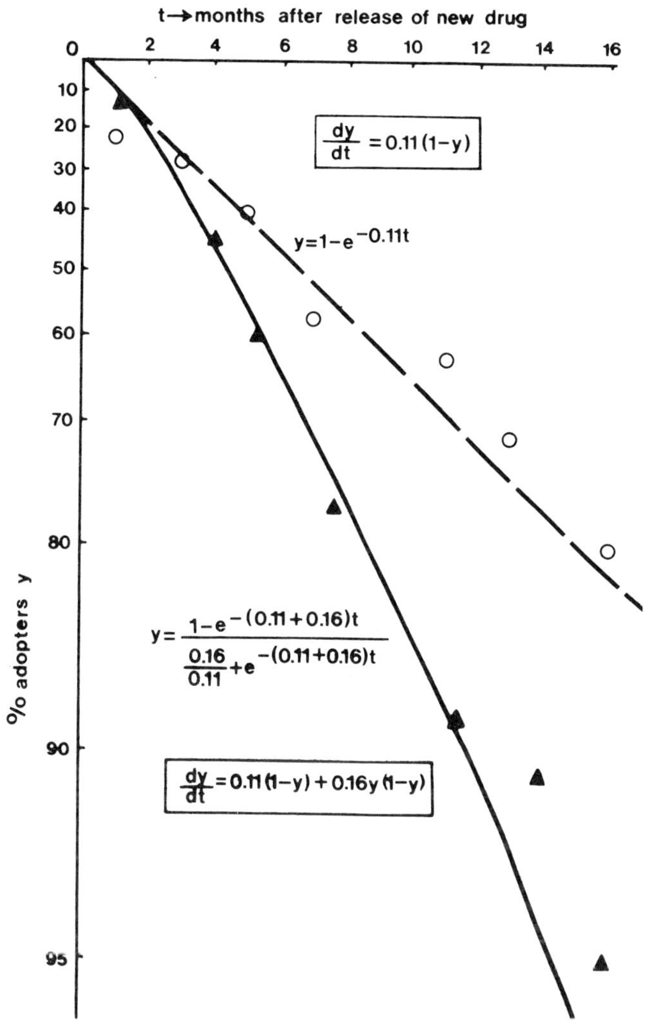

FIGURE 1.9.4. Data from Coleman et al. on adoption of new drug among 53 physicians with low (circles) and 43 physicians with high (triangles) social participation. The percentage of adopters, y, is indicated on the ordinate on a $\log_e y$ scale. The hypothesis fitted to the group with low participation is that of independent, constant probability of adoption. The hypothesis fitted to the group with high participation is that, in addition to the same process of independent adoption, with the same rate as the other group, there is a logistic, socially conditioned process of acceptance. The fitted curves are the integrals of the boxed differential equations.

in the addition to the rate of a term expressing the autocatalytic logistic growth curve

$$\frac{dy}{dt} = \beta(n - y) + \gamma y(n - y). \tag{1.9.12}$$

On integration this produces a slightly more complicated curve, shown in Fig. 1.9.4.

(iv) *The proportionality constant must not only be valid over time, but must also apply for all individuals.* This requirement may pose a major problem, for in some cases there is a clear indication that it is not true. In fact, in their analysis, Rogers and Schoemaker (1971) choose an interpretation that is completely different from the one underlying the logistic. They propose that the reason for variation in time of adoption is not at all a consequence of the interactions between members of the population that happen according to a specified rule of contact, as in the logistic. Rather, they suggest that it is due entirely to the variation in individual tendencies. The early adopters are a peculiar class of individuals, the "pioneers," who like to try new things. They are closely watched by the next in line, who will adopt if convinced that the results obtained by pioneers are worth it. The mass of adoptions then occur and the tail is formed by "laggards." Thus the S-shaped curve of adoption reflects an underlying variation in the ability to accept. It is nothing but the cumulative probability curve of the tendency to adopt. Thus, according to Rogers and Schoemaker, the kinetics of adoption is entirely reducible to the variation in the tendency to learn of or accept an innovation (Figure 1.9.5).

The choice between the "statistical" interpretation, based entirely on individual differences, and the kinetic model that produced the logistic curve, cannot be made from the shapes of empirical curves alone. The bell shaped, Gaussian (or normal) curve of variation, if used to fit the variation of individual proneness to learn of or accept an innovation, would lead to an

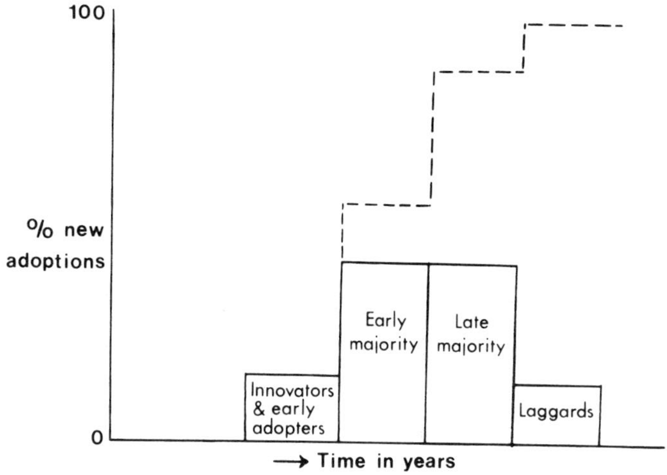

FIGURE 1.9.5. The interpretation of the kinetics of adoption on the basis of individual variation in willingness to adopt. The cumulative curve of adoptions is dotted. The frequency distribution of *new* adoptions is the derivative of the cumulative curve of adoptions. (Modified from Rogers and Schoemaker 1971.)

S-shaped cumulative frequency distribution of acceptance very difficult to distinguish from the logistic. Now there are very many types of distributions of individual variation that could be invoked, and could explain any observed curve of adoption. The choice between these hypotheses must be made in more subtle ways than by mere fitting of curves, perhaps even to the extent of an analysis of the process at the individual level.

It seems very likely, *a priori*, that there is variation between individuals in their capacity both to learn of an innovation and to decide for adoption. Many factors contribute to such variation, including social and economic stratification, geographic conditions such as means of transportation, availability of communication networks, and, last but not least, individual differences in the behavioral characteristics that govern both awareness and eventual adoption. The capacities to become aware of and to adopt an innovation may even be correlated. Ordinarily such a correlation would be expected to be positive, but it might turn out to be negative. For instance, if during the process of adoption by the population there is increased pressure to adopt on the later adopters, the correlation may well be negative.

(v) *The population is assumed to be spatially homogeneous.* One source of heterogeneity that has received some special attention is geographical ("spatial") partitioning of the population. The purely logistic model, and in general the epidemic models, assume the homogeneity of the population, or "perfect mixing," so that every individual in the population has an equal chance of eventually contacting every other individual. The spatial distribution and movement pattern of a population may be serious limitations to the validity of the hypothesis of perfect mixing. Contacts are ordinarily more probable between individuals closer in space, and it is often reasonable to assume that movement of individuals over short intervals of time is not very pronounced. An important treatment that takes account of spatial spread under simple conditions was suggested by R. A. Fisher (1937) and might apply to any of the following situations: (i) the diffusion across space of an advantageous gene; (ii) the spread across space of an epidemic or (iii) rumor or (iv) innovation; (v) the growth and spread of a population from some point of origin. In all cases it is assumed that migration is isotropic, that is, equal in all directions (see also Kendall 1965; Skellam 1951).

In Fisher's model individuals migrate according to a Gaussian law. This means that the change in the concentration variable p is given by

$$\frac{dp}{dt} = M \frac{d^2p}{dx^2} \tag{1.9.13}$$

where in case (i), p is the local gene frequency; in case (ii), p is the local frequency of people affected by the disease, or in (iii), having heard the rumor, or in (iv) having adopted the innovation. In case (v), p is the local population density. The parameter M is a "diffusion" or "migration" coefficient, and can be identified with the variance of the distance migrated per individual in unit time. Note that in each case p is a function of time, t, and the spatial variable x, where the latter might be a vector if more than a single dimension is studied.

If the only population change were Gaussian diffusion from an initial source, then in one dimension the curve $p(t, x)$ would represent, for changing t, a series of Gaussian curves that became flatter as t increased. (See Figure 1.9.6A.) But in the examples mentioned above there is an additional process causing growth of p. In (i), this comes from carriers of the advantageous gene at each location, in (ii) it represents the growth of the proportion of the

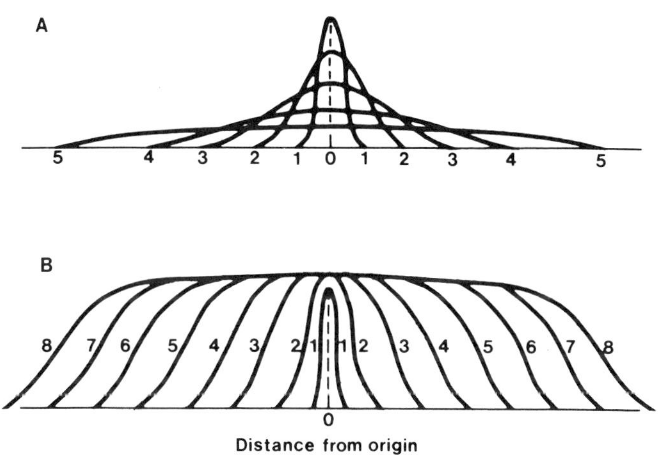

FIGURE 1.9.6. Diffusion processes from a source in position labeled at zero in the abscissa. Ordinate is population density at distance x from source, at time indicated by numbers written near respective curves. In A there is no growth accompanying diffusion: near center density decreases as diffusion proceeds with time. In B there is growth in addition to diffusion, and local saturation (the maximum local height) is always reached eventually. This is the process modeled by R. A. Fisher, and others (see equation 1.9.11).

infected, and so on. If the logistic law were assumed in the perfectly mixing population, $dp/dt = \alpha p(1 - p)$, then the combination of the growth and diffusion would be described by

$$\frac{dp}{dt} = M\frac{d^2p}{dx^2} + \alpha p(1 - p). \tag{1.9.14}$$

R. A. Fisher (1937) showed that a solution of the differential equation (1.9.11) is the equation of a wave moving at speed from the origin as in Figure 1.9.6B. The asymptotic speed with which the wave moves across space is $\rho = (2\alpha M)^{1/2}$.

There have been successful applications of this principle in ecology, archaeology, and epidemiology. The archaeological example given by Ammerman and Cavalli-Sforza (1971, 1977) is the spread of an important complex of innovations in food production, namely the domestication of plants and animals; these innovations involved replacement of the age-old hunting-gathering food economy with that of food production by agriculture and animal breeding. The complex set of innovations involved here spread from cities of origin, not so much as an idea, but, at least in part, as a wave of people, the farmers themselves. This process was of course a slow one; it took over 3,000 years (from about 9,000 to 5,500 years before the present) for agriculture to spread from an area of origin in the Middle East to Europe. (See Figure 1.9.7). Other archaeological processes of diffusion where the innovation itself spread—for example, the use of copper or of iron—were much faster.

It is remarkable that, in spite of the nonisotropy of the European area of diffusion with its many rivers, seas, and mountains, the process of spread of farming appears to have been relatively smooth and homogeneous. Nonisotropy can be detected in the slightly faster spread westward via the Mediterranean than northward, and in the delayed arrival of agriculture to the mountainous regions of Switzerland and to Scandinavia. The spread of many infectious diseases, ranging from the plague in

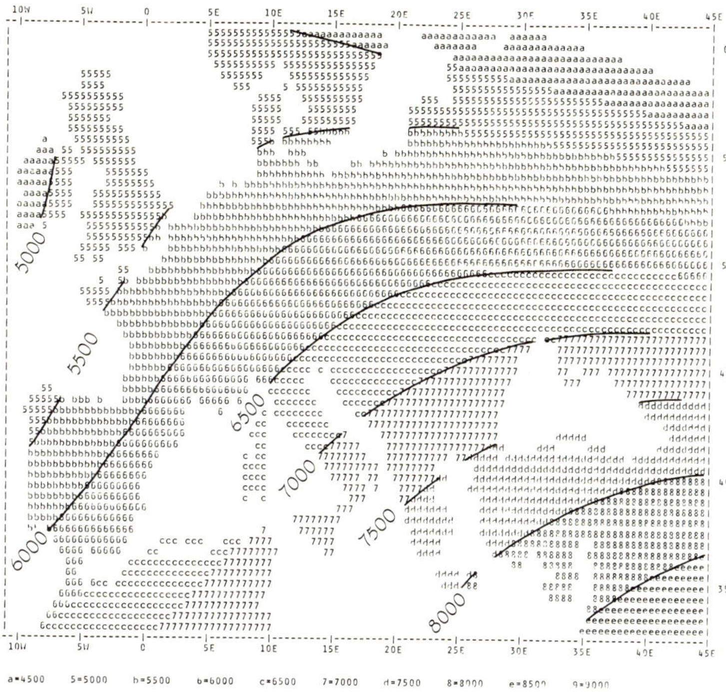

FIGURE 1.9.7. The spread of early farming from the Near East to Europe. A surface has been interpolated by least squares to first-arrival dates obtained by radiocarbon (uncorrected). The lines indicated are isochrons (i.e. lines of equal time of arrival), for the years before present, given in the figure. Data are those used by A. Ammerman and L. Cavalli-Sforza (1971) but some new sites have been added and a more general type of surface has been fitted that does not assume the hypothesis of the constant rate of spread. The isochrons are limited to the range for which archaeological data are available. Numbers are years before present.

Europe during the sixteenth century to many plant infections, was also relatively smooth and isotropic (see Figure 1.9.8 and Zadoks and Kampmeijer, 1977).

In more modern situations the spread follows less direct contact routes. An extreme example is the spread of TV stations in the United States, which were first located in the larger cities, New

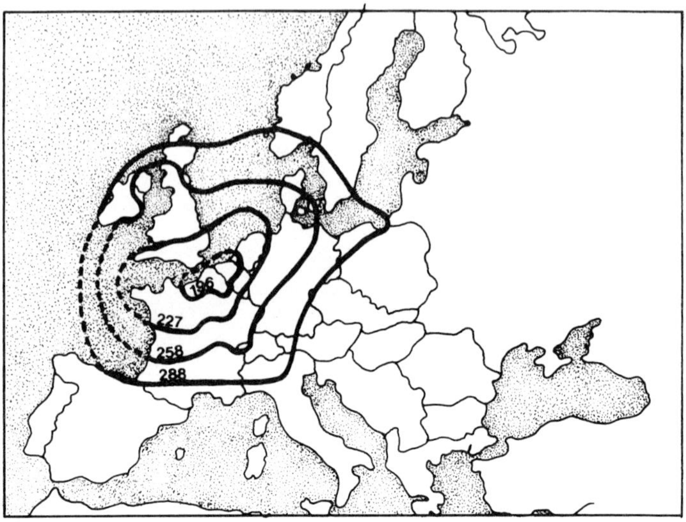

FIGURE 1.9.8. The spread of potato blight in Europe in 1845 (after Bourke, in Zadoks and Kampmeijer 1977). The numbers are days since the beginning of year 1845.

York, Los Angeles, Chicago, then in progressively smaller cities. Here economic reasons have dictated the spread by city size rather than by geographical location.

A classic study of the spatial spread of innovations was made by the Swedish geographer Hägerstrand (1967). In an analysis of a small rural area in Sweden, Hägerstrand followed the diffusion of a variety of agricultural, technological and administrative innovations, including the antitubercular vaccination of cattle, and the use of tractors, telephones and postal checking accounts. The spread was followed jointly in time and space, and various models were constructed to test it. An interpretive device basic to Hägerstrand's analysis is the "information field" (see Figure 1.9.9). Using data from migrations or from telephone communications, the probability of contact between individuals living in different locations can be computed as a function of distance between

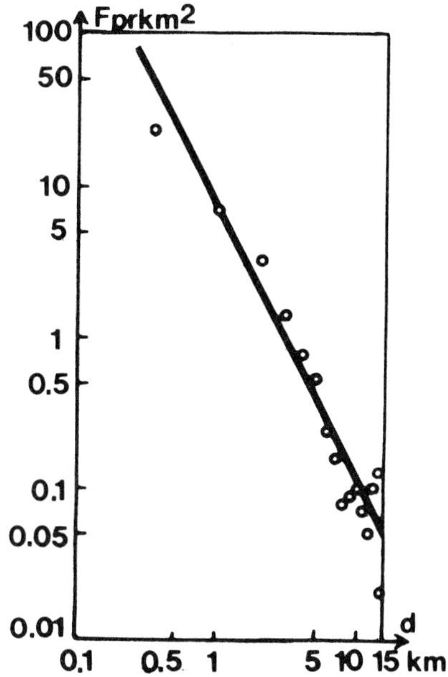

FIGURE 1.9.9. Research by Hägerstrand (1967) has helped establish the concept of "private information field." This is studied indirectly by ascertaining individual migration distances, range of communication by telephone, and so on, for a specific population. It is found that distance has an important effect on all types of contacts. The curve after Hägerstrand represents the number of migratory "units" (migrant individuals or families) per km^2, F, in the area of Kisa, Sweden, who have moved to another area at distance d, indicated in the abscissa (log-log plot). The ordinate represents the number of migratory units per unit area of destination (calculated as a ring of one km width). The curve is represented adequately by the empirical expression $y = 6.35\ d^{-1.8}$. Similar curves are obtained for telephone calls, etc. The numerical constants differ for area and phenomenon studied but the exponent is ordinarily not far from 2.

permanent residents. The district under analysis, or more usually a suitable small fraction of it, was subdivided into squares, and the number of migrants moving from the center of the area, that is, the central square, to any other square, is recorded in the cells of a matrix. The same can be done for telephone communications, using incoming or outgoing calls, or some combination of both. The matrix so constructed is specific to that particular place located at the center of the matrix. Matrices from many different places in the district under study can be employed in order to obtain a representative picture of the study area. Migration data and telephone data give similar information fields, though the modes of dependence on distance in the two cases are different.

The information fields thus obtained serve to predict how far, in a given time unit, news of an innovation can travel. The adoption curves can then be predicted under a variety of assumptions. It is of interest that for those innovations of exclusively agricultural interest, it was necessary to introduce variation in the individual "resistance" to adoption in order to explain the observed curves. Individual variation is thus again found to be of importance, even though other limitations to the process of perfect mixing, namely the spatial distribution of the population, have been taken into account.

1.10 EPIDEMICS

The mathematical study of epidemics has produced an actively developing theory that has, perhaps, not received all the attention at the practical level that it deserves. The theory has been developed in two parts, one involving infinite populations (deterministic theory), the other finite populations subject to random effects (stochastic theory). We limit our considerations here to some major aspects of the former. A recent fuller exposition is by N.T.J. Bailey (1975), and the major deterministic models are reviewed in Waltman (1974). Most of the theory is couched in the form of differential equations in continuous time rather than the discrete time-recursion systems used in most of our treatments.

The simplest theoretical formulation of an epidemic assumes a population of N susceptible individuals, of which, at the commencement of the process, one or more becomes infectious.

The disease is propagated on contact between a susceptible and an infectious individual. Not all contacts are necessarily successful; a contact that is successful is termed *effective*. The rate of effective contact is α; it is the product of the probability that a contact occurs and that, having occurred, it results in propagation. The first of these probabilities is clearly a property of the nature of social interactions in the population. The second depends on the type of trait considered. It is assumed that any two individuals have the same probability of contact, and thus there is "perfect, or homogeneous mixing." This is analogous to "mass action" in chemical reaction theory.

In the simplest case the population consists of N individuals divided into two classes S and I as follows:

S is the number of susceptibles who have not contracted the disease but are capable of doing so.

I is the number of infectious individuals who have contracted the disease and can pass it on to susceptibles. In the simplest model an infected individual is also infectious.

Under the above hypotheses, the increase in the number of infectious individuals ΔI, in the small time ΔT, is

$$\Delta I = \alpha SI\Delta t, \qquad (1.10.1)$$

giving rise to the differential equation

$$\frac{dI}{dt} = \alpha I(N - I), \qquad (1.10.2)$$

since $I + S = N$. This equation (1.10.2) describes the logistic curve given earlier for the number of infectious individuals at one time t.

This two-class model is too simple to describe accurately the course of any infectious disease in a population. Such complexities as the existence of an incubation period after contagion must be recognized, and there are treatments that take this into account.

More important is the fact that the disease may last for a relatively short time, at the end of which the patient either dies or recovers. If the patient recovers and remains immune to the disease, then from the point of view of susceptibility an individual is lost, just as if the patient were dead. Thus we should include a third class of individuals in our total N, namely R, the number of either dead or immune individuals together designated as removed. Then $S + I + R = N$. The sequence of events is then $S \to I \to R$.

If the process of removal occurs at the constant rate β (an obvious oversimplification as it does not take account of variation in severity of the disease, relapses, etc.), then we have

$$\frac{dI}{dt} = \alpha SI - \beta I \tag{1.10.3a}$$

$$\frac{dR}{dt} = \beta I \tag{1.10.3b}$$

with the third equation fixed by these two since $S + I + R = N$ is constant:

$$\frac{dS}{dt} = -\alpha SI \tag{1.10.3c}$$

The following are the main qualitative properties of the system, in terms of S_0, the initial number of susceptibles.

(1) There is a threshold for the epidemic to start. In fact, for $dI/dt > 0$ it is required that $[\alpha S_0 - \beta] > 0$. Thus there must initially be a number S_0 of susceptibles not less than β/α, that is,

$$S_0 > \beta/\alpha = \rho. \tag{1.10.4}$$

The quantity ρ, the ratio of the removal rate β to the rate of infection α, is therefore responsible for setting a threshold to the initial number S_0 of susceptibles in the population. If $S_0 < \rho$, then it can be proved that $I(t)$ decreases monotonically to zero.

(2) If $S_0 > \rho$, the number of infectives increases as t increases,

and then tends monotonically to zero. $S(t)$ decreases and converges to the positive limit which is the unique root, S_∞, of the transcendental equation in Z:

$$S_0 \exp\left[\frac{-1}{\rho}(N - Z)\right] - Z = 0 \qquad (1.10.5)$$

(3) R, the total size of the epidemic, is monotonically increasing and converges to $N - S_\infty$. An approximate estimate of R_∞ is $2A$, where $A = S_0 - \rho$.

(4) The epidemic curve is bell-shaped, but not exactly symmetric like its simple counterpart (1.10.2). It is, however, close to symmetric unless the initial number of infectives is high or unless the total number of susceptibles that eventually contract the disease is large. A classic example is given in Figure 1.10.1.

More elaborate models with more categories of individuals or with varying dependence of sensitivity to contagion and degree of infection have been developed. The major application of these models in epidemiology is potentially to predict optimal methods for the control of infectious diseases. Choices must be made between expenditure on vaccination prophylaxis, vector eradication, and the like. It is too early to estimate how much practical gain can be achieved from the theoretical framework, but in some empirical circles models like those described above and relatively simple extensions of them are starting to receive attention.

The machinery necessary to describe the course of events in the spread of customs or innovations in a quantitative way is at least as complex as that required for epidemics. As stated in the previous section, translation into terms of cultural change involves substituting for infection the state of awareness and for removal the state of adoption. (Indeed some epidemiologists refer to the study of "epidemics or rumors," meaning that their treatment covers both phenomena equally well.) The sequence of states may be more complex than just these two, for instance awareness →

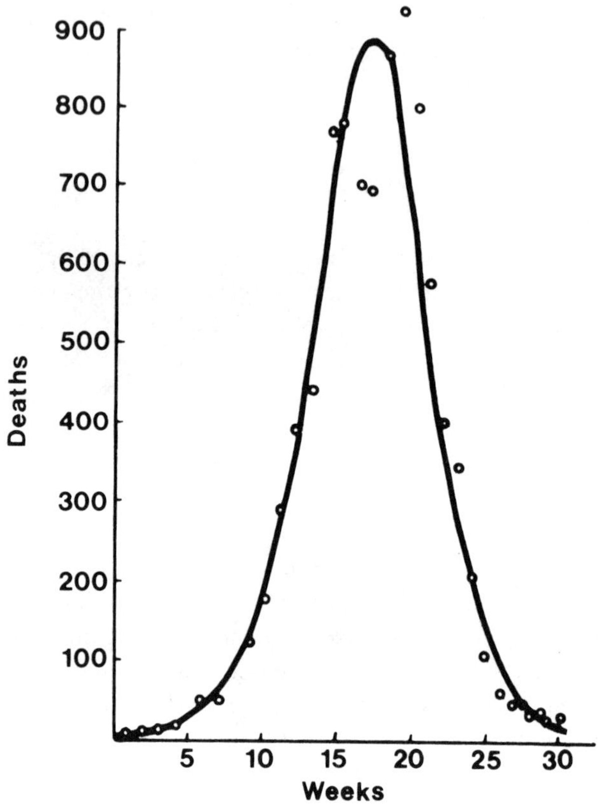

FIGURE 1.10.1. An epidemic of plague in Bombay: number of new deaths per week. Fit obtained by Kermack and McKendrick (1929) using approximate solution of the system (1.10.6) $dR/dt = 890 \operatorname{sech}^2 (0.2t\text{-}3.4)$ (after Waltman 1974).

trial → adoption → rejection. It should be remembered that this theory does not involve any variation in susceptibility at any stage and that this may be a severe practical limitation.

There is another implicit assumption that can be removed, and already has been in the study of epidemic diseases. When the duration of the disease is long, the study of the process may demand consideration of changes in the population due to birth and death. In this case it may be that the epidemic disease will tend to persist in the long run, with the population reaching a state of equilibrium in which there is a constant proportion of diseased. This is the case with *endemic* diseases (habitually prevalent in a country, for example tuberculosis in Europe until World War I, malaria in many parts of the world, leprosy in central Africa). The theory of endemic diseases was first studied analytically in 1760 by Daniel Bernoulli (although the work was actually published in 1765), who was concerned with the efficacy of vaccination against smallpox, which was then endemic in Europe.

The major difference between the mathematical models of endemic and epidemic diseases is that, in the latter, the population is closed. In the former, although the population size N may remain constant, new individuals arrive in the population through a birth process, but are balanced by an equal death rate, denoted by r. In the simplest model $S \rightarrow I \rightarrow R$. With $S + I + R = N$ we now replace (1.10.3) by

$$\frac{dS}{dt} = rN - \alpha SI - rS$$

$$\frac{dI}{dt} = \alpha SI - \beta I - rI$$

$$\frac{dR}{dt} = \beta I - rR, \tag{1.10.6}$$

All individuals are born susceptible, and therefore dS/dt contains the positive term rN. It is assumed that S, I, and R all die at the same rate r, which explains the negative r-terms. The other terms are exactly as in the closed population epidemic model. It should be noted that the identity of the death rate, r, in the three categories S, I, and R, makes the model applicable only to diseases that have a negligible effect on survival. The above model, originally due to Dietz (1972, 1976), is in this respect well suited for use in the diffusion of innovations, rumors or habits, at least because it is probably the case that these do not often have an important effect on the Darwinian fitness of the individual. Following Dietz's treatment, let us set

$$x = S/N, \quad y = I/N, \quad z = R/N \tag{1.10.7}$$

so that $x + y + z = 1$. Apart from the trivial solution $x = 1$, $y = z = 0$, where all individuals are susceptible and the infection is absent, the equilibrium of (1.10.6) (obtained by setting all the derivatives equal to zero) is

$$\begin{aligned} \hat{x} &= 1/A, \\ \hat{y} &= (1 - 1/A)/M, \text{ and} \\ \hat{z} &= (1 - 1/A)(1 - 1/M) \end{aligned} \tag{1.10.8}$$

where

$$A = \alpha N/(r + \beta), \quad M = 1 + \beta/r,$$

and is valid for

$$A = \alpha N/(r + \beta) > 1. \tag{1.10.9}$$

The condition (1.10.9) for the validity of the nontrivial equilibrium (1.10.8) is merely the condition that the number of infectious individuals increases when one is introduced into a totally susceptible population. Indeed, A is the number of infected individuals introduced per unit time under these conditions, and unless A exceeds 1, the disease, or rumor, will not establish itself. As an example, in an application to the study of the endemic disease rubella, A was found to be 7.65.

The system of equations (1.10.6), like (1.10.3), has no age structure built into it. The age at death follows a simple negative exponential distribution, as a consequence of the constant mortality rate r. The average length of life is $1/r$. Now if D is the average age at which disease is contracted, then it can be shown (N.T.J. Bailey, 1975; Dietz, 1976) that

$$A = 1 + 1/(rD). \tag{1.10.10}$$

The above mentioned estimate for rubella was derived by taking an estimate of average age at infection, $D = 10.5$ years and $1/r = 70$ (if normal life terminates on average at that age). Naturally, the distribution of the human lifespan, especially today in western society, is far from a negative exponential, so the above estimate of A is probably rough. It is, however, larger than 1, as would be expected, given the endemic nature of rubella in most populations. The relation (1.10.10) should be used with caution, since the approximations are severe (e.g., it gives $A > 1$ for any positive age at contracting the disease).

The theory of endemics is still in its infancy. It may be of importance not only for explaining the maintenance of endemic disease, and the effects of prophylactic measures, but also for improving our understanding of the maintenance of social customs and habits that are spread like infectious diseases and, like the endemic ones, are present over long periods of time.

1.11 CULTURAL TRANSMISSION

From the history of biology it is clear that the theories of evolution and of transmission can be developed to some extent independently of each other. Darwin's (1859) theory of evolution by natural selection was constructed in the absence of knowledge about genetic mechanisms and, indeed, in spite of incorrect postulates concerning biological transmission. The correct mechanism was discovered by Mendel (1865), who did not use evolutionary knowledge. The theory of biological evolution, however, could develop a rigorous and quantitative framework only after Mendel's transmission laws were incorporated, and a mechanism of variation (by mutation) was postulated.

Similarly, theories of cultural transmission and evolution can, to some extent, be developed independently of each other, although for a complete theory of cultural evolution rules of cultural transmission are essential. We will begin our study of cultural transmission with a classification of different mechanisms from which laws might be derived.

As outlined in the previous section, the theory of epidemics has many parallels with that of the diffusion of innovations. The theory of endemics may be even more relevant, especially for the study of the maintenance of beliefs, values, and the like. Thus, epidemiology is an important source of inspiration for the study of cultural transmission, and in this section some of the terminology used in epidemiology will be adopted for use in our theory of cultural evolution. *Vertical* transmission is used to denote transmission from parent to offspring and *horizontal* transmission denotes transmission between any two (usually unrelated) individuals. Here the same adjectives will be used to apply to cultural traits. We will, however, use the term *horizontal* as restricted to members (related or not) of the *same* generation, and in addition we introduce the word *oblique* to describe transmission from a member of a given generation to a member of the next (or later) generation who is not his or her child or direct descendant. Clearly genetic (chromosomal) transmission is strictly vertical (and in addition has a very precise mechanism). A few examples of vertical nonchromosomal transmission are known in animals, and a few have been described in man. As a human example, the Kuru virus has been transmitted mostly from parents to children in ceremonial contacts with dead relatives (including cannibalism) in the Fore tribe of the New Guinea highlands. Figure 1.11.1 shows an example of a pedigree of Kuru. The resemblance with chromosomal transmission is clear. Until it was shown that the disease can be transmitted to chimpanzees by inoculation of the brain from dead Kuru patients it was not unreasonable,

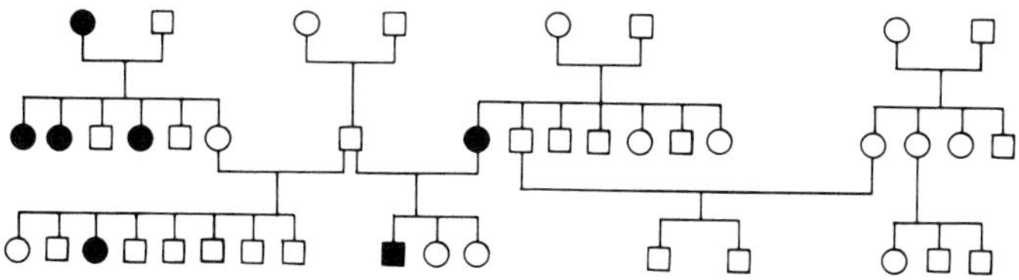

FIGURE 1.11.1. A pedigree of kuru. Circles: females; squares: males. Filled circles or squares: deaths from kuru. The individuals at right come from tribes free from the disease. After Bennett et al. (1959).

because of its behavior in pedigrees, to consider it as a potentially genetic disease.

Cultural transmission must have been primarily vertical for much of human evolution. For more than 99% of human evolution social groups were of small size on average, and we might speculate that during this period the cultural influence of members of the social group other than the parents or members of the group on a given individual was probably small. Following the introduction of agricultural practices some 10,000 years ago, social groups started increasing in size and complexity. The resultant social stratification into social classes, castes, age groups and other hierarchies must have continuously increased the importance of oblique and horizontal transmission for the coordination and maintenance of the group as a community.

The following is a survey of the major modes by which cultural transmission can occur.

(1) *Vertical,* that is, parent to offspring. Unquestionably there is even today cultural transmission from parent to child and in earlier periods of human evolution it probably was a major component of cultural transmission. The high specialization of activities and division of labor among the sexes may, for some traits, make the transmission *uniparental,* though not necessarily always unisexual. More generally, both parents may contribute,

sometimes in unequal proportions (say p_1 and p_2), to the trait of the child (*biparental* transmission).

(2) *Foster parent–foster child* transmission may be different from the above when there is biological differentiation in the capacity to learn. In such cases, the comparison of foster parent–foster child pairs with natural parent–natural child and natural parent–foster child (raised in another family) pairs may be the only way to disentangle the effects of biological and cultural transmission.

(3) *Family members other than parents* may contribute to the education of the children. They usually belong to the parental generation, in which case we have an example of oblique (intrafamilial) transmission.

(4) *Members of the social group other than the family,* also belonging to the parental generation, may serve as models for children. This is another example of oblique (extrafamilial) transmission.

(5) *More remote generations* can contribute. In the absence of writing or of a strong oral tradition, this contribution will be mostly made by grandparents.

(6) *Sib-sib* interactions are certainly of importance and may complicate the outcome of the previous forms. This is purely horizontal (intrafamilial) transmission, but might interact with vertical transmission.

(7) *Age peers* constitute the group with whom much contact occurs, especially during the formative years. In (4) and (6) above there would usually be a substantial difference, if not in age, then in experience between the teacher and the learner. Among age peers the differences in age and experience tend to be small, but the relationship of psychological influences may be highly asymmetric. This is a most common type of horizontal (extrafamilial) transmission.

(8) Wherever schools exist, the *teacher-pupil relationship* centralizes the source of information into one or a few individuals,

usually of an earlier generation (and therefore oblique), who transmit to many recipients. Also, wherever there is a political organization (even at the simplest tribal level), decisions are usually made and the spread of information is directed by one or few people, again to a potentially larger group.

If there is a *social leader* or *teacher* of the group, then interactions between this individual and others can be represented by a bunch of arrows departing from this individual toward all others.

1 teacher or leader

↙ ↓ ↓ ↘

2 3 4 5

This can also be written as a matrix, in which columns represent the communicators, and rows the recipients of information:

		Communicators				
		1	2	3	4	5
	1	1	0	0	0	0
	2	1	0	0	0	0
Recipients	3	1	0	0	0	0
	4	1	0	0	0	0
	5	1	0	0	0	0

We have chosen to set the first column at 1 for reasons that will be clearer later. Individual 1 transmits to everyone in the group at the next generation. He may be from an earlier generation, or from the age peer group.

(9) With the increase in political organization, there will be a *social hierarchy* that further complicates patterns of transmission. A complex, multilayer hierarchy is necessary in a large group; in the absence of mass communication, the immediate audience of a

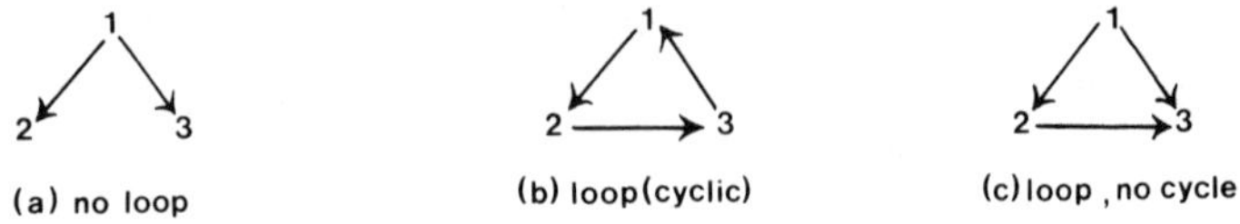

FIGURE 1.11.2. Examples of relationships among three individuals, represented as graphs and corresponding to the three matrices shown in the text.

speaker is limited in size, and hierarchies are an efficient way of spreading information. The flow of information or influence can be represented as a graph with arrows indicating the direction of flow or influence, as in Figure 1.11.2. This can be formally represented by the following matrices, in which every arrow between individuals i and j is represented by 1 in the ij cell, all other cells being zero.

$$\text{(a)}\begin{pmatrix}1 & 0 & 0\\ 1 & 0 & 0\\ 1 & 0 & 0\end{pmatrix}\quad \text{(b)}\begin{pmatrix}0 & 0 & 1\\ 1 & 0 & 0\\ 0 & 1 & 0\end{pmatrix}\quad \text{(c)}\begin{pmatrix}1 & 0 & 0\\ 1 & 0 & 0\\ 1 & 1 & 0\end{pmatrix}$$

When loops are absent and transmission occurs in one layer only, the transmission matrix can be represented in a generalized "teacher" form as given in the matrix (a) above. If leaders exist, they are assumed to have self determination, that is, to be influenced only by themselves, as in matrix (a) and (c) above. The graphs in Figure 1.11.2 are extremely simple. Typically the hierarchical model is likely to have several layers, for example,

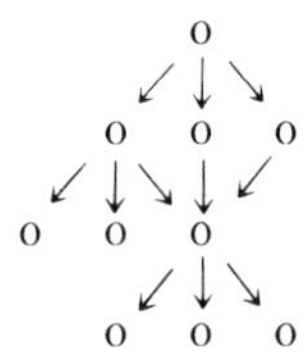

The simplest hierarchical matrix is

$$\begin{array}{c} 1 \\ \downarrow \\ 2 \\ \downarrow \\ 3 \end{array} \quad \text{or} \quad \begin{pmatrix} 1 & 0 & 0 \\ 1 & 0 & 0 \\ 0 & 1 & 0 \end{pmatrix}$$

For more details on the use of graphs and matrices in social networks, see Roberts (1976).

(10) *Telecommunications* have had a double effect. By increasing the range of direct two-way communication between any two individuals, telecommunications have increased the size of the social network enormously, and potentially produced a transmission matrix between all living individuals. A $4 \cdot 10^9 \times 4 \cdot 10^9$ matrix is a real challenge even for modern computers, so this representation would require great simplification for numerical analysis. Of course the number of two-way interactions for a given individual would usually be very small. The second effect of telecommunication has been to increase the range of one-way communication, so that a single individual can communicate in *one* direction with huge numbers of individuals. The advent of printing, the availability of newspapers and of the mass media in general greatly increases the audience of a single speaker and may eliminate many of the intermediate steps which would otherwise be essential in a complex society.

(11) All societies are organized in groups that partially overlap, or communicate. Between some groups, however, overlap or communication may be minimal. Separation into groups with poor or absent intercommunication creates opportunities for independent evolution, both biological and cultural.

These modes of transmission can obviously interact and produce transmission matrices of great complexity. In the next chapters we shall consider each major mode of transmission by itself, and some of the simpler interactions between them, that might describe real situations. A well-developed concept that can usefully be applied to the study of social transmission is that of social networks. Social networks are commonly employed to describe the relations between members of a selected group of individuals. Normally the directionality of the relationship between members of a given pair is not indicated in social networks. In principle, therefore, matrices corresponding to social networks should be symmetric. But it is possible that the data may not be symmetric even when no directionality is considered in the model. Thus person *a* may report contacts with *b*, *c*, and *d*, but *b* may report contacts with *e* and *f* but not with *a*.

For the purposes of cultural transmission, the elements a_{ij} of the matrix should refer to either the probability that person i is made aware by person j of a given fact, or that person i is influenced by person j to decide on adoption of a given behavior; or they may refer to the joint probability of these two events. The matrix corresponding to a given social network can then be written as a transition matrix for a Markov process (as exemplified and discussed in more detail later), that is, the sum of the elements of each row must be one. Each coefficient a_{ij} thus represents the probability that the ith individual will be like the jth individual at the next time point. The principal diagonal represents the probabilities a_{ii} that the ith individual will not change spontaneously in a time unit. Such a notation is appropriate if the trait being studied can be in one of two states. For a metric trait the elements in the row may take the significance of weights to be used in computing the trait value of the ith individual in the next time interval as a weighted average of the trait values of all individuals in the current population. We shall call the elements of the stochastic matrix a_{ij} $(\Sigma_j a_{ij} = 1)$ the coefficients of cultural transmission from individual j to individual i, and the matrix, the *transmission matrix*.

Under these conditions the matrix may have a number of rows

and columns equal to the number of individuals in the group between whom cultural transmission is possible. A limitation of this representation is that the matrix would have the addition or deletion of a row and column with the birth of a new individual or the death of an old one. This is not easily incorporated into the discussion. Thus this kind of matrix representation is useful especially for short term phenomena in which addition or deletion of individuals can be neglected. At most, individuals who die in the observation period (or who were dead at the beginning but were still influential in some way) may be represented as columns to which there corresponds no row. Vice versa, individuals who do not participate in the teaching process but only at the learning level may be present as rows and not as columns. As a result the relevant transmission matrices may be rectangular. When long term change over several generations is considered, and transmission is only vertical or oblique, but not horizontal, columns can represent individuals of the older generation, and rows their offspring.

With simple linear assumptions, the use of transmission matrices allows the cultural transmission process of the whole population to be iterated easily over generations, so that projections can be made as to the long-term evolution of the system. The population size here is the size of the square transmission matrix and is constant, as is the matrix itself. These assumptions can be relaxed, but are satisfactory for our initial exploration. The diagonal elements then have the role of coefficients of cultural transmission from parent to child in a unisexual scheme, for example, father to son or mother to daughter. When bisexual transmission or mixed systems of horizontal, vertical, and oblique transmission are considered, the coefficients have to be defined in a more complicated way and different recursive processes result.

There are thus at least four criteria that are useful in the classification of transmission rules.

(i) Relationship of teacher and taught;

(ii) age differences, which will be modeled as generational differences between teacher and taught (remembering that a cultural generation may not be the same as a biological one);

(iii) numerical relations between teachers and taught (one to one, one to many, and so on);

(iv) complexity of society. This may involve social structure (several hierarchical layers) as discussed above. It may also involve subdivision into castes or classes that are to some extent segregated.

Many of the following chapters will be dedicated to a formal study of transmission and the evolution expected under the various conditions which we have briefly summarized in this section.

1.12 TRANSMISSION AS A TWO-STAGE PROCESS

Cultural transmission—the acquisition by one individual of a trait from another individual—may involve long and complex learning processes. These processes may in practice be wholly or partially reversible. Our models deal with traits that do not change after the process of learning is complete. This can be accomplished by studying the population at the same age in every generation—an age at which all individuals are mature for the trait under study.

There are two stages in this transmission process, analogous to those referred to above in the remarks on diffusion of innovations. The first stage is awareness, which requires the existence of a signal (via teaching or observation), and the second is acceptance (or learning). It is at the second stage that cultural selection may occur.

For many traits the two stages, awareness and adoption, are not clearly distinct. We do not refuse to learn our native language. The very few individuals who are entirely unwilling to communicate are very unlikely to reproduce and are eliminated by Darwinian selection. Some of the routines imposed by the socialization process may be rejected, but there is clear danger involved in nonconforming, in that individuals who do not accept a significant proportion of these routines, for example, personal hygiene or standard rules of conduct, may be discriminated against and therefore may have a lower chance of finding mates and reproducing. They may be segregated from the rest of society if their behavior is considered inadequate or dangerous, or they may die before reproducing for one reason or another. But these are relatively rare cases. The tendency to conform may have very many different origins; a genetic origin is conceivable, but difficult to demonstrate. Whatever the origins, individuals "deviant" beyond a certain degree are heavily penalized by society. The majority of people do accept most of the teachings that constitute the norms of society. Some of these are so important that natural selection may well have favored their acceptance: The acquisition of language may be one example for which the distinction between awareness and acceptance is dim. Imprinting is another learning situation in which the distinction is hard to make. In other situations, society conditions us to learn certain things, and the predisposition or imposition thus determined is so effective that we do not even notice that we have been conditioned. *The outcome is that many cultural traits are accepted with limited or no choice. In these cases awareness and acceptance are difficult or impossible to distinguish*, and transmission may be truly a one stage process.

A further complication that strengthens the confounding between awareness and acceptance is the possible existence of spontaneous tendencies or drives to teach and thus impart knowledge. Thus, in the transmission process, both the teaching and the learning may be deeply motivated. Ideally a partition of transmission coefficients into three parts might be sought: the drive to teach, which could be expressed as a probability that the act of teaching is carried out under appropriate circumstances; the drive to learn, similarly defined; and the probability that the conditions for the appropriate teacher-learner interactions actually occur. But it is rare that transmission coefficients can be broken down into these components. The confounding is especially deep in situations in which some knowledge or behavior is spread in the population at the command of a religious or political leader (for example, the acceptance of the Fascist salute, or of a new religious dogma, or a decision to pay a fixed tithe or fine). Here acceptance could be called obedience and is determined not by true cultural selection (that is, a true persuasion that the requested behavior is good or adaptive), but by acceptance of the authority of the leaders who directly or indirectly request the act.

Another way in which choice may be circumscribed is through the economic or marketing system. If a certain product or process produces a consistent and acceptable profit, there may eventually be no choice in the market place. Even if white eggs are preferred in Massachusetts, or brown eggs in California, the individual may not realize this preference. Certain eating habits have been transformed as a result of most or all of a certain agricultural product being processed in specific ways. The spectrum of food from which the individual chooses might be determined solely by packaging, transportation, and esthetic constraints, rather than by intrinsic desirability. All of these are ways in which there has been an imposed restriction of the range of alternatives among which a decision is to be made. Such cases of imposition may well destroy or at least minimize the amount of variability in cultural traits, and lead to the adoption of trait values not by choice but by default.

Further constraints on cultural selection result from the social system. Earlier work has stressed the importance of conforming to certain standards. Custom is defined as "the name we give to uniformities, regularities, continuities, etc., in cultural social systems" (L. A. White, 1959). In our terms a custom is any behavioral trait that is transmitted with little individual variation. In White's words, custom is always the absence of novelty, which is avoided because it is disruptive and costly. As such, custom is a powerful means of social integration and regulation, serving as a social badge, a means of identifying societies or classes. A custom is to be distinguished according to whether it refers to ethics or etiquette. The first is a "set of rules designed to regulate the behavior of individuals so as to promote the general welfare, the welfare of the group. Etiquette is a set of rules which recognizes classes within society, etc." (White, 1959). Whether the distinction into ethics and etiquette is necessary and sufficient is not our major point. Rather, we stress the pervasiveness of cultural transmission, and the fact that it seems geared to promote social laws, which fix customs and are thus made for keeping "novelty" (or, in general, variation) within certain bounds. This is achieved by the combined and often conflicting forces of cultural transmission, selection, and cultural mutation.

1.13 A SUMMARY OF EVOLUTIONARY FACTORS IN CULTURE

In the discussion of evolution two evolutionary factors have been singled out: *mutation* and *selection,* which can be regarded as the major contributors to biological and, with the appropriate substitutions in definitions discussed above, to cultural evolution. Given the necessary chemical material and environment, living beings produce offspring via a genetic blueprint that is almost exactly the same as that which gave rise to the parents. For cultural objects, the mechanisms of transmission are quite different. But it is also important to emphasize the distinction between *mutation* in the biological and in the cultural processes. In the former, mutation is typically an error in the copy, which as a result is not identical to the master, or it is a chemical change in the master to be copied. In both cases *randomness* seems to be the rule. On the other hand, in the cultural process, the change is not necessarily a copying error, but can often be directed innovation, that is, innovation with a purpose, and might therefore appear to be nonrandom. This would be a crucial issue only if there were a high chance that a cultural innovation could be truly adaptive as a consequence of being purposeful. All evidence points to this being impossible in biology: biological "innovation" by mutation is apparently truly random. If cultural innovations are not truly random, but are designed to solve specific problems, they may increase the rate of the corresponding adaptation in evolution over that expected for a truly random process. We might speculate, however, that whatever the good faith and insight of the proponents of innovations of any kind, the chance that the innovations will prove truly adaptive in the long run is not 100%, so that many innovations, however purposeful and intelligent they may seem to their proponents and first adopters, may not turn out to be highly adaptive, at least on a long term basis. Because of this, and because some cultural mutation is simply copy error, a significant proportion of new cultural mutations might be truly random without any semblance of adaptiveness.

We now turn to the other major evolutionary factor, *selection*. Darwinian selection could in principle affect any trait of an organism, physical or behavioral, innate or learned, "biological" or "cultural." But a cultural trait is selected at another level before being subject to Darwinian selection. The trait must be culturally "accepted" first. In the discussion of the diffusion of innovation two stages, awareness and adoption, were distinguished with the first stage involving communication and at least some of the learning. However, the final test of fitness is whether the learned trait will be really incorporated into the final permanent phenotype of the individual, or alternatively forgotten, rejected or replaced. In the latter cases the trait evidently does not pass the test of cultural selection.

Apart from mutation (in the forms of copy error or directed

innovation) and selection (both cultural and Darwinian), other factors of cultural evolution can be recognized as important and found to be analogous to those already described in biological evolution. Two such factors are drift and migration. The similarities between analogous factors of biological and cultural evolution have already been noted by various authors (e.g., Gerard et al., 1956; Cavalli-Sforza, 1971). We will add a short description of them for the readers who are unfamiliar with the biological counterparts.

The word "drift" as used in evolutionary biology has a somewhat different, in fact almost opposite, meaning to the same word as used in physics or in linguistics (Sapir, 1921), but the same as in archaeology (Binford, 1963). (This will be discussed in greater detail in Chapter 5.) In biology the term, introduced by S. Wright, expresses the changes due to random sampling processes in populations that are of finite size. It has been suggested that it might be defined as "random genetic drift" or "random sampling drift." In the case of discrete traits, as we discuss in later chapters, much of the theory developed for drift in biological evolution can be used, if some redefinition of constants to account for vertical transmission is made. Certain types of horizontal transmission can also be treated with the same theory.

In general, the effect of random drift alone on small populations is to make each population homogeneous and, with high probability, different from every other population. Being the consequence of random fluctuations, the process is expected to be faster, the smaller the population. In other words, if a trait is variable in a population, it will tend to "fix" under drift, that is, only one type of the trait will survive, and all others will become extinct in that population. But another form of that trait may fix in another population.

A new theoretical treatment is necessary for some types of completely cultural transmission. A continuous trait under selection also demands a different approach. In the chapters dealing with the various types of traits and their transmission, we give an interpretation of classical genetic theory relevant to cultural drift. The other conflicting uses of the term "drift" in cultural evolution will also be discussed.

The process of spontaneous homogenization of individuals of a population and of simultaneous heterogenification of the populations under drift is opposed by recurrent mutation, and also by migration. Migration can play the same role for cultural traits that it does for biological evolution. However, it can take various forms which should be distinguished. In cultural evolution there may be migration of people (with their ideas) or migration of ideas (customs and the like). These migrations might be referred to as *demic* and *cultural.* The latter phenomenon is sometimes called "borrowing" in anthropology, or sometimes referred to as "diffusion," a term which in fact is often employed unspecifically to cover a number of different processes ranging from transmission to spread. Before the existence of writing and telecommunications, cultural migration necessitated the migration of at least one individual who could carry the idea from one place to another. Even today, however, demic migration contributes in an important way to the transportation of ideas, customs, values, and inventions.

One other interesting analogy between biological and cultural evolution is found in comparing migration and mutation. Mutation of gene A to gene A' in a population of all A genes may be indistinguishable, in terms of its evolutionary consequences, from the introduction of an A' gene into a population of only A genes by immigration from another population in which the A' gene is present. More generally, in most formal treatments of evolution, one can replace mutation by migration. As an example, if it is impossible to distinguish by our means of observation the sickle-cell mutation found in Africa from that found in India, it may be impossible to say whether they arose as two independent mutations or as a single mutation that spread from one place to the other by migration. (It happens that, very recently, differences have been found between the two types that make it likely that they arose from independent mutations.) Similarly, it is often difficult to say whether an invention, for example pottery or the wheel, found at different times in different parts of the globe, spread from one part to the other or arose independently. And without knowledge of neighboring languages it may be impossible to say if a new word entering a language is the result of invention or is borrowed from another language.

In summary, our approach to cultural evolution will be through discussion of specific traits, rather than through some overall description of culture. Biological evolutionary theory has generally developed along the same lines: specific features of an organism, rather than the whole organism, have been the focus of attention. The forces of mutation, selection, migration, and drift, which are central to the theory of biological evolution under genetic transmission, have analogues in the specific-trait approach to cultural evolution. The mode of transmission strongly differentiates genetic from cultural evolution. Among the many different mechanisms of cultural transmission, it might be anticipated that some can be extremely conservative, whereas others permit very rapid change. The differences between evolutionary rates and equilibria of biological and cultural traits should be explained, at least in part, by differences in their modes of transmission.

1.14 SOME CAVEATS AND PROBLEMS

Progress in evolutionary biology during this century has been due to the convergence of lines of thought from many different disciplines. Mathematical formulations have often antedated observations and have helped to promote the discovery of interesting phenomena. The interaction between theory and observation has followed a path somewhat similar to that it has taken in physics. Quantitative observations of processes are generally much more difficult to make in biology than in physics, and in areas relevant to evolutionary theory, gross estimates are all that can be expected. Even so, theory has guided both the interpretation of macroevolution—major changes as evidenced from the fossil record—and microevolution—smaller evolutionary changes that can take place over shorter periods and have therefore, on occasion, been amenable to experimentation in organisms with short generation times.

In our view, the possibility of reducing the analysis of cultural change to terms of a few evolutionary factors of general validity can greatly add to the understanding of the phenomenon. Such a formulation does, however, require the limitation of the objects of study to traits that are possibly quantifiable and easily observable. This simplification carries with it the danger often ascribed to "reductionism." Here the reduction is from an undefined "whole" to specific traits, and allows a description of the evolution in relatively few parameters. It is not intended to suggest that the mechanisms of transmission, learning, and choice are entirely reducible to this level. As long as the limitations of the oversimplification are kept in mind, and as long as general, qualitative conclusions are sought from the quantitative analysis, this is a reasonable procedure.

It is perhaps a pity that the objects of cultural evolution lack an appropriate collective name. Terms like "cultural traits," or more specific words, such as behaviors, skills, values, rules, tools, technologies, connote such different phenomena that any essential similarities are lost in the absence of more detailed discussion. Dawkins (1976) has suggested the word "memes" (an abbreviation of "mimeme" from a Greek root for imitation) for "units of imitation." Specific units, such as memes were intended to represent, have meaning when there is essential discontinuity between categories. Such convenient discontinuities are found in atoms, elementary particles, genes, and DNA. It is not clear how commonly such discrete changes occur in cultural traits. In linguistic evolution the replacement of a word by another that is phonetically unrelated has some superficial similarity to the process of nucleotide substitution in DNA and of amino acid substitution in proteins. Here units of change can be discerned. For other cultural traits, including phonetic changes, the level of description or measurement is made discontinuous by, for instance, perceptual thresholds.

The recognition of discontinuity, when it exists and can be quantified, will be important for the elucidation of genuine qualitative principles. But many traits are more naturally described on a continuous scale. Also, cultural objects are often so complex that they have to be dissected into simpler components or aspects for an analysis to be useful, and very often these are most simply evaluated on a quantitative scale.

The choice between discrete and continuous values depends on the nature of the phenomenon, but it may also depend on the method of observation and the aim of the study. The choice of the level of observation may sometimes determine the success of a study. Gregor Mendel chose a qualitative level of observation: his peas were classified according to seed color (green or yellow), height of stem (normal or dwarf), flower color, and so on. The chosen traits suggested the mode of classification: observable differences were sufficiently sharp that they indicated a basic discontinuity. Even though some peas may have been difficult to classify as green or yellow, the great majority fell unambiguously into one or other class. By following up this very tenuous hint, Mendel made one of the most significant discoveries in the history of science. The macroscopic discontinuity was actually dictated by a microscopic one, and the inference of the latter is one of the great advances made by Mendel. At almost the same time, Francis Galton and Karl Pearson, who perhaps were not lesser scientists, but who were apparently unaware of Mendel's work, chose to study the inheritance of measurable traits like stature, and weight. Although their work led to the development of important statistical methods it made little or no contribution to the understanding of biological inheritance, which was their primary aim.

For our studies here there is usually no greater value intrinsic to the study of qualitative (or discrete-valued) traits over continuously varying traits. They are both important for evolutionary theory. The distinction between them is made because they require different methods of analysis and different statistical treatments. But intermediate situations arise. Thus qualitative traits with many states can sometimes be ordered unambiguously. Others are more clearly enumerable (e.g. number of teeth, number of vertebrae). Traits that cannot be measured but can be ranked (e.g. preferences for style or food) occupy an intermediate position. In some cases it is useful to consider these as qualitative, and classify them by adjectives, while in others measurement can be attempted. An example of this could be intelligence, which can be described for some purposes by such categories as genius, bright, normal, dull, imbecile, and for others by IQ measurements.

In some investigations of cultural evolution the choice of classification schemes may be crucial for the success of the endeavor, as was true in Mendel's case. But in the present discussion the choice may depend on the ease of collection of reliable quantitative data and on the fact that some mathematical problems are easier to pose and solve in one context than another. Wherever possible, the same problem will be discussed in reference to both continuous and qualitative traits.

As an alternative to the dissection of cultural entities that we propose, some kind of mathematical formulation might be achieved by representing a given state of culture as an element in some abstract space subject to formal mathematical rules that describe its change over time. But this approach is not likely to be qualitatively useful because, first, the mathematical relationships describing the transformation of culture (in this holistic sense) over time are difficult to relate to real phenomena and, second, even if some general transformation of this form is written down, its evolutionary consequences, that is, the way in which the mathematical definition of culture changes over long periods of time, will be difficult to determine. For these reasons specific aspects of culture are chosen for our evolutionary studies. These allow the construction of direct analogies with mutation, selection and other concepts from evolutionary biology.

We accept as the cultural unit, or trait, *the result of any cultural action* (by transmission from other individuals) *that can be clearly observed or measured on a discontinuous or continuous scale.* Thus, if the observation is made at the discrete level, we require that an individual can be unambiguously assigned to just one of a set of categories. For continuous traits we require that a reproducible measurement can be made. A detailed and complete classification of all possible cultural traits is obviously impossible within this definition.

Starting with this desire to treat single, easily defined, easily observed and measured cultural traits, we have found it useful to dichotomize our theoretical treatment into two classes, depending on whether the trait is discrete or continuous (which ordinarily means measurable). Often it is possible to transform one type of trait into the other; it is obvious that any continuous trait can be transformed into a discrete, dichotomous, or polychotomous one by the introduction of thresholds along the continuous scale of measurement. The reverse is occasionally also possible. Thus the dichotomy between the theories of discrete and continuous traits can be conceived as largely due to limitations inherent in the methodology of model building and data analysis. This formal subdivision should not, therefore, be given excessive weight.

Of greater importance is the fact that for some traits it may be legitimate to invoke the existence of a biological component of the variation. Here we are interested in phenomena, all of which are, by definition, learned. Thus biological variation must refer to variation in the ability to learn (or to teach). Such variation need not be innate, but it is conceivable that it has innate components; such questions of mechanism and process are frequently emotionally loaded, and encroach upon areas in which prejudice has a strong foothold. The difficulty of the problems, the emotional overtones, and the willingness of some to tie them into the public policy arena, make some of these areas of research particularly difficult.

Transmission can be studied on a short-term basis, over one or a few generations, and sometimes even over fractions of generations. Transformation of individuals over time is best studied longitudinally over their lifespans. This is difficult, and is often replaced by retrospective studies, whose shortcomings are well known, or by simultaneous studies of different people of various ages. In the latter case, observed age differences may be due either to physiological changes accompanying maturation and aging, or to evolutionary changes (when evolution is very fast). Thus, to take a biological example, the smaller stature of older people relative to younger ones is not entirely due to the secular trend in the increase in stature, which took place in the last 100 years or so, but also to modifications of the vertebral column with age; it is thus partly a phenomenon accompanying aging. In other cases there also may be effects of differential mortality. Some traits may also be short lived, or fluctuate to some extent. With appropriate followups, all these components can be isolated from secular trends.

The study of transmission may center on several aspects. One is the contribution of the various modes of transmission, of which we have classified almost a dozen. They can undoubtedly be further subdivided and the relative importance of each is likely to vary greatly among the characters studied. The transmission mechanism may be confounded with biological phenomena, namely, the existence of "critical" periods, such as are known, for example, in the learning of pronunciation and language. But there must exist a host of other age-dependent sensitivities in acquiring specific behaviors. For a collection of articles on critical periods in man and animals, see Scott (1978). If the most important age for learning pronunciation is, say, between 6 and 12 years of age, and social conditions dictate that this period should be spent with parents or members of the older generations, then the transmission will be mostly vertical or oblique. But if social conditions, for example obligatory schooling, favor spending most time at that age with age peers, then the transmission will be horizontal between age peers. The result is that the mode of transmission is age specific. Cross-cultural studies may be especially useful in exhibiting the importance of such life-cycle phenomena.

Whenever a cultural trait is acquired predominantly by copying a single model or a few models, and there is a potential choice between models to be copied, a most intriguing problem arises concerning the mechanism of choice of model. Another important problem in the study of transmission is that of dissecting the elements that enter into it, for example, communication and cultural selection (or the equivalent stages of awareness and decision). As we have discussed previously, a coefficient of transmission includes the chance and quality of teaching, the chance and quality of learning, and the decision to retain and practice whatever was learned. The teaching may be so persuasive and efficient that there is no room for variation of learning. At the other extreme, even with poor "quality of teaching," the need for homogeneity may be so great (e.g. mutual understanding in communication by language) that it will almost always be obtained. Society generates between its members communication channels that ensure the necessary degree of cohesiveness.

It will be convenient to refer to some concrete examples. To this end we have collected some information pertinent to our theoretical investigations—information concerning, for instance, the relative effects of some types of transmission (Cavalli-Sforza, Feldman, Chen and Dornbusch, 1981). The data have been collected from samples of students (we are well aware that these are hardly representative of any population other than themselves). We have found it useful to report here some of the analyses of the set of questionnaires filled out by Stanford students, their parents and friends. This was designed as a preliminary survey of values, beliefs, and everyday life activities in which the questions were chosen so as to have simple answers that were likely to be honest, while questions that were found to cause difficulties were avoided. The problems relating to errors of observation are beyond the present discussion. Suffice it to mention that the data on the students and their parents are obtained at different ages for the two generations and that it is difficult to say whether the differences between the students and their parents are due to a secular trend or to individual change with age. In the same way similarity between husband and wife is measured after twenty years of life together, and so could be due to convergence of interests or of conceptions, rather than to true homogamy (assortative mating). There is no way to clearly differentiate here between these two explanations. Similarly correlations obtained between friends could be due to influence in either of two directions, or alternatively they could be the result of reciprocal choice due to preexisting affinity. None of these problems can be settled by the existing material. The result of the questionnaire has in general indicated that vertical (and perhaps oblique) transmission is more important than was anticipated. Data collected in this way obviously cannot distinguish between biological and cultural (vertical) transmission. But the nature of most of the questions, and evidence from other sources, make the contributions of innate factors likely to be small or negligible.

CHAPTER TWO

Vertical Transmission

2.1. INTRODUCTION

This chapter is devoted to the study of transmission from parent to child and its evolutionary consequences for the simplest case, namely that of a binary (dichotomous) trait. In conformity with epidemiological usage this mode of transmission is called *vertical.* A model of vertical transmission of cultural traits is introduced that includes, as special cases, certain Mendelian examples. We concentrate here on transmission from one generation to another and ignore intragenerational differences and age structure. A major objective of this study is the determination of stationary-trait frequency configurations, that is, those frequencies from which no change in frequency occurs under the evolutionary forces assumed. If a trait exists in only one state, that form of the trait is said to be fixed—the other forms are lost. If more than one trait exists at the stationary state it is called polymorphic. We shall use the term *equilibrium* to mean any stationary state, and the context will clarify what sort is meant. We shall also try to determine whether the population can attain such an equilibrium frequency configuration, and if so which initial frequencies lead to this equilibrium. Such properties are collectively termed the *stability* of the equilibrium. We are also interested in the time-dependent behavior of the trait, whenever possible, and in the *correlations between relatives* that arise under the mechanisms of transmission envisaged. Of particular interest will be the evolutionary roles of natural and cultural selection, mutation, migration, and the stochastic effects of population size, as well as the interactions of these factors with the mode of transmission. Much of the conceptual framework of classical population genetics can usefully be extended to cultural evolution by making appropriate changes.

2.2 VERTICAL TRANSMISSION

The simplest discrete-valued trait takes on one of two possible values or, alternatively expressed, the trait exists in one of two possible states. For example, one is either a medical doctor, or is not. One is a political conservative (perhaps measured by voting Republican), or is not. Obviously, by increasing the number of professions or political parties these traits can become multi-valued. In both of these examples cultural inheritance may well be assumed to be of importance. In the simplest case of vertical transmission the coefficients of transmission are constant from generation to generation. Let H and h represent the two states that the trait can take. There are then four possible mother-father pairs. In biology these are usually called *mating types,* a term which we shall also adopt here. It is necessary to specify for each mating type the probability that the progeny be of type H or h. Denote by b_0, b_1, b_2, b_3 the probabilities (assumed constant) that an H child results from the matings $h \times h$, $h \times H$, $H \times h$, and $H \times H$ respectively, where the mating specifies the mother $\times$ father pair with mother listed first. The frequencies with which these matings occur among adults at generation t in the population are $p_0^{(t)}$, $p_1^{(t)}$, $p_2^{(t)}$, and $p_3^{(t)}$, respectively, with $\Sigma_{i=0}^3 p_i^{(t)} = 1$. If

TABLE 2.2.1. Vertical transmission for a 2-state trait; adults at generation t.

Mating Type		Probability of		Frequency of Matings	
Mother	Father	H Child	h Child	General	Random Mating
H	H	b_3	$1 - b_3$	$p_3^{(t)}$	u_t^2
H	h	b_2	$1 - b_2$	$p_2^{(t)}$	$u_t v_t$
h	H	b_1	$1 - b_1$	$p_1^{(t)}$	$u_t v_t$
h	h	b_0	$1 - b_0$	$p_0^{(t)}$	v_t^2

the frequency of H in the adult individuals at generation t is u_t, with $v_t = 1 - u_t$ that of h, then under random mating we have $p_0 = v_t^2, p_1 = u_t v_t = p_2, p_3 = u_t^2$. The whole structure is depicted in Table 2.2.1.

The frequency of H in the offspring generation $t + 1$ is

$$u_{t+1} = \sum_{i=0}^{3} p_i^{(t)} b_i, \tag{2.2.1}$$

which under random mating is

$$u_{t+1} = u_t(b_3 u_t + b_2 v_t) + v_t(b_1 u_t + b_0 v_t) \tag{2.2.2}$$

$$= u_t^2 B + u_t C + b_0, \tag{2.2.3}$$

where

$$B = b_3 + b_0 - b_1 - b_2 \tag{2.2.4}$$

$$C = b_2 + b_1 - 2b_0. \tag{2.2.5}$$

The dynamic properties of this quadratic recursion system can be ascertained according to methods described in detail, for example, in Roughgarden (1979, pp. 577–581). They were discussed in Feldman and Cavalli-Sforza (1976). At equilibrium $u_{t+1} = u_t$. Any value satisfying the equation

$$u = u^2 B + uC + b_0 \tag{2.2.6}$$

is an equilibrium of (2.2.3). If $B = 0$ then the equilibrium exists if $C < 1$ and is

$$\hat{u} = b_0/(1 - C). \tag{2.2.7}$$

If $B \neq 0$, then from the fact that an equilibrium must be real, and between zero and one to be valid, the only admissible root of (2.2.6) is

$$\hat{u} = (1 - C - \sqrt{A})/2B \tag{2.2.8}$$

where $A = (1 - C)^2 - 4b_0 B$. Note that if $C < 1$ then $u \geq 0$, while $\hat{u} \leq 1$ always. If $A < 4$, then u_t converges to $\hat{u}$ and the equilibrium is stable. If $A < 1$, then the convergence of u_t to $\hat{u}$ is monotonic. If $1 < A < 4$, the convergence is oscillatory. If $A > 4$, there is no stable equilibrium, but a stable two-point cycle is established such that the population eventually oscillates between two different values. The way in which the values u_t change in time, that is, the transient behavior of u_t, is depicted in Figure 2.2.1. The parameter ranges under which the various convergence behaviors are observed are shown in Fig. 2.2.2. Notice that the conditions under which cycling occurs are extreme, requiring $4 < A < 5$ and that $A \leq 5$ always. These are transmission coefficients that model the situation in which a child has a behavior essentially opposite to that of its parents, since b_0 must be large and b_3 small. In other words, the offspring of $H \times H$ matings must be nearly all h and those of $h \times h$ matings nearly all H. This explains the oscillatory behavior of (2.2.3) with these parameters. There are some traits, for example type of clothing, whose fashion shows oscillatory

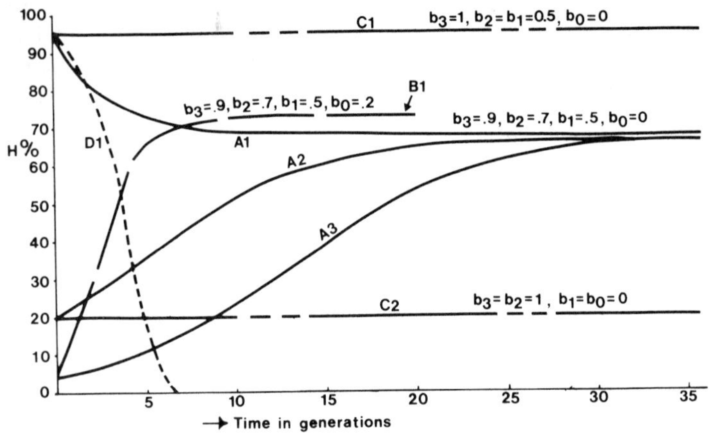

FIGURE 2.2.1. Examples of trajectories for vertical transmission. A1,2,3(b_3 = .9, b_2 = .7, b_1 = .5, b_0 = 0) stable equilibrium at u = 2/3 irrespective of initial values. $B1$: effect of changing b_0(b_0 = .2) in A_3, equilibrium shifts to 0.73. $C1$,2: neutral equilbria, for $B = 0$, $C = 1$, b_0 = 0. $D1$: Rapid loss of trait under vertical transmission alone (b_3 = 1, b_2 = b_1 = b_0 = 0).

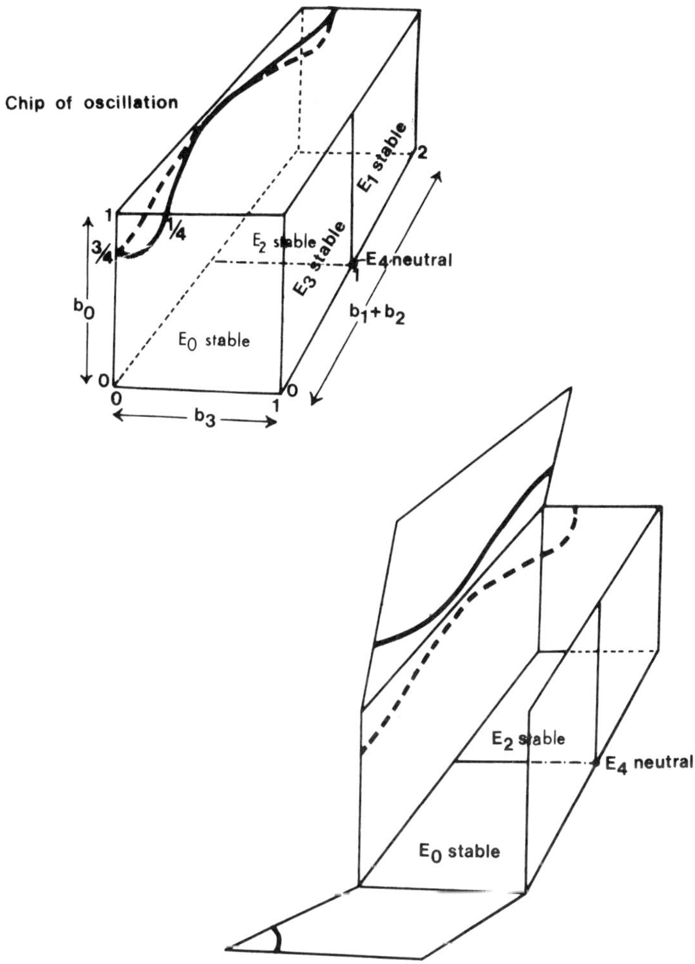

FIGURE 2.2.2. On the labeled surfaces, the $E_0(\hat{u} = 0)$, $E_1(\hat{u} = 1)$, $E_2((1 - C)/B)$, $E_3(b_0/B)$, are stable. The E_4 point is a neutral equilibrium. In the chip of oscillation (top left) there is a two-point limit cycle. In the internal 3-dimensional area, (2.2.7) or (2.2.8) are stable but may be reached in an oscillatory fashion, if $A > 1$. (From Feldman and Cavalli-Sforza, 1976.)

behavior. The period of the oscillation in such cases is often not very far from one generation. Negative reactions to examples set by parents may perhaps contribute to explaining this state of affairs, although it would be naive to suppose that this is a complete explanation.

From the Stanford survey of beliefs and values, described in section 1.14, Table 2.2.2 presents some examples of data on vertical transmission. The five items chosen represent areas from among the spectrum spanned by the survey. Included is a question about dietary habits (item 1, on frequency of salt usage), one on religious habits (item 2, frequency of prayers to God), one on sports participation (item 3, frequency of swimming), one (item 4) on belief about the relative roles of ability versus luck in determining success in life, and one (item 5) on political interest (registration with some party). The first three items were originally scored as frequencies (for example, of salt usage) and have been dichotomized by subtracting average frequencies of the students and taking positive values as H and negative values as h. The χ^2 test, applied in Table 2.2.2 to all items, each considered as a 4×2 table, is used to ask the question of whether there is an association between a parent's trait and that of the child. An affirmative answer is statistical evidence for heterogeneity between the individuals of trait H in the progeny of the four types of mating, which may be due to vertical transmission. The answer is clearly positive for the first three items and the last one. The frequencies of the traits in the parental generation and in the offspring, and the expected equilibrium, are shown in the three bottom lines of Table 2.2.2. The last line gives the equilibrium value, computed on the basis of (2.2.8), to which the population frequency of the trait would move if there were continued evolution under constant transmission coefficients (b_3, b_2, b_1, b_0) equal to those determined from this set of data. The transmission coefficients and their standard errors are shown in the upper part of the table. Note that the standard errors can be fairly large

TABLE 2.2.2. Vertical transmission for five items from the Stanford survey of values and beliefs (Item 1, habitual salt usage; Item 2, frequency of praying to God; Item 3, frequency of swimming; Item 4, belief in ability versus luck; Item 5, political interest). Trait value H, h is determined on the basis of self rating. n_i is the number of matings of the type shown at left as a binary trait. The frequencies of H individuals in the offspring of these matings are approximated to two digits by b_i, and are followed by their standard errors: $\pm [b_i(1 - b_i)/n_i]^{1/2}$ In the fifth line of the table the χ^2 value for heterogeneity among the four progeny frequencies is recorded. The last three lines present the frequencies of H observed in parents, offspring and at equilibrium (using the estimated b_i above).

	b_i	Item 1		Item 2		Item 3		Item 4		Item 5	
Father × Mother		n_i	b_i	n_i	b_i	n_i	b_i	n_i	b_i	n_i	b_i
$H \times H$	b_3	20	60 ± .11	87	68 ± .05	17	65 ± .12	29	48 ± .09	130	72 ± .04
$H \times h$	b_2	67	36 ± .06	10	20 ± .13	19	58 ± .11	45	29 ± .07	18	44 ± .12
$h \times H$	b_1	17	65 ± .12	53	57 ± .07	24	42 ± .10	36	28 ± .07	25	40 ± .10
$h \times h$	b_0	99	26 ± .04	52	19 ± .02	141	29 ± .04	91	29 ± .05	28	25 ± .08
χ^2 with 3 *d.f.*		15.16		35.26		13.24		4.53		28.26	
Significance[(a)]		$p < 1\%$		$p < 0.1\%$		$p < 1\%$		$p > 5\%$		$p < 0.1\%$	
Obs. H in parents		30.5%		58.7%		19.1%		34.6%		75.4%	
Obs. H in progeny		36.0%		50.0%		36.3%		31.3%		59.2%	
Exp. H at equilibrium		45.0%		32.9%		47.4%		30.3%		41.5%	

(a) χ^2 is generally computed using the likelihood ratio method.

when the numbers of matings are small. Error in the estimates of the coefficients, changes in the coefficients over time (which is difficult to predict or assess), assortative mating (discussed later), and perhaps most important, the fact that the observations on the progeny (which determine the b_i values) are not made at the same age as those of the parents, all affect the amount of faith to be placed in the equilibrium frequencies of H. Note, however, that these estimated equilibria appear to continue the trend of change observed from parent to offspring in these data.

2.3 SPECIAL CASES OF VERTICAL TRANSMISSION

If the coefficients of transmission in the vertical case are constant, as in Table 2.2.1., they could, in principle, be estimated from observations of families. If random mating applies, then the evolutionary behavior could be predicted from equation (2.2.3). In this paragraph some special cases of (2.2.3) are considered. The parameter specifications given in Table 2.3.1 are simple cases of the general scheme in Table 2.2.1. Case 1 of Table 2.3.1 illustrates a situation in which the sex of the parent has a strong influence on the transmission rule, to the point that the transmis-

TABLE 2.3.1. Some simple cases of vertical transmission with constant coefficients and random mating derived from the general array of Table 2.2.1.

Mating ♂ ♀	General Case	Case 1 (Uniparental) Father-dep.	Mother-dep.	Case 1a Extreme Uniparental	Case 1b Degenerate	Case 2 Genetic (Haploid)
$H \times H$	b_3	b	b	1	b	1
$H \times h$	b_2	b	c	1	b	1/2
$h \times H$	b_1	c	b	0	b	1/2
$h \times h$	b_0	c	c	0	b	0
Equilibrium frequency of trait ($\hat{u}$)	see formula (2.2.8)	$\frac{c}{1-b+c}$		u_0 (no change)	b	u_0 (no change)

sion is effectively *uniparental.* For case 1, the recursion equation (2.2.3) is now linear, and the full transient behavior is easy to state:

$$u_t = (b - c)^t u_0 + c[1 - (b - c)^t]/[1 - (b - c)]. \quad (2.3.1)$$

Hence, u_t converges to

$$\hat{u} = \frac{c}{1 - (b - c)}, \quad (2.3.2)$$

unless, of course, $b = 1$, and $c = 0$, in which case u_t retains its original value, u_0, throughout the evolution; this is case 1a in Table 2.3.1. From equation (2.3.1) it is obvious that, if $b = c$, so that all transmission coefficients are equal, u_t becomes equal to b after the first generation and forever more, as in case 1b.

In case 2 (genetic haploid), as with 1a, u_t does not change over time. Case 2 is not only an example of cultural transmission, it is also an example of a class of genetic systems, namely those in which the organism is *haploid,* that is, in which every gene is represented only once. The *diploid* phase results from the fusion of two parents, which are haploids, and lasts a very short time. Haploidy is restored by a process of separation of the two parental contributions. This cycle occurs in fungi, for example.

TABLE 2.3.1. (cont'd)

Case 3 Symmetric	Case 4 Additive	Case 5 Multiplicative	Cases 6 Infectious	Cases 7 Infectious
$1 - b_0$	$b_0 + \alpha_p + \alpha_m$	$1 - (1 - b_0)(1 - \alpha_p)(1 - \alpha_m)$	1	1
$1 - b_1$	$b_0 + \alpha_m$	$1 - (1 - b_0)(1 - \alpha_m)$	1	0
b_1	$b_0 + \alpha_p$	$1 - (1 - b_0)(1 - \alpha_p)$	1	0
b_0	b_0	b_0	0	0
1/2	$\dfrac{b_0}{1 - \alpha_p - \alpha_m}$	$[(\alpha_p + \alpha_m)(1 - b_0) - 1 + \{[(\alpha_p + \alpha_m)(1 - b_0) - 1]^2 + 4\alpha_p\alpha_m(1 - b_0)\}^{1/2} \div 2\alpha_p\alpha_m(1 - b_0)$	1,0	0,1

Most higher organisms have a prolonged diploid phase: for these, the representation in terms of Table 2.2.1 is much more complicated, as we will later show (Chapter 4.1). This is because type H may include representatives of the two genetic types AA or Aa, while h individuals may belong only to the genetic class aa. To parallel the equation (2.2.1), the transmission coefficients b_0, b_1, b_2 and b_3 would have to be gene-frequency dependent.

Case 2 is itself a special case of 3 (with $b_1 = 1/2$, $b_0 = 0$) and of 4 (with $b_0 = 0$, $\alpha_p = \alpha_m = 1/2$), all from Table 2.3.1. The latter two transmission models we term "symmetric" and "additive," respectively. In contrast to case 4, case 5 involves multiplication of the probabilities of transmission. Cases 4 and 5 will be compared in more detail later. Cases 6 and 7 are termed "infectious." In the latter, both parents must be affected for the progeny to be so, while in case 6 it is enough that one parent is affected. We named the latter case after the viral disease "Kuru" in an earlier paper (Feldman and Cavalli-Sforza, 1976) in which the detailed kinetics of such a case were examined. Later we will consider sex-influenced transmission, that is, a different set of transmission coefficients for male and female offspring. Genetic differences in susceptibility may also exist but these will be discussed in the next volume. A constant value $b_0 \neq 0$ allows contagion from unrelated individuals at a constant rate: frequency dependence (see Chapter 2.3) may be more realistic for this type of transmission. Finally, the model represented by cases 6 and 7 ignores mortality due to the infection, as if all deaths due to the virus occurred post-reproductively. This is not the case in reality, but we postpone a more detailed analysis until our general discussion of natural selection in section 2.6.

The discussion above has shown that certain special choices of the transmission parameters maintain a constant frequency of H. Among these is the haploid genetic transmission scheme $b_0 = 0$, $b_3 = 1$, $b_1 + b_2 = C = 1$. Other cases involving perfect transmission by one sex, and none by the other ($b_1 = 1$, $b_2 = 0$ or

$b_1 = 0, b_2 = 1$) also allow $b_1 + b_2 = 1$, and result in constant H frequency. When $b_3 = 1, b_0 = 0$, but $C = b_1 + b_2 \neq 1$, only $\hat{u} = 0$ and $\hat{u} = 1$ are the possible equilibria. If $C < 1$ and $b_0 = 0$, then starting from any intermediate frequency there will be convergence to $\hat{u} = 0$; if $C > 1$, then convergence is to $\hat{u} = 1$.

In most practical cases, b_0 is different from zero, and C is unlikely to be exactly equal to 1. The b_0 value may represent contributions from types of transmission other than parental, which will be examined later. When $b_0 \neq 0$, then if $C = b_1 + b_2 - 2b_0 > 1$, convergence occurs to $\hat{u} = 1$, while if $C < 1$, there is convergence to an interior equilibrium. In forty-one examples of traits from the Stanford survey, C was less than unity in forty cases, the only exception being nonsignificant (there were very few matings of the $h \times h$ type). Thus in practically all cases the estimated transmission coefficients were compatible with an internal stable equilibrium. All of the traits are assumed to be equally expressed in both sexes. It is useful to ask if some of the simpler models from Table 2.3.1 can apply, in particular, to uniparental transmission (model 1) or to biparental additive transmission (model 4). The two can be fitted by maximum likelihood. For model 1, and paternal inheritance, for instance, the estimates of b and c are then simply obtained by pooling frequencies of H in the offspring of $H \times H$ and $H \times h$ matings, and of $h \times H$ and $h \times h$, respectively. To fit the biparental model, an iterative approach is necessary, yielding α_p, α_m values and their standard errors. The goodness of fit of the models can then be tested.

Of the forty-one traits examined in the Stanford survey, twenty do not show a significant χ^2 of heterogeneity between the matings (as in the fifth line of Table 2.2.2) and therefore show no statistically significant evidence of vertical transmission (see, for instance, item 4 in Table 2.2.2). Of the traits in Table 2.2.2, we see that for items 1 and 2 there is maternal transmission (model 1). In item 1, the original data were

father × mother			number of pairs	observed progeny	
				H	h
H	×	H	20	12	8
H	×	h	67	24	43
h	×	H	17	11	6
h	×	h	99	26	73
Total			203		

The estimate of b is $\hat{b} = (12 + 11)/(20 + 17) = 0.63$ ($\pm$ .08) and that of c is $\hat{c} = (24 + 26)/(67 + 99) = 0.30$ ($\pm$ .04). The fit of the model can be tested by computing expectations, as shown below.

father × mother			expected progeny	
			H	h
H	×	H	12.43	7.57
H	×	h	20.18	46.82
h	×	H	10.57	6.43
h	×	h	29.82	69.18

Goodness of fit is then tested by comparing observed and expected numbers of progeny, giving $\chi^2 = 1.86$ with two degrees of freedom, which is not significant. Similarly, item 2 is also in good agreement with the hypothesis of maternal transmission.

Item 4 does not require testing, since there is no evidence for vertical transmission. In item 5, both parents seem to influence children, and the biparental additive model 4 can be tried. Estimation by maximum likelihood yielded the following numerical values for the paternal and maternal contribution and for the baseline b_0 (with standard errors):

$$\hat{\alpha}_p = 0.27 \pm .09$$

$$\hat{\alpha}_m = 0.21 \pm .09$$

$$\hat{b}_0 = 0.23 \pm .07$$

The expected relative frequencies are computed from these and given below, near the corresponding observed ones:

father × mother	observed	expected
$H \times H$	.72	.72
$H \times h$	.44	.50
$h \times H$	.40	.44
$h \times h$	.25	.23

Since $\chi^2 = 0.46$ with 1 d.f., the agreement is very good.

Details of the analysis can be found in the original paper. Twenty-one traits showed evidence for vertical transmission, as follows:

Maternal component only significant—frequency of listening to classical music; liking to camp; frequency of attending movies; frequency of reading horoscopes; believing margarine better than butter for health; frequency of praying to God; religious preference Protestant, Catholic, Jewish; salt usage high; Republican party preference.

Paternal component only significant—frequency of watching football; frequency of watching baseball; liking to visit art museums; liking big parties; belief in ESP; frequency of going swimming; political position conservative.

Both paternal and maternal component significant—frequency attending church; registered with a party; Democratic party preference.

Significance was tested also with a statistical technique known as log-linear analysis: the results suggested that three-way interactions were absent, and were in good agreement with the additive model. The numbers of matings involved were not large enough to guarantee that relatively small parental effects would not be missed. The analysis above should not be interpreted to exclude, for instance, a paternal component where only the maternal one is significant, but rather to show that in such cases there is likely to be a maternal component and that it is likely to be more important than the paternal one.

It should be remembered that this analysis cannot distinguish other sources of transmission, which may be confounded with vertical cultural transmission. In particular, we cannot distinguish cultural from vertical, biological (genetic) transmission, for which discrimination the analysis of adopted children would be necessary. However, evidence from twin data (Loehlin and Nichols, 1976) for a large number of similar traits similar to the above, suggests that the biological component is likely to be negligible, since correlations between identical and fraternal twins are similar. These data alone cannot distinguish vertical from oblique cultural transmission. For this purpose, partial correlation techniques can be used to assess the effect of socioeconomic variables. The effect of this conditioning on correlations between parent and offspring was negligible. This test is only suggestive and cannot exclude the possibility of other modes of transmission, which might be discovered through a trait-by-trait longitudinal analysis of the social contacts and influences to which individuals are subjected.

When there is deviation from additivity, the four coefficients of transmission in the additive case may be rewritten as:

$$\begin{aligned} b_3 &= b_0 + \alpha_m + \alpha_p + \alpha_{mp} \\ b_2 &= b_0 + \alpha_p \\ b_1 &= b_0 + \alpha_m \\ b_0 &\text{ unchanged} \end{aligned} \tag{2.3.3}$$

with α_{mp} positive or negative depending on whether there is synergism or antagonism between the effects of the two parents. From (2.2.4), $B = \alpha_{mp}$ measures *deviation from additivity*. Similarly, from (2.2.5), $C = b_2 + b_1 - 2b_0 = \alpha_p + \alpha_m$ measures the sum of the *separate parental effects*.

In all of our numerical computations the additive model was quite satisfactory. Its goodness of fit was fairly close to that estimated for the absence of triple interaction progeny × father × mother in log-linear analysis.. For completeness, the multiplicative model was also tested, but no improvement on average over the fit of the additive model was obtained. It is possible, however, to think of situations in which the multiplicative model should prove preferable.

2.4 CORRELATIONS BETWEEN RELATIVES

Transmission probabilities, which are the basic parameters of the above model of cultural evolution, may sometimes be difficult to estimate. Correlations between relatives, on the other hand, are easier to obtain and may provide useful information about the importance of cultural effects. In purely cultural inheritance it is foster parents of adopted children that are important. For the same reason, the various types of twin pairs need not be distinguished, at least as a first approximation. For the present discussion three types of relationship will be studied: parent-offspring, sib-sib, and half-sib. The last defines two children who have only one parent in common, as a result of widowhood, divorce, polygamy, or the like. More remote degrees of relationship could be treated in the same way if desired.

For the values in the 2 × 2 Table 2.4.1 the value of the correlation coefficient, r, is defined to be

$$r = \frac{q_1 q_4 - q_2 q_3}{\sqrt{q(1-q)q'(1-q')}} \tag{2.4.1}$$

In practice r is evaluated as $\chi/N^{-1/2}$, where χ is the square root of

TABLE 2.4.1. Frequencies among relatives.

		One member of related pair		
		H	h	Total
Other member of related pair	H	q_1	q_2	q
	h	q_3	q_4	$1-q$
	Total	q'	$1-q'$	1

TABLE 2.4.2. Parent-offspring relationship.

		Father		
		H	h	Total
Children	H	$p_3^{(t)}\, b_3 + p_2^{(t)}\, b_2$	$p_1^{(t)}\, b_1 + p_0^{(t)}\, b_0$	u_{t+1}
	h	$p_3^{(t)}\,(1 - b_3) + p_2^{(t)}(1 - b_2)$	$p_1^{(t)}\,(1 - b_1) + p_0^{(t)}(1 - b_0)$	$1 - u_{t+1}$
	Total	$p_3^{(t)} + p_2^{(t)} = u_t$	$p_1^{(t)} + p_0^{(t)} = 1 - u_t$	1

χ^2 computed in the classical way for the 2 × 2 contingency table corresponding to Table 2.4.1, and N is the total number of observations, that is, the sum of the absolute observed frequencies. The same result would be obtained by computing the Pearson product moment correlation coefficient as if the trait were metric, taking the values of H and h as 1 and 0 respectively. Table 2.4.2 exhibits the parent-offspring relationship using the basic transmission system, Table 2.2.1, in Table 2.4.1, where the parents are members of generation t and the children belong to the next generation. The value of the correlation coefficient r_{PO} in the case of random mating is written, for fathers (for mothers, replace b_2 with b_1):

$$r_{FO} = R_{t,t+1}[b_2 - b_0 + Bu_t] \tag{2.4.2}$$

where B is as given in (2.2.4), and

$$R_{t,t+1} = \sqrt{u_t(1 - u_t)/u_{t+1}(1 - u_{t+1})}, \tag{2.4.3}$$

which is equal to unity if the population is at equilibrium.

As a numerical example, we use item 5 of Table 2.2.2. The original numbers from which the b_i values were computed are given below:

father × mother	H progeny	h progeny
$H \times H$	94	36
$H \times h$	8	10
$h \times H$	10	15
$h \times h$	7	21

For computing the father-offspring correlation, we arrange them in the following table:

		Father H	Father h	totals
Offspring	H	102	17	119
	h	46	36	82
totals		148	53	201

The correlation is $r_{FO} = 0.330$. Similarly, that between mother and offspring is $r_{MO} = 0.295$.

For sib-sib correlations Table 2.4.2 is used. This is constructed by allotting to each of the four cells the four possible sib pair frequencies, b_i^2, $b_i(1 - b_i)$, $b_i(1 - b_i)$, $(1 - b_i)^2$, and adding over all mating pairs after weighting for the appropriate frequency $p_i^{(t)}$.

Table 2.4.3 gives a sib-sib correlation of

$$r_{SS} = \frac{\Sigma p_i^{(t)} b_i^2 \, \Sigma p_i^{(t)} (1 - b_i)^2 - [\Sigma p_i^{(t)} b_i (1 - b_i)]^2}{u_{t+1}(1 - u_{t+1})} = \frac{\Sigma p_i^{(t)} b_i^2 - (\Sigma p_i^{(t)} b_i)^2}{u_{t+1} v_{t+1}}, \tag{2.4.4}$$

which for random mating reduces to

$$r_{SS} = R_{t,t+1} \{u_t^2[(b_1 - b_3)^2 + (b_2 - b_3)^2] + u_t v_t[(b_0 - b_3)^2 + (b_1 - b_2)^2] + v_t^2[(b_0 - b_1)^2 + (b_0 - b_2)^2]\}, \tag{2.4.5}$$

where R is as in (2.4.3).

TABLE 2.4.3. Sib-pair frequencies*

		One sib H	One sib h	Total
Other sib	H	$\Sigma p_i^{(t)} b_i^2$	$\Sigma p_i^{(t)} b_i(1 - b_i)$	u_{t+1}
	h	$\Sigma p_i^{(t)} b_i(1 - b_i)$	$\Sigma p_i^{(t)} (1 - b_i)^2$	$1 - u_{t+1}$
	Total	u_{t+1}	$1 - u_{t+1}$	1

*sums are for i from 0 to 3.

An alternative derivation of r_{SS} is possible by considering a random variable X which takes on the value 1 if an individual's phenotype is H and 0 if it is h. At time $t+1$ there are u_{t+1} individuals with H and $1 - u_{t+1}$ with h, so that the mean value of X is $M = \Sigma p_i^{(t)} b_i = u_{t+1}$, while its variance is $V = u_{t+1} v_{t+1}$. The variance between families is computed by taking the variance between family means, of which there are four (see Table 2.2.1). These means are just the b_i values. The variance between family means is therefore

$$V_B = \Sigma p_i^{(t)} b_i^2 - M^2, \tag{2.4.6}$$

which is equal to the numerator of (2.4.4). The intraclass correlation between sibs (the Pearson product moment correlation, entering every pair of sibs twice in the table, with one sib in rows and the other in columns and vice-versa) is easily checked and found to be

$$r_{SS} = V_B / V, \tag{2.4.7}$$

equal to (2.4.4).

For half-sibs a table similar to those of the previous examples can be constructed. With random mating the correlation coefficient reduces to

$$r_{HS} = R_{t,t+1}^2 \{u_t(b_3 - b_1) + v_t(b_2 - b_0)\}^2. \tag{2.4.8}$$

The correlations between relatives are dynamic variables which change as u_t does. For estimation it is usually assumed that the system has come to equilibrium, in which case our theory, which predicts u, can be applied to predict equilibrium correlations. $R_{t,t+1} = 1$ at equilibrium.

In Table 2.4.4 some of the equilibrium correlations from the simple cases in Table 2.3.1 are presented. These are the cases for which an equilibrium distinct from $u = 0$ or $u = 1$ is attained. The values in Table 2.4.4 apply to the random mating case.

In summary, correlation coefficients between relatives can be

TABLE 2.4.4. Correlations between relatives for some simple cases of vertical transmission from Table 2.3.1 with random mating, at equilibrium.

Case		r_{PO}	r_{SS}	r_{HS}
1	Uniparental	$b - c$	$(b - c)^2$	$(b - c)^2$
1a	Extreme uniparental	1 or 0**	1	1
2	Genetic, haploid	1/2	1/2	1/4
3	Symmetric	$1 - b_1 - b_0$*	$1 - 2(b_0 + b_1) + 2(b_0^2 + b_1^2)$	$(1 - b_1 - b_0)^{2}$*
4	Additive	α_m or α_p**	$\alpha_p^2 + \alpha_m^2$	α_m^2 or α_p^2†

*this is for mother-offspring; the father offspring value is obtained replacing $b_1 - b_0$ for $1 - b_1 - b_0$.
**for mother-offspring and father-offspring respectively.
†for maternal and paternal half-sibs respectively.

expressed, at equilibrium, in terms of the transmission probabilities. The reverse would be possible provided enough correlation coefficients were given. The correlation coefficients expected between parent and offspring, between sib and sib and between half-sibs, presented here, all depend on the trait frequency. A more serious limitation of the above formulas is that they use the assumption of random mating, which is frequently false as we shall see in the next section.*

2.5 ASSORTATIVE MATING

The assumption of random mating clearly simplifies the transmission problem, but is not necessarily valid. Moreover, the formulation of Table 2.2.1 is independent of this assumption, and one example of non-randomness, namely the phenomenon of assortative mating, is of particular interest to the study of population dynamics. Assortative mating occurs when the matings between similar phenotypes are more frequent (or less) than would be the case under random mating. Positive (negative) assortative mating is the term used to describe the situation where like (unlike) phenotypes mate more frequently than random mating would suggest. In this definition of assortative mating the intention is not to introduce selection by excluding some types from the mating pool. The formulation by Sewall Wright (1921), which we shall adopt here, accomplishes this as follows. The population is conceived of as containing a fraction $(1 - m)$ that mates at random and a complementary fraction m which mates assortatively. That is, a fraction m of matings involving H and h are obliged to be with H and h, respectively, with $(1 - m)$ of the matings at random. Thus $(1 - m)u_t^2$ is the fraction of random matings at generation t of type $H \times H$, while mu_t is the fraction of $H \times H$ matings due to obligate assortative mating. The total fraction of $H \times H$ matings is the sum

$$(1 - m)u_t^2 + mu_t = p_3^{(t)} = u_t^2 + mu_t v_t,$$

and so on. In this way the mating Table 2.5.1 is constructed.

Note here that the marginal sums are the same for the two sexes, and are equal to the population frequencies; this was a key ingredient of Wright's formulation. According to our previous formula (2.4.1) for the correlation from a 2×2 table, this system of assortative mating is equivalent to assuming a correlation between husband and wife

$$r_{HW} = m. \tag{2.5.1}$$

This "mating correlation" m can now be incorporated into the general recursion connecting the frequencies of H in successive

TABLE 2.5.1. Frequencies of mating pairs with assortative mating of m of the population.

		Husband		
		H	h	Totals
Wife	H	$u_t^2 + mu_t v_t$	$u_t v_t (1 - m)$	u_t
	h	$u_t v_t (1 - m)$	$v_t^2 + mu_t v_t$	v_t
Totals		u_t	v_t	1

generations. We now have

$$u_{t+1} = u_t^2 B + u_t C + b_0 + m u_t v_t B \tag{2.5.2}$$

Thus assortative mating affects the evolution of H frequencies only if $B \neq 0$. The possible equilibria of (2.5.2) are roots of the quadratic equation obtained by deleting the time subscripts. When $B \neq 0$ and $m < 1$, the equilibrium, if it exists, is given by (2.5.3). An analysis of the existence conditions similar to that made for (2.2.8) can be made but is omitted here.

$$\hat{u} = \frac{1 - C - mB - \sqrt{(1 - C - mB)^2 - 4b_0 B(1 - m)}}{2B(1 - m)} \tag{2.5.3}$$

for $m = 1$, $\hat{u} = 1 + b_0 - b_3$.

Among the simple examples of Table 2.3.1, cases 1–4 have the same equilibrium with or without assorting, because $B = 0$ in these cases. In cases 6 and 7, although the transient behavior of (2.5.2) is affected by m the equilibria are not; both 0 and 1 are still the only equilibria.

The correlations between relatives may be affected by assortative mating. The values from Table 2.5.1 can be substituted into Table 2.4.2 to obtain the parent offspring correlation with assortative mating as

$$r_{PO}^{(m)} = r_{PO} + m[b_3 - b_2 - u_t B] \tag{2.5.4}$$

where r_{PO} is given in (2.4.2). Similarly the sib-sib correlation turns out to be

$$r_{SS}^{(m)} = r_{SS} + m(b_3^2 + b_0^2 \quad b_1^2 - b_0^2 - B) \tag{2.5.5}$$

where r_{SS} is as in (2.4.5). The equilibrium values for u_t in the special cases of Table 2.3.1 can be substituted into equations (2.5.4) and (2.5.5) to produce the equilibrium correlations that are presented in Table 2.5.2

For the case of a polymorphic equilibrium, Table 2.5.2 shows that assortative mating increases r_{PO} and r_{SS} when both parents have an effect on the progeny. In the genetic haploid and additive

TABLE 2.5.2. Correlations between relatives under assortative mating for cases in Table 2.3.1 at equilibrium of the trait frequency**.

	Case 1 Sex Dependent	Case 1a Extreme Sex Dependence	Case 2 Genetic Haploid	Case 3 Symmetric	Case 4 Additive
$r_{PO}^{(m)}$	r_{PO}*	r_{PO}*	$r_{PO}(1+m)$	$r_{FO} = 1 - b_0 - b_1 + (b_1 - b_0)m$	$\frac{\alpha_p + \alpha_m}{2}(1+m)$
(correlation) between parent and offspring)			$= \frac{1}{2}(1+m)$	$r_{MO} = b_1 - b_0 + m(1 - b_1 - b_0)$	
$r_{SS}^{(m)}$	r_{SS}*	r_{SS}*	$r_{SS}(1+m)$	$r_{SS} + 2m(b_0^2 - b_1^2)$	$\alpha_m^2 + \alpha_p^2 + 2m\,\alpha_m\alpha_p$
(correlation between sibs)			$= \frac{1}{2}(1+m)$		

*No change with respect to random mating.

**r_{PO} is the correlation between parent and offspring.
r_{FO} is the correlation between father and offspring.
r_{MO} is the correlation between mother and offspring.

case the increase is by a factor $1 + m$ for both r_{PO} and r_{SS}. For the symmetric case the effect of assorting is more complex.

On the basis of the survey of activities and beliefs, the extent to which mating is assortative in cultural transmission is likely to be great. Consider, for example, item 5 of Table 2.2.2. The n_i's can be arranged in the following 2 × 2 contingency table.

		Father		
		H	h	Total
Mother	H	130	25	155
	h	18	28	46
	Total	148	53	201

Here $\chi^2 = 36.57$, which is highly significant and shows that the hypothesis of random mating is untenable. The correlation coefficient between mates is $r = \chi(201^{-1/2}) = 0.427$. Almost all 41 items in the survey have positive assortative mating significantly different from zero.

Most items showing vertical transmission in the survey are uniparental, and all items that show biparental vertical transmission have no significant deviation from additivity. This is tantamount to saying that observed B values are not significantly different from zero. Under these circumstances, assortative mating has no significant effect on equilibrium values, such as were calculated using observed b_i's in the last row of Table 2.2.2.

Assortative mating does, however, tend to increase the correlation between one parent and the offspring, and the sib-sib correlation, by a factor of the order of $(1 + m)$. Since m is large, raw r_{PO} and r_{MO} values may overestimate the influence of each parent in a nontrivial way. The effect may be especially misleading if the correlation between father and mother is at least in part the result of convergence of the traits of spouses during a long life in common. It is not usually clear at what stage of life the influence of parents on children takes place. Changes in parental values over time would be especially difficult to model. In such cases, in the absence of appropriate longitudinal studies, it may be especially difficult to evaluate parental effects in an unbiased way. When the additive model holds, it seems more satisfactory to make direct recourse to the estimates of α_m and α_p. In the example of item 5, correlation between father and mother is $m = 0.427$, and correlations between father and offspring, and mother and offspring, are $r_{FO} = 0.330$ and $r_{MO} = 0.295$, respectively. Here, as for practically all items in the survey, m is greater than the parent-offspring correlations and hence the bias is not negligible. The α_p and α_m values are, correspondingly, $\alpha_p = .27$ and $\alpha_m = .21$. The partial correlation coefficients between offspring and one parent, eliminating the effect of the other parent, are $r_{FO.M} = 0.235$ and $r_{MO.F} = 0.181$: not very different from α_p and α_m. Over most items we found a good correspondence between partial correlations and separate parental effects estimated by the additive model.

Summarizing, if mating is not random, the frequencies of four mating types p_0, p_1, p_2, p_3 can be expressed as functions of the trait frequency u (assumed to be the same in males and females) and a quantity m, which is equal to the correlation coefficient between mates. The latter is zero if mating is random, positive if mates are like each other (maximum value of $m = 1$), negative if they tend to differ (minimum value $m = -1$). Assortative mating ($m \neq 0$) affects the equilibrium frequency if $B = b_3 + b_0 - b_1 - b_2$ is different from zero. The correlations between relatives (r_{PO} and r_{SS}) are unaffected by assortative mating if transmission is from one parental sex only; they are multiplied by $(1 + m)$ in the genetic haploid and in the additive case, and show a slightly more complex behavior for symmetric transmission. In the presence of assortative mating, the correlation coefficients between one parent and offspring, therefore, supply a biased representation of the effect of that parent. The parental contributions obtained by fitting the additive model, or partial correlation coefficients eliminating the effect of the other parent, are ordinarily preferable.

2.6 NATURAL SELECTION

Individuals having a vertically transmitted trait H may have a probability of survival different from others in the population. Alternatively, H individuals may produce a different number of offspring from those with phenotype h. Either of these will result in differential contributions of the phenotypes to the next generation, even when the transmission rules for all matings are the same. The manifestation of these different contributions from generation to generation due to altered viability or fertility, or ability to mate is the process of *natural selection*. The relative abilities of H and h to contribute to the next generation are their *fitnesses*, as discussed in §1.6. Ordinarily we assume that the fertility of a mating can be computed as the arithmetic mean of the fertility of the two mates. When this is not the case, the fertility of each mating has to be defined separately. We shall ignore this complication, and, as is customary in almost all studies of natural selection in genetic evolution, our discussion will be valid strictly for selection due to differential survival and additive behavior of the fertilities.

We call the relative fitnesses of the two types:

$$1 + s \text{ for } H$$

$$1 \text{ for } h.$$

As these are relative values, they can express higher (or lower) fertility of H versus h, and higher (or lower) mortality of H versus h.

When the trait value H is the presence of a skill that is absent in h, s is probably positive, so that H individuals have, on average, an advantage over h. If the trait value H corresponds to having an infectious disease, s is likely to be negative, with a lower bound of $s = -1$: this would be the case if the disease killed all its carriers before reproduction. Such a trait is called *lethal*. In the earlier sections of this chapter, the traits have been assumed to be *selectively neutral*, that is, $s = 0$.

As this is the first time that we introduce natural selection in a model of cultural transmission we should reemphasize its difference from cultural selection. The latter is determined by the acceptance or refusal by the individual after his or her exposure to the trait. Thus cultural selection is assumed to have already occurred at the time the Darwinian fitness of the trait is tested. According to this treatment, cultural selection has been incorporated into the transmission step. In fact, the parameters b_i should be considered as composed of the two steps described in Chapter 1: signaling by the transmitter and acceptance by the receiver of what is being transmitted.

The complexity of the mathematical treatment is usually increased by the presence of natural selection. The reason is that the relative proportion of H individuals in the tth generation, u_t, must now be multiplied by $1 + s$ to obtain those who survive to reproduce to form the $(t + 1)^{\text{st}}$ generation. Similarly, the relative proportion of h must be multiplied by its relative fitness, in this case, unity. Then the relative proportion of H individuals after selection is obtained by relating the number of H individuals after selection to the sum of expected H and h individuals modified by selection:

$$u_t^* = \frac{u_t(1 + s)}{(1 + s)u_t + 1 - u_t} = \frac{u_t(1 + s)}{1 + su_t}. \tag{2.6.1}$$

The star has been added to indicate that the census is after selection. Now matings occur between adults: $H \times H, H \times h, h \times H$, and $h \times h$ with frequencies $p_3^{(t)*}$, $p_2^{(t)*}$, $p_1^{(t)*}$, and $p_0^{(t)*}$, respectively. In the case of random mating, these are just $(u_t^*)^2$,

$u_t^*(1 - u_t^*)$, $u_t^*(1 - u_t^*)$, $(1 - u_t^*)^2$, respectively. The offspring are H or h according to the transmission rules of Table 2.2.1. With random mating this process results in a frequency u_{t+1} of juveniles in the next generation, with

$$u_{t+1} = b_3(u_t^*)^2 + (b_1 + b_2)u_t^*(1 - u_t^*) + b_0(1 - u_t^*)^2. \quad (2.6.2)$$

The frequencies after selection in generation $t + 1$ are then

$$u_{t+1}^* = (1 + s)u_{t+1}/(1 + su_{t+1}), \quad (2.6.3)$$

which, using (2.6.2) gives a one-step recursion connecting successive adult frequencies after selection but before mating. At equilibrium the values of u^* (after selection) and u (before selection) are connected by

$$u^* = \frac{u(1 + s)}{1 + su}. \quad (2.6.4)$$

The complexity due to selection is revealed in (2.6.3), where the transformation relating successive frequencies is now a ratio of quadratic functions. Analytical iteration of such a recursion is, in general, impossible. However, some progress can be made in the analysis of equilibrium properties of the transformation. At equilibrium $u_{t+1}^* = u_t^*$, so that equilibria are the solutions of the cubic equation

$$sB(u^*)^3 + (u^*)^2[sC - B(1 + s)] + u^*[1 + sb_0 - C(1 + s)] - b_0(1 + s) = 0. \quad (2.6.5)$$

In (2.6.2) to (2.6.5) $b_1 + b_2$ appears as a single parameter. If cultural transmission is for an advantageous trait (i.e. $s > 0$) such as a skill, a reasonable assumption might be

$$b_3 > \frac{b_1 + b_2}{2} > b_0. \quad (2.6.6)$$

If (2.6.6) holds, then with $s > 0$ the right side of (2.6.4) is a monotonically increasing function of u_t^*. If, in addition, $b_0 > 0$

and $b_3 < 1$, then there must be either one or three valid equilibria of (2.6.6), (valid in the sense that the equilibria $\hat{u}$ satisfy $0 \leq \hat{u} \leq 1$). If there is a single equilibrium, then the monotonicity ensures its stability. If there are three valid equilibria, say $\hat{u}_1 \leq \hat{u}_2 \leq \hat{u}_3$, then $\hat{u}_1$ and $\hat{u}_3$ must be stable, with $\hat{u}_2$ unstable. When (2.6.6) does not hold and the transformation is not monotonic, more complicated trajectories can arise.

For the special cases in Table 2.3.1 more can be said. In case 1 (sex dependent), (2.6.3) reduces to

$$u^*_{t+1} = \frac{(1 + s)[bu^*_t + c(1 - u^*_t)]}{1 + s[bu^*_t + c(1 - u^*_t)]}. \tag{2.6.7}$$

This is a linear fractional transformation and can be iterated explicitly; the valid equilibrium to which global convergence occurs is

$$\hat{u}^* = \frac{(1 + s)(b - c) - (1 + sc) + \{[(1 + s)(b - c) - (1 + sc)]^2 + 4cs(1 + s)(b - c)\}^{1/2}}{2s(b - c)} \tag{2.6.8}$$

With $b = 1, c = 0$ (case 1a) the only equilibria are $\hat{u} = 0$, and $\hat{u} = 1$, and, if $s < 0$, the former is stable, while if $s > 0$ the latter is stable. The same conclusions hold for case 2 (genetic haploid). The symmetric case does not reduce to a quadratic, while the additive model does, and has a stable equilibrium

$$\hat{u}^* = \frac{2\alpha(1 + s) - 1 - sb_0 + \{[2\alpha(1 + s) - 1 - sb_0]^2 + 8b_0\alpha s(1 + s)\}^{1/2}}{4s\alpha}. \tag{2.6.9}$$

This simplifies to

$$\hat{u} = [2\alpha(1 + s) - 1]/2s\alpha, \tag{2.6.10}$$

when $b_0 = 0$.

The infectious case 6, which we called "Kuru," has a number of interesting features. Case 6 allows three equilibria:

$$\hat{u}_1 = 0,\ \hat{u}_2^* = \frac{1 + 2s}{s},\ \text{and } \hat{u}_3 = 1, \tag{2.6.11}$$

all of which are valid if $s < -1/2$. $\hat{u}_3^* = 1$ is always stable and $\hat{u}_1^* = 0$ is stable if $\hat{u}_2^*$ is valid, i.e., if $s < -1/2$. Thus the infection can kill up to (but not more than) 50% of its carriers and still spread through the population from an initially low frequency. Of course if the population is small enough this may result in its extinction, a problem that the present model does not address. If $s < -1/2$, then $\hat{u}_2^*$ merely bounds the domains of those initial u values that result in convergence to $\hat{u}_1^* = 0$ and $\hat{u}_3^* = 1$. More details and discussion of this model can be found in Feldman and Cavalli-Sforza (1976).

When $b_3 = 1$ and $b_0 = 0$, but $b_1 + b_2 = C < 2$, we have a less restrictive version of the Kuru model. Here (2.6.3) simplifies to

$$u_{t+1} = \frac{(1 + s)u_t[C + (1 - C)u_t]}{1 + su_t[C + (1 - C)u_t]}, \tag{2.6.12}$$

which has $\hat{u}_1^* = 0$, $\hat{u}_3^* = 1$ as equilibria, and in addition the interior equilibrium:

$$\hat{u}_2^* = \frac{1 - C(1 + s)}{s(1 - C)}. \tag{2.6.13}$$

The last is valid only in the range

$$1 - s < C < \frac{1}{1 + s}. \tag{2.6.14}$$

The equilibrium $\hat{u} = 0$ is stable if $C < 1/(1 + s)$ and $\hat{u} = 1$ is stable if $C > 1 - s$, so that in the range (2.6.14) of validity of $\hat{u}_2^*$ both $\hat{u}^* = 0$ and $\hat{u}^* = 1$ are stable (see Fig. 2.6.1), and $\hat{u}_2^*$ separates the domains of attraction to these two.

In summary, natural selection for or against carriers of trait H can be measured by a selection coefficient s that is positive if H is

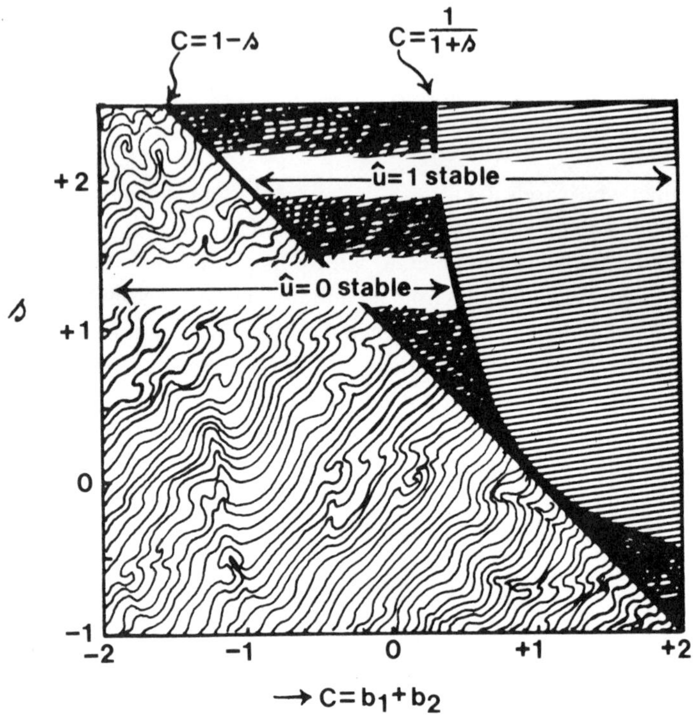

FIGURE 2.6.1. Equilibrium for vertical transmission $b_3 = 1$, $b_0 = 0$, $b_1 + b_2 = C$. In the presence of selection (s) in the striped area, $\hat{u} = 1$ is stable; in the area of wavy lines, $\hat{u} = 0$ is stable. In the area between $C = 1 - s$ and $C = 1/(1 + s)$ both equilibria are stable; the trait frequency u will turn to 0 or 1 depending on whether the initial trait frequency is smaller or greater than the unstable equilibrium value, $\hat{u} = [1 - C(1 + s)]/s(1 - C)$.

advantageous and negative if H is deleterious. The incorporation of s increases the complexity of the analysis; for instance, the equilibria are now roots of a cubic equation. A selectively disadvantageous trait could be a maladaptive social custom (e.g., one creating some degree of danger to life that is not compensated for by other advantages in Darwinian fitness) or a custom decreasing fertility (see Section 3.9), or an infectious disease. In the simple situations where $b_3 = 1$ and $b_0 = 0$, but $b_1 + b_2 = C$

($0 \le C \le 2$), then depending on the relationship between C and s, even a disadvantageous ($s < 0$) trait can spread to the whole population from an initially very low frequency. If $C = 2$, the extreme case, this can happen with selection as strong as $s = -0.5$ (i.e. half the carriers die before reproduction). For large initial frequencies even stronger selective disadvantages can be overcome.

2.7 MUTATION

The various models discussed so far have produced equilibria with single phenotypes represented (either H or h), or with a polymorphism (that is, both H and h types are present at equilibrium; $\hat{u} \neq 0, 1$). These processes of evolution require variability in the population before natural selection or the variety of transmission rules can operate. It is natural then to inquire as to the origin of the variability. That is, in a population that was originally all h, how did H arise, whether these phenotypes define skills, beliefs, customs, or states of a disease. The formal answer familiar to biologists for a genetically determined trait is that H arises either by mutation of one or more h individuals to H, or by immigration of H from some outside source into the monomorphic h population. As we have already mentioned, these two explanations may be difficult to distinguish in both genetic and cultural evolution. Independent inventions of a process or object are the cultural equivalent of recurrent mutation in genetics.

By analogy with genetic evolution, we can expect population size to determine many important properties of cultural evolution in small populations. In particular, the interaction between population size and rate of innovation should play a key role, as it does in genetics, and we shall discuss this in a later section. Here we consider the phenomenon at a deterministic level.

In the formulation of Table 2.2.1, H individuals can be produced by $h \times h$ matings if $b_0 > 0$ and h individuals from $H \times H$ matings if $b_3 < 1$. Such offspring can arise by: (i) mutation,

(ii) immigration from other populations and appearance in this group of progeny because of some form of adoption, or (iii) by transmission from H members of the group (other than the parents, of course). The third alternative is the equivalent of nonvertical transmission and is frequency-dependent. It is the subject of the next chapter. We will here consider mutation, and discuss special aspects of migration in Chapter 3, keeping in mind the practical difficulty of a distinction between mutation and migration from the outside when there is not sufficient historical information. The problem of correct interpretation of transmission parameters in terms of the three alternatives above is not a mathematical one although it is qualitatively important in the reconstruction of evolutionary history.

In the *homotypic* matings, $H \times H$ and $h \times h$, the occurrence of progeny unlike either parent may be interpreted as a mutation. Call the mutation rates in the two directions μ for H to h, and ν from h to H. The probability of a parent H not having a child h by mutation is then $1 - \mu$, and the probability that in mating $H \times H$ the progeny is H, that is the transmission parameter b_3, is then $(1 - \mu)^2$. In the $h \times h$ mating the probability of h progeny is similarly $(1 - \nu)^2$, and parameter b_0 is then $1 - (1 - \nu)^2$. Assuming the mutation rates to hold in the heterotypic matings, $H \times h$ matings will have progeny H half of the time with probability $(1 - \mu)$ (from the H mother), and half of the time with probability (ν) (from the h father), or on average $(1 - \mu)/2 + \nu/2$. Similarly, for the other heterotypic mating the probability is $\mu/2 + (1 - \nu)/2$. The transmission parameters, reinterpreted as mutation rates, will then be as in the table below:

$$
\begin{array}{ll}
H \times H & b_3 = (1 - \mu)^2 \\
H \times h & b_2 = (1 - \mu + \nu)/2 \\
h \times H & b_1 = (1 - \nu + \mu)/2 \\
h \times h & b_0 = 1 - (1 - \nu)^2
\end{array}
\qquad (2.7.1)
$$

Under random mating with no selection, the recursion for the trait frequency will then be:

$$u_{t+1} = u_t^2(1-\mu)^2 + u_t(1-u_t) + (1-u_t)^2[1-(1-\nu)^2]. \quad (2.7.2)$$

Calling $z = u/(1-u)$, at equilibrium

$$z^2[1-(1-\mu)^2] = 1-(1-\nu)^2,$$

or

$$z^2 = \frac{\nu(2-\nu)}{\mu(2-\mu)} \simeq \frac{\nu}{\mu}, \text{ if } \mu \text{ and } \nu \text{ are small,}$$

and therefore

$$\frac{\hat{u}}{1-\hat{u}} \sim \sqrt{\nu/\mu}. \quad (2.7.3)$$

If migration rather than mutation is the cause of a new H phenotype, then we can introduce a parameter m, the proportion of immigrants into the population per generation. If P_0 is the frequency of H in the source population, then the fraction mP_O takes the same role as the mutation rate. In certain situations the treatment of mutation and migration is formally identical, and it is not too surprising that frequently they cannot be distinguished in empirical situations. This is discussed in detail in Chapter 3.

2.8 RANDOM-SAMPLING DRIFT

The analysis of trait frequency change has so far been deterministic. Populations have been assumed to be so large that, for example, every $H \times H$ mating produces exactly the proportions $1 - b_3$ of H and b_3 of h offspring. Within this class of matings no variation in offspring proportions is allowed. In the Mendelian example of a mating between the diploid genotypes $Aa \times Aa$, among the offspring, AA, Aa, and aa occur with probability 1/4,

1/2 and 1/4, respectively, but if there are few offspring, each $Aa \times Aa$ mating will realize different offspring frequency arrays that should vary around the expected array of 1/4, 1/2, 1/4. These variations are *sampling* fluctuations due to the finite size of the population. In this section, the investigation concerns the consequences of finiteness of population size for the evolution of a culturally inherited trait. The highly developed corresponding theory for genetic characters will play a central role in the analysis.

It is widely accepted that much of hominid evolution has taken place in small groups. These may have been extended family or larger tribal aggregations. It is in this small population setting that the precursors of what is now called culture evolved. The groups were, presumably, not entirely isolated, nor were they all permanently identified. The formation and extinction of subgroups must have played an important evolutionary role, but the focus here will be on a single group of individuals, N in number, isolated from all other groups. There are two phenotypes, H and h, and the transmission probabilities are b_3, b_2, b_1 and b_0, as before. N and b_i are constant in time.

There are many ways to model the change in frequency of H over time in this finite population setting. The primary obstacle is to define the sampling procedure that enables N offspring to be produced by a total of N male and female parents while at the same time using the b_i to assign the offspring into H and h classes. The model used most widely in population genetics to represent evolution in finite populations is due to Fisher (1922, 1930) and Wright (1931b). It has the following structure: suppose that the allele A occurs with frequency x_t in the tth generation and its alternative form a with frequency $1 - x_t$. That is, the number of A is $Nx_t = i$, say. Suppose also that if the population were infinitely large, so that deterministic analysis were valid, then the frequency x_{t+1} of A in the next generation would be

$$x_{t+1} = f(\theta, x_t), \tag{2.8.1}$$

where θ is a vector of parameters describing all evolutionary forces under study. [For instance, if there is mutation from A to a at rate μ_1 and from a to A at rate μ_2 and the relative fitnesses of genotypes AA, Aa, and aa are $1 - s_1$, 1, and $1 - s_2$, then θ is the vector (μ_1, μ_2, s_1, s_2).] The model then postulates that the N offspring are produced by binomial sampling using the probability x_{t+1} in (2.8.1). Thus the probability that there are j of A in generation $t + 1$ is

$$P_{ij} = \binom{N}{j} x_{t+1}^j (1 - x_{t+1})^{N-j}. \tag{2.8.2}$$

P_{ij} are the entries of a matrix $\mathbf{P}$ of size $N \times N$ specifying the probabilities of transitions from the various values of i in one generation to j in the next. The model is now fully specified.

The evolution of the population is expressed in terms of probabilities of passage from i_n copies of A in the population at generation n_n to j_{n+m} copies of A m generations later. These probabilities are the entries $P_{ij}^{(m)}$ of the powered matrix $\mathbf{P}^m$.

$$P_{ij}^{(m)} = (\mathbf{P}^m)_{ij}. \tag{2.8.3}$$

The study of the fate of A then entails the mathematical analysis of powers of $\mathbf{P}$, each of whose rows $\mathbf{P}$ is expressed in terms of a probability density.

The possible numbers 0, 1, 2, . . . , N taken by i and j are the *states* of the population expressing how many A there are. The random process that occurs by the transition from state i in one generation to state j in the next does not depend on any generation earlier than the former. Such a process is called a *Markov* process, and, in particular when the transition probabilities are specified by a matrix such as (2.8.2), the process is called a Markov chain.

The highly developed general theory of Markov chains has been successfully applied to the Wright-Fisher model outlined above. The mathematical details and applications may be found in the books by Karlin (1966), Moran (1962), Ewens (1979), Crow and Kimura (1970) and Cavalli-Sforza and Bodmer

(1971). Rather than present this population genetic theory, we shall discuss the results for a corresponding cultural evolution scheme and show how the finite population genetic models relate to those with cultural transmission.

The first case we consider is a stochastic version of the deterministic model of section 2.1. The sequence of events is as follows. First there is cultural transmission from the parents in generation t, according to Table 2.2.1, producing an expected H frequency in the offspring of u_{t+1}. Then there is binomial sampling among these offspring to produce the N individuals of the next generation. The frequency after sampling is a random variable which we write U_{t+1}. According to the sampling procedure just outlined,

$$P_{ij} = P[NU_{t+1} = j \,|\, NU_t = i] = \binom{N}{j} u_{t+1}^j (1 - u_{t+1})^{N-j} \tag{2.8.4}$$

is the probability of the transition from i H individuals in generation t to j H individuals in generation $t + 1$, where, in (2.8.4),

$$u_{t+1} = b_3 u_t^2 + (b_1 + b_2) u_t (1 - u_t) + b_0 (1 - u_t)^2 \tag{2.8.5}$$

as in (2.2.2); u_{t+1} takes the role of x_{t+1} in (2.8.3) and the vector $\theta = (b_3, b_2, b_1, b_0)$.

The matrix $\mathbf{P} = \|P_{ij}\|$ completely specifies the process of change in numbers of H individuals among the N in the population. The values of i and j are in the set $\{0, 1, 2, \ldots, N\}$.

Obviously in terms of real mating and families the process specified by (2.8.4) and (2.8.5) must be regarded as a simplification. However, if males and females are equally frequent, if the frequencies of H in the two sexes are equal, if random mating takes place among the N parents, and if the number of offspring per family has a Poisson distribution, then the resulting changes in overall H frequency (keeping the population size fixed at N) follow the binomial sampling rules. Some other offspring number distributions produce qualitatively similar results. In genetics the offspring generation is said to be sampled from a "gene pool," in which the frequencies are those among the parents. Here the idealized equivalent is a "phenotype pool." Neither exist in reality; both are devices constructed to formalize evolutionary theory in finite populations. In genetics, as a conceptual device, the binomial sampling model has been extremely successful.

In the haploid genetic model $b_3 = 1$, $b_1 = b_2 = 1/2$ and $b_0 = 0$. This is the model that has been exhaustively studied in genetics, and is the classical Wright-Fisher model. Since $b_3 = 1$, a population with only H individuals cannot change. Similarly, since $b_0 = 0$, a population of only h individuals cannot change. These statements are expressed quantitatively in the **P** matrix by the relations $P_{Ni} = 0$ for all $i \neq N$ with $P_{NN} = 1$ and $P_{0i} = 0$ for $i \neq 0$ with $P_{00} = 1$. The two states 0 and N are special in this regard. They are called *absorbing* states. With probability one the process ends up in one of them, that is, in fixation of H or h. The probability of fixation in H is the frequency of H in the initial generation, u_0, while $1 - u_0$ is the probability of fixation in h.

Our interest here focuses on similar questions in cases where the transmission is *not* genetic. First, if $b_1 \neq b_2$ but $b_1 + b_2 = 1$, then the situation is indistinguishable from the genetic haploid model. Otherwise there are three qualitatively different cases. When $b_3 = 1$, $b_0 = 0$ but $b_1 + b_2 \neq 1$, then 0 and N are still absorbing states but the model is *not* the haploid genetic one. The first class of cases examines the role of $b_1 + b_2 \neq 1$. In the second class either $b_3 \neq 1$, $b_0 = 0$ or $b_3 = 1$, $b_0 \neq 0$, so that a population of all H can produce h offspring but the opposite cannot occur and only 0 is an absorbing state. Finally, if $b_3 \neq 1$, $b_0 \neq 0$, then there are no absorbing states. The states 0 and N are "reflecting barriers" and in the long term the population remains polymorphic.

The analysis of these three classes of models requires that we pass from the discrete time representation U_t, U_{t+1}, and so on, to a continuous time approximation. Let $f(u_0, u;t)$ be the probability density that the frequency of H becomes u at time t given that the initial H frequency is u_0. Then the continuous approximation to the discrete process specified by (2.8.4) produces $f(u_0, u;t)$ as the solution of what is called the Kolmogorov forward equation:

$$\frac{\partial f}{\partial t} = \frac{1}{2}\frac{\partial^2}{\partial u^2}(V_{\delta u} f) - \frac{\partial}{\partial u}(M_{\delta u} f), \tag{2.8.6}$$

where $V_{\delta u}$ and $M_{\delta u}$ are the mean and variance of the changes in U per time unit. The way in which this approximation, usually called the diffusion approximation, is achieved as follows. The changes per generation in the discrete process are assumed to be so small that the mean and variance of the change per N generations (N is the population size) remain constant as N becomes large. By changing the time scale to units of N generations, each single generation in the original time scale is $1/N$ time units in the new time scale. As N becomes large, the process specified by (2.8.6), with t measured in units of N generations, approximates the original discrete process.

This process is presented rigorously in textbooks (e.g. Breiman, 1968; Karlin and Taylor, 1980). We shall be content here to demonstrate how $M_{\delta u}$ and $V_{\delta u}$ are obtained. The definitions are

$$\begin{aligned} M_{\delta u} &= \lim_{\Delta t \to 0} \frac{1}{\Delta t} E(U_{t+1} - U_t \mid u_t = u) \\ &= \lim_{N \to \infty} NE(U_{t+1} - U_t \mid u_t = u), \end{aligned} \tag{2.8.7}$$

and

$$V_{\delta u} = \lim_{N \to \infty} NE\{(U_{t+1} - U_t)^2 \mid U_t = u\}. \tag{2.8.8}$$

To calculate $M_{\delta u}$, return to (2.8.4), where we see that

$$E(U_{t+1} - U_t | U_t = u)$$
$$= b_3u^2 + (b_1 + b_2)u(1 - u) + b_0(1 - u)^2 - u \quad (2.8.9)$$
$$= (b_3 - 1)u^2 + (b_1 + b_2 - 1)u(1 - u) + b_0(1 - u)^2. \quad (2.8.10)$$

Now for the limit in (2.8.6) to exist, the three parameters $(b_3 - 1)$, $(b_1 + b_2 - 1)$, and b_0, which appear in (2.8.10), should satisfy

$$\lim_{N\to\infty} N(b_3 - 1) = -\beta_3, \lim_{N\to\infty} N(b_1 + b_2 - 1) = \beta_{12},$$
$$\lim_{N\to\infty} Nb_0 = \beta_0, \quad (2.8.11)$$

where β_3, β_{12}, β_0 are independent of N. To obtain a diffusion equation approximation of the form (2.8.8), the assumptions (2.8.11) must be made. (Qualitatively the conditions (2.8.11) indicate that for the limiting process specified by (2.8.6) to be valid, the departure of b_3 from 1, for example, must be sufficiently small in each generation that in the new time scale defined by (2.8.7) and (2.8.8) the accumulation of such departures remain bounded.) It should be emphasized that this parameterization is not unique. We have already written (2.8.9) as (c.f. 2.2.3)

$$Bu^2 + (C - 1)u + b_0,$$

so that an alternative to (2.8.11) might be

$$\lim_{N\to\infty} NB = \beta, \quad \lim_{N\to\infty} N(C - 1) = \gamma, \quad \lim_{N\to\infty} Nb_0 = \beta_0. \quad (2.8.12)$$

Note that $\beta = 0$ under additivity of parental effects. The assumptions (2.8.11) and (2.8.12) are interconvertible since

$$\gamma - \lim_{N\to\infty} N(C - 1) = \lim_{N\to\infty} N(b_1 + b_2 - 2b_0 - 1)$$
$$= \lim_{N\to\infty} N(b_1 + b_2 - 1) - 2\beta_0,$$

so that $\beta_{12} = \gamma + 2\beta_0$. Similarly, $-\beta_3 = \gamma + \beta + \beta_0$. Using (2.8.11) we have

$$M_{\delta u} = -\beta_3u^2 + \beta_{12}u(1 - u) + \beta_0(1 - u)^2. \quad (2.8.13)$$

To calculate $V_{\delta u}$, write

$$E\{(U_{t+1} - U_t)^2 | U_t = u\} = E\{[U_{t+1} - EU_{t+1}]^2 - 2U_tEU_{t+1} + U_t^2 + (E(U_{t+1}))^2 | U_t = u\}$$
$$= \text{Var}\,(U_{t+1} | u_t = u) + \{E[(U_{t+1} - U_t) | U_t = u]\}^2. \quad (2.8.14)$$

Now the second term in (2.8.14) is the square of (2.8.10), and under the assumptions (2.8.11) is of the order $1/N^2$. The first term in (2.8.14) is the variance of the proportion in a binomial sample, and by use of (2.8.7) and (2.8.4) is shown to be

$$\text{Var}\,(U_{t+1} | U_t = u) = (M_{\delta u}/N + u)(1 - M_{\delta u}/N - u)/N. \quad (2.8.15)$$

The only term of order higher than $1/N^2$ in (2.8.15) is $u(1 - u)/N$. Now apply the definition (2.8.8); only this last term makes a contribution as $N \to \infty$, so

$$V_{\delta u} = u(1 - u). \quad (2.8.16)$$

Higher infinitesimal moments make no contribution, so that the continous approximation of (2.8.4) is (2.8.6) with $M_{\delta u}$ and $V_{\delta u}$ as in (2.8.13) and (2.8.16). With the diffusion equation (2.8.6) fully specified we can turn our attention to the three types of parameter configurations.

Case I. $b_3 = 1, b_0 = 0$. From (2.8.11), $\beta_3 = 0, \beta_0 = 0$. This case is analogous to the genetic case without mutation. The difference is that in the genetic case there is the additional stipulation that $\beta_1 + \beta_2 = 1$, so that $\beta_{12} = 0$ and $m_{\delta u} = 0$. In the present more general case we have

$$M_{\delta u} = \beta_{12}u(1 - u). \quad (2.8.17)$$

Using well-known results from diffusion theory (e.g., Crow and Kimura 1970, p. 424), $F(u_0)$, the probability of eventual fixation in H starting from an initial H frequency of u_0 can then be calculated:

$$F(u_0) = \int_0^{u_0} G(u)du \Big/ \int_0^1 G(u)du, \qquad (2.8.18)$$

where

$$G(u) = \exp\{-2\int (M_{\delta u}/V_{\delta u})du\}. \qquad (2.8.19)$$

Since $M_{\delta u}/V_{\delta u} = \beta_{12}$, it is easy to see that

$$F(u_0) = \frac{1 - e^{-2\beta_{12}u_0}}{1 - e^{-2\beta_{12}}} \qquad (2.8.20)$$

The initial frequency u_0 would be $1/N$ if the trait H arose by a single non-recurrent cultural mutation, that is, by a unique vertically transmitted innovation from a single individual. u_0 would be expected to be strongly tied to the population size in this case. If there were a massive cultural takeover, then u_0 would be much larger.

It is mathematically true that

$$\begin{aligned} F(u_0) &> u_0 \text{ if } \beta_{12} > 0 \\ F(u_0) &< u_0 \text{ if } \beta_{12} < 0. \end{aligned} \qquad (2.8.21)$$

These equations state the intuitively obvious fact that, if the transmission of H is more frequent than genetic, the probability of fixation in H is greater than in the genetic case, and vice versa.

It should be noted that in the haploid genetic model, with types A and a, where the fitnesses are $(1 + s)$:1 exactly, the same functions $M_{\delta u}$ and $V_{\delta u}$ arise with s in the role of β_{12}. For this reason we can translate the familiar results for the genetic case into those above for cultural transmission. Another such result is the expected time until fixation of H. Suppose that there is initially a fraction u_0 of H. How long does it take for the whole population to become of phenotype H, assuming that H has not been lost from the population? For the haploid genetic model this was investigated by Kimura and Ohta (1969). The answer in the case at hand is

$$T_H(u_0) = J_1 + \frac{1 - F(u_0)}{F(u_0)} J_2, \tag{2.8.22}$$

where $F(u_0)$ is given in (2.8.20),

$$J_1 = \frac{1}{\beta_{12}(1 - e^{-2\beta_{12}})} \int_{u_0}^{1} \frac{(e^{-2\beta_{12}\xi} - 1)(e^{-2\beta_{12}\xi} - e^{-2\beta_{12}})}{\xi(1 - \xi)} d\xi \tag{2.8.23}$$

and

$$J_2 = \frac{1}{\beta_{12}(1 - e^{-2\beta_{12}})} \int_{0}^{u_0} \frac{(e^{2\beta_{12}\xi} - 1)(1 - e^{-2\beta_{12}\xi})}{\xi(1 - \xi)} d\xi. \tag{2.8.24}$$

In Figure 2.8.1 curves are drawn showing the theoretical relationship between $T_H(u_0)$ and u_0 for $\beta_{12} = 0.5$ and $\beta_{12} = 5$. The curves were obtained by numerical integration of (2.8.22). There is a strong decrease in T_H as β_{12} increases. In view of the diffusion limit procedure, time in (2.8.22) is measured in units of N generations where N is the population size.

Case II. $b_3 = 1, b_0 \neq 0$ ($\beta_3 = 0, \beta_0 \neq 0$). We discuss only this case, since the corresponding results when $b_3 \neq 1, b_0 = 0$ are completely analogous to those presented. As mentioned above, if $b_3 = 1$, then fixation of phenotype H is an absorbing state, whereas with $b_0 \neq 0$, fixation in h represents a reflecting barrier. The general theory of Markov chains informs us that in this situation fixation of H is inevitable from any starting frequency, that is, the probability that the population is ultimately all H is one. From (2.8.13) in this case,

$$M_{\delta u} = \beta_{12}u(1 - u) + \beta_0(1 - u)^2 \tag{2.8.25}$$

while

$$V_{\delta u} = u(1 - u). \tag{2.8.26}$$

The application of a result from general diffusion theory, whose proof is not within the scope of this presentation, produces the average time for the fixation of H from an initial state u_0 as

$$\int_{u_0} d\eta g(\eta) \int_0^{\eta} [g(x)\,x\,(1 - x)]^{-1} dx, \tag{2.8.27}$$

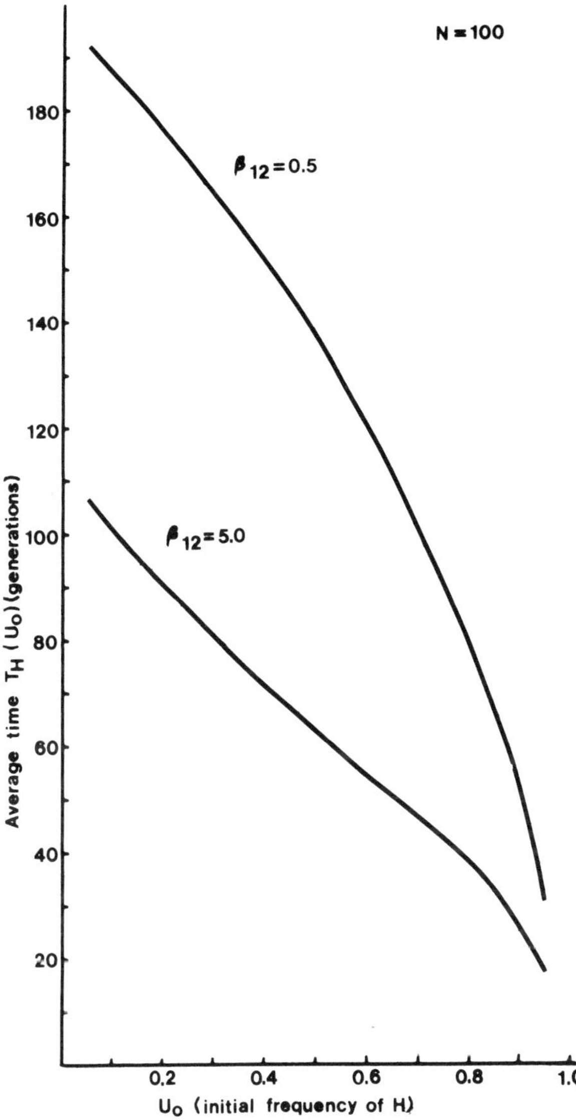

FIGURE 2.8.1. Conditional expected times to fixation in H as a function of initial frequency. The curves are obtained by numerical integration of (2.8.22). (Courtesy of K. H. Chen.)

where

$$g(y) = e^{-2(\beta_{12}-\beta_0)y} y^{-2\beta_0}. \tag{2.8.28}$$

This integral is considerably more complicated than the analogous result from population genetics. The reason for this is that, when there is no genic selection, irreversible mutation in the corresponding genetic model produces an $M_{\delta u}$ function that is linear. From (2.8.25), the present case gives a quadratic function. Even if $\beta_{12} = 0$, the form of (2.8.27) is still too complicated to integrate in closed form, while with the linear form of $M_{\delta u}$, the expected time to fixation with unidirectional mutation is no simpler than an integral of a hypergeometric function. In fact, the model of two alleles A and a, with mutation only in the direction of a to A at the rate $\alpha_2 = (\beta_2/N)$, produces $M_{\delta u} = \beta_2(1 - u)$, $V_{\delta u} = u(1 - u)$. The expected time to fixation starting from an A frequency of u_0

$$2\int_{u_0}^{1} d\eta\, \eta^{-2\beta_2} \int_0^{\eta} \xi^{2\beta_2 - 1}(1 - \xi)^{-1} d\xi,$$

which can be expressed in terms of the hypergeometric function (Abramowitz and Stegun 1965, p. 558).

Case III. $b_0 \neq 0$, $b_3 \neq 1$. Here both $u = 0$ and $u = 1$ are reflecting barriers, and the general theory of Markov chains informs us that a stationary distribution should be attained. There are standard methods for obtaining the approximate form of this distribution from the diffusion equation approximation. $M_{\delta u}$ is given by (2.8.13), and $V_{\delta u} = u(1 - u)$. The stationary distribution is

$$\phi(x) = C\, e^{-2\beta x} x^{2\beta_0 - 1}(1 - x)^{2\beta_3 - 1}. \tag{2.8.29}$$

This is completely analogous to the haploid genetic model with allele A having a selective advantage s over a and where the mutation rates from A to a and a to A are $N\beta_3$ and $N\beta_0$, respectively. In (2.8.29), the factor analogous to s is

$$-\beta = -NB = -N[b_3 - b_1 - b_2 + b_0]$$
$$= \beta_3 + \beta_{12} - \beta_0. \quad (2.8.30)$$

The selection term here is *confounded* with the "mutation" terms. Indeed we can speculate that the models of Table 2.3.1 are easy to study because with $B = 0$ there is no "effective selection" on H.

In conclusion, models of cultural drift *without* natural selection, when compared with analogous models from haploid population genetics, are formally similar to the genetic models *with* natural selection. This form of selection can even more profitably be viewed as a generalized "segregation distortion," to use a genetic term. That is, in genetics it occasionally happens that the Mendelian transmission rules are violated. In certain cases this results in a profound effect which can overcome a selective disadvantage and a deleterious gene (or genotype) in the population. The ultimate distortion is to have no preassigned rules, that is, to have b_i arbitrary. This is what we have in cultural vertical transmission, and this is why the formulas are formally similar to the genetic ones with natural selection. The correspondence is difficult to make complete because of confounding with the corresponding "mutation" factors $1 - b_3$ and b_0, as can be seen in (2.8.29) and (2.8.30). Thus the term β which is the "effective selection coefficient" includes β_3 and β_0.

Other models of random-sampling drift more general than the binomial can be constructed and analyzed (see for instance Section 3.9). Our purpose here was to introduce a few key results rather than present an exhaustive analysis of the stochastic problem. In the next section we consider the addition of natural selection to drift for cultural vertical transmission.

2.9 DRIFT AND NATURAL SELECTION

In section 2.6 we introduced natural selection on the phenotype H, h through the transformation

$$u_{t+1}^* = \frac{(1 + s)u_{t+1}}{1 + su_{t+1}} \tag{2.9.1}$$

with [(2.6.1) – (2.6.3)]

$$u_{t+1} = Bu_t^{*2} + Cu_t^* + b_0.$$

To transform this deterministic model into a stochastic finite population model we consider the sequence of random variables $\{U_t^*\}$ where NU_{t+1}^* is binomially distributed with parameters N and u_{t+1}^*. Hence

$$E[U_{t+1}^* \mid U_t^* = u_t^*] = u_{t+1}^*, \tag{2.9.2}$$

and further

$$E[U_{t+1}^* - U_t^* \mid U_t^* = u_t^*] = \frac{(1 + s)u_{t+1}}{1 + su_{t+1}} - u_t^*$$

$$= [1 + su_{t+1}]^{-1}\{M_{\delta u^*} + s(1 - u_t^*)(M_{\delta u^*} + u_t^*)\}, \tag{2.9.3}$$

where

$$M_{\delta u^*} = Bu_t^* + Cu_t^* + b_0 - u_t^*,$$

exactly as in (2.8.9). Now, in the same way that we assumed $(1 - b_3)N$, $(b_1 + b_2 - 1)N$, and b_0N all tend to 0 as N tends to ∞, to validate a diffusion approximation, we must assume in addition that the selection coefficient s satisfies

$$s = \sigma/N. \tag{2.9.4}$$

With this assumption, the term $[1 + su_{t+1}]$ in the denominator of (2.9.3) can be ignored. In addition, the product $sM_{\delta u^*}$ can be ignored, since it is of the order $1/N^2$. We are left with

$$E[U_{t+1}^* - U_t^* \mid U_t^* = u_t^*]$$

$$= \{M_{\delta u^*} + \sigma[u_t^*(1 - u_t^*)]\}/N. \tag{2.9.5}$$

Similarly,

$$E[(U_{t+1}^* - U_t^*)^2 \mid U_t^* = u_t^*] = \frac{u_t^*(1 - u_t^*)}{N}, \tag{2.9.6}$$

with smaller order terms neglected. The moments (2.9.5) and

(2.9.6) are both correct to the order N^{-1}, that is, terms neglected are like N^{-2} or smaller. Now following the limiting procedure in (2.8.7) and (2.8.8) we write

$$\begin{aligned} M^*_{\delta u} &= M_{\delta u^*} + \sigma[u^*(1 - u^*)] \\ &= -\beta_3 u^{*2} + (\beta_{12} + \sigma)u^*(1 - u^*) + \beta_0(1 - u^*)^2 \quad (2.9.7) \end{aligned}$$

$$V^*_{\delta u} = u^*(1 - u^*). \quad (2.9.8)$$

The only difference between these expressions including selection and (2.8.13) and (2.8.14) is the inclusion of $\sigma u^*(1 - u^*)$ in (2.8.7). Nevertheless, it should be remembered that there are assumptions on s which enable this convenient representation.

The inclusion of σ as a coefficient of $u^*(1 - u^*)$ enables us to substitute the combined term $\beta_s = \beta_{12} + \sigma$ for β_{12} in all analyses of section 2.8. Thus, for example, if $\beta_3 = 0, \beta_0 = 0$ the probability of fixation from an initial frequency u_0 is

$$F^*(u_0) = \frac{1 - e^{-2\beta_s u_0}}{1 - e^{-2\beta_s}}. \quad (2.9.9)$$

What is interesting about this expression is its implication for (2.8.21). If

$$\beta_{12} + \sigma > 0, \text{ then } F(u_0) > u_0,$$

so that the selection-transmission process increases the fixation probability of H over the genetic haploid case if one of two events occurs: 1) the transmission favors H to the extent that, even if H is disadvantageous ($\sigma < 0$), the sum $\beta_{12} + \sigma > 0$. 2) Selection favors H to the extent that even if transmission favors h, (that is, $\beta_{12} < 0$), $\beta_{12} + \sigma > 0$. To see how this relates to the deterministic model with $b_3 = 1$, $b_0 = 0$, $b_1 + b_2 = C$, return to the inequality (2.6.14). If s is small we can write the right hand inequality

$$C < 1/(1 + s) \simeq 1 - s$$

or

$$C - 1 + s < 0. \quad (2.9.10)$$

If this inequality holds, then in the deterministic case, $\hat{u} = 0$ is stable. If the converse inequality holds, $\hat{u} = 1$ is stable. The carry-over to the stochastic case is obviously in terms of $N[C - 1 + s] = \beta_{12} + \sigma$; with $\beta_{12} + \sigma > 0$, H has a net selective advantage, while $\beta_{12} + \sigma < 0$ entails a net advantage to h.

To summarize, the inclusion of viability selection, if it is of the correct order of magnitude relative to the population size, affects only the parameter from the heterotypic matings. As in the deterministic models, even if selection is against H ($s < 0$), transmission can favor it ($C > 1$) to the extent that there is a net positive effective selection. This is measured by an increase in the probability of fixation of H over the haploid genetic model without selection. Similarly selection for H ($s > 0$) can be overcome by transmission favoring h ($C < 1$). All other results from the previous section 2.8 pass over to the selection case by substituting $\beta_s = \beta_{12} + \sigma$ for β_{12}.

2.10 CONCLUDING REMARKS ON VERTICAL TRANSMISSION

The analysis above has been based on the transmission coefficients that are the probabilities of H progeny from matings $h \times h$, $h \times H$, $H \times h$ and $H \times H$. There is too little data in the literature to allow the estimation of these coefficients in concrete situations. This paucity of empirical information makes it inevitable that doubts about the validity and interest of these analyses are not easily dispelled. Thus it can be asked whether these parameters can ever be constant in a real situation. Can they be estimated satisfactorily, and if so, how? Even if it were possible to separate vertical cultural transmission from other types of transmission, would it ever be nontrivial relative to those other modes of cultural transmission that do not involve the parent-child relationship? Or can it be separated from the powerful mechanism of biological transmission from parent to child? Is it not likely that most cultural traits change with development, making analysis of parent to child transmission even more difficult? None of these questions can be fully answered at present. But we can make some suggestions as to potentially profitable avenues of study.

Here we summarize the results for the two phenotype cases and indicate possible extensions and ways in which the conclusions might be tested. We have described the conditions under which transmission laws produce fixation or loss of H, or equilibrium with H and h present. Assortative mating must be included for certain cases to be realistic, but the analysis is simpler with random mating. The equilibria and the rates at which they are achieved depend on the extent of natural selection. The transmission scheme can result in the preservation of a trait that is disadvantageous to the carriers, or, vice versa, an advantageous trait can be prevented from becoming established. These questions involve the delicate balance between different evolutionary forces.

In the absence of natural selection, from (2.2.2) and (2.4.8) it can easily be proven that there is an equilibrium with 100% H if $b_3 = 1$ while it must be less than 100% if $b_3 < 1$. Similarly, there will be an equilibrium with 100% h if $b_0 = 0$, while if $b_0 > 0$ some H must be present. Equilibria near these states of fixation may not be stable: a more even configuration is possible. But the conditions for a neutral equilibrium to exist, namely $b_0 = 0$, $b_3 = 1$, $b_1 + b_2 = 1$ are stringent. They are satisfied under genetic haploid transmission, or under cultural transmission when only one parental sex can teach (but does it so perfectly that $b = 1$ in Table 2.3.1). The coefficients $1 - b_3$ and b_0 can be compared to migration or mutation rates. When b_3 and b_0 are near 1 and 0, respectively, then $b_1 + b_2$ largely determines the equilibrium behavior. If $b_1 + b_2 < 1$, the frequency of H will be less than 1/2. The asymptotic rate of approach to the equilibrium will also be well approximated by $b_1 + b_2$. If, on the other hand, $b_1 + b_2 > 1$, then at equilibrium H will predominate and the rate of approach will be close to $2 - b_1 - b_2$. In sections 2.8 and 2.9 the role of $b_1 + b_2$ in the stochastic random drift model is elaborated.

Throughout the above chapter we have used a discrete or non-overlapping generation approach. This has been chosen as a matter of convenience; under some circumstances overlapping generations may be more realistic. One way to translate the discrete to a continous time treatment is to take the basic recursion

$$u_{t+1} = Bu_t^2 + Cu_t + b_0, \tag{2.2.3}$$

and write it as

$$u_{t+1} - u_t = \Delta u_t = Bu_t^2 + (C - 1)u_t + b_0. \tag{2.10.1}$$

Then assuming (2.10.1) is valid over ever-decreasing time intervals, we rewrite it as

$$\frac{du}{dt} = Bu^2 + (C - 1)u + b_0. \tag{2.10.2}$$

Write $\hat{u}_1$, $\hat{u}_2$ as the roots of the quadratic $u^2 + [(C - 1)/B]u + b_0/B$. Then the time course of the variable $u(\cdot)$ is given by the solution of (2.10.2), namely

$$u(t) = \frac{\hat{u}_1 - K\hat{u}_2 \exp\{Bt\sqrt{A}\}}{1 - K\exp\{Bt\sqrt{A}\}}, \tag{2.10.3}$$

where $A = (1 - C)^2 - 4b_0B$, $K = (u_{(0)} - \hat{u}_1)/(u_{(0)} - \hat{u}_2)$ with $u_{(0)}$ the initial frequency.

In the cases of Table 2.3.1 with $B = 0$ we have the recursion

$$u_{t+1} = Cu_t + b_0,$$

since $B = 0$, $C = b_3 - b_0 < 1$ in the nonneutral cases. The corresponding differential equation is

$$\frac{du}{dt} = (C - 1)u + b_0, \tag{2.10.4}$$

so that

$$u(t) = \hat{u} + (u_0 - \hat{u}) \exp[(C - 1)t], \tag{2.10.5}$$

where $\hat{u} = b_0/(1 - C)$. Thus $u(t) \rightarrow \hat{u}$ as $t \rightarrow \infty$. Other relationships between the discrete-time models such as (2.2.3) and continuous time formulations have been suggested that in many respects can be superior to (2.10.2). The latter should be regarded as merely one among many alternatives.

The testing of models such as the ones we have proposed here can only be done when a suitable and substantial body of data is available. In the empirical study of transmission of beliefs and values described earlier, substantial correlations between parents and offspring exist for certain types of traits. The chosen traits were selected because we felt they were likely to be transmitted in an entirely cultural manner. It has not been excluded that some are subject to innate influences. In many widely studied traits the confounding of innate and learned effects is complete. A carefully designed study including adopted children could contribute to the resolution of the two, especially if the adoptees include separated identical twins, a very rare occurrence. Some of the questions pertaining to religion and politics in our survey are similar to those asked by Loehlin and Nichols (1976) in a survey of twins. From his analysis the correlations between twins were practically the same for identical and fraternal twins. This strongly supports the notion that the innate contribution to variation in these traits is not significant.

Another difficulty is caused by social stratification, which may create positive correlations between parents and offspring that might occasionally be mistaken for direct vertical transmission. If there are differences among parent-child correlations according to the sex of the parent (or the offspring), this is an indication of cultural transmission. But, again, sex-linked inheritance may be difficult to distinguish from specific sex-role transmission (see e.g. in Chapter 3.3, the analysis of Kuru).

Among other sources of complication which must be considered in such a study, the first is the change of the trait with age. This may be overcome by studying age cohorts separately or investigating the age effect specifically. A second problem could be lack of information about one or other parent, making it impossible to use the separation into four mating types. Here one can make recourse to ordinary correlation coefficients, for example, between father-offspring, mother-offspring, and the like. It is not impossible, under auxiliary assumptions, to convert the estimates obtained by this more indirect approach into the transmission coefficients in which we are interested.

Third (and perhaps most important and interesting) is that observed incidence rates of H from the four matings may not necessarily represent the b coefficients. Other types of cultural transmission (in particular, oblique and horizontal, to be defined in the next chapter) may also operate. These are frequency dependent and can substantially alter the dynamics of the process. Especially b_0 is likely to appear inflated by components that depend on the presence of the trait in members of the population other than parents. A high value for b_0 is suspicious and could be taken as suggesting that b_1 and b_2 and perhaps b_3 are also inflated by such nonvertical transmission. The recognition of frequency dependent components is not simple and would involve an accurate study of who the teacher was or what models were imitated.

Many important customs, attitudes or tastes are determined by an early imprinting (a term used here to cover a more extensive basis of imitation and attachment than is customary in ethology). Such early imprinting is most likely to be due to parents or close relatives. Perhaps the earliest learnt are the least reversible behaviors. Whatever is ultimately shown to be true of these conjectures, we should not dismiss vertical cultural transmission as unimportant merely because of our present ignorance about it. For some of the topics addressed by our survey, like religion or politics, substantial vertical transmission was expected from earlier data, which, however, were not published in a form that can be directly transferred to our analytical framework. Our results indicated that religious preferences and attitudes seem to be highly influenced by maternal transmission. For politics, the effect of the family was only slightly less, with fathers almost as important as mothers. In sports and entertainments, there was some effect of parents. The correlations of parent and offspring were smaller for beliefs, but correction for possible sources of error increased them considerably. Only for habits was the correlation of parents and offspring practically negligible.

That parental contributions to the traits studied are large has interesting evolutionary consequences. Of particular interest are rates of evolution: it is natural to ask whether traits that are strongly influenced by parents are more conservative in evolution than others. We will be in a better position to answer this question when other modes of transmission have been introduced. But this seems to be a useful point at which to indicate how an assessment of "conservation" might be made. The simplest evaluation involves rates of passage toward equilibrium, whether it be polymorphic or trait fixation. Such rates can be evaluated locally, that is, by using the linear approximation to the transformation in the neighborhood of the equilibrium. If a trait is bound for extinction, then the closer the leading eigenvalue of the local linear transformation near extinction is to unity, then the longer will be its time to extinction and the more conservative the evolution. Similarly, the change near a polymorphic equilibrium will be slow if the leading eigenvalue near the equilibrium is close to unity. For example, if the transformation is given by (2.2.3), then the rate of return to a polymorphic equilibrium, $\hat{u}$ is $[2\hat{u}B + C]$. For additive biparental transmission, $B = 0$, $C = b_1 + b_2 - 2b_0 = \alpha_p + \alpha_m$. Thus the higher the sum $\alpha_p + \alpha_m$, that is, the greater the additive parental contribution, the slower is the return to the equilibrium—in other words, the more conservative is the evolution. Rates of cultural change are often thought to be fast relative to biological evolution. Yet some cultural traits are extremely conservative; perhaps the latter are often transmitted vertically.

CHAPTER THREE

Oblique and Horizontal Transmission

3.1 OBLIQUE TRANSMISSION

In the preceding chapter cultural transmission was assumed to be only vertical, that is, from parent to child, though not necessarily with the rules of vertical transmission in biology. The models we have studied show that evolution under vertical transmission may have any of a number of possible outcomes. Some examples result in a *polymorphic equilibrium*, while others allow the trait in question, say H, to take over the total population (*fixation* of H), or to disappear (*extinction*).

Here the question is asked as to the evolutionary importance of transmission from individuals other than the parents but who belong to the same generation as the parents. Teaching by individuals of earlier generations or even by individuals of the same generation would be among the possible modes of non-vertical transmission in a polymorphic population. If in addition to the parents other people of the *older* generation serve as teachers we call it *oblique* transmission, to contrast it with transmission from people of the offspring generation, which we call *horizontal*. In the absence of precise knowledge we can only rely on the impression that in traditional societies many of the social customs are probably transmitted in this way: parents teach and other adults reinforce the teaching. The elders may sometimes assume this task; formally this amounts to generation $t - 1$ teaching generation $t + 1$. It seems intuitively correct that teaching by elders should make a society more conservative on average, a point which shall be formalized later when linear models of transmission for continuous traits are discussed. In other cases, one person may be charged with teaching a special custom. This teacher-type transmission (see section 1.1) is an important variant of oblique transmission. The distinction between intra- and extra-familial oblique transmission would be important in situations where the nature of the trait dictated that these occur at different rates.

In a group of N individuals, assume that the number of potential teachers (other than parents) that an average person can contact is n. Of these at time t, a proportion u_t is of type H. Then, if g $(0 \leq g \leq 1)$ is the chance that an h individual of generation $t + 1$ is transformed to H by contact with an H individual who is a potential teacher, the probability that he will be transformed after n contacts is

$$1 - (1 - g)^{nu_t},$$

which, if g is small, simplifies to ngu_t. The constant $ng = f$ expresses the probability that an h individual of generation $t + 1$ is modified by contact with individuals of generation t. Another way of looking at this calculation entails that the number of contacts an individual makes has a Poisson distribution with mean n. That is, the probability that r contacts are made is $p_r = e^{-n}n^r/r!$. Then the probability of transformation to H of h individuals is

$$\begin{aligned}
&\sum_{r=0}^{\infty} (1 - (1 - g)^{ru_t}) \frac{e^{-n} n^r}{r!} \\
&\quad = 1 - e^{-n} \sum_{r=0}^{\infty} \frac{[n(1 - g)^{u_t}]^r}{r!} \\
&\quad = 1 - e^{-n} e^{n(1-g)^{u_t}} \\
&\quad \simeq 1 - e^{-ngu_t} \simeq ngu_t = fu_t \qquad (3.1.1)
\end{aligned}$$

as before, as long as g and ngu_t are small. The value fu_t derived in (3.1.1) might more generally be regarded as a *rate*, rather than a probability, whose sign could also be negative. More generally

TABLE 3.1.1. Transmission rules for vertical and oblique transmission of a 2-state trait. Frequencies u_t and $v_t = 1 - u_t$ refer to the tth generation. Assortative mating occurs with correlation m between mates.

Mating	Frequency of Mating	Probability of H Progeny	Probability of h Progeny
♀ ♂	(p_i)		
$H \times H$	$u_t^2 + mu_tv_t$	$1 - (1 - b_3)(1 - fu_t)$	$(1 - b_3)(1 - fu_t)$
$H \times h$	$u_tv_t(1 - m)$	$1 - (1 - b_2)(1 - fu_t)$	$(1 - b_2)(1 - fu_t)$
$h \times H$	$u_tv_t(1 - m)$	$1 - (1 - b_1)(1 - fu_t)$	$(1 - b_1)(1 - fu_t)$
$h \times h$	$v_t^2 + mu_tv_t$	$1 - (1 - b_0)(1 - fu_t)$	$(1 - b_0)(1 - fu_t)$

still, oblique transmission will be indicated by the function $\varnothing(u_t)$.

Now suppose that the chance that an individual obtains the phenotype H from its parents is b, with b not necessarily small. Then $(1 - b)(1 - \varnothing(u_t))$ is the chance that H is not obtained vertically from the parents nor horizontally from the others in the population. The complementary value $1 - (1 - b)(1 - \varnothing(u_t))$ is then the probability that H is transmitted by one or other (or both) modes of transmission. In what follows we shall usually assume

$$\varnothing(u_t) = fu_t, \tag{3.1.2}$$

as in (3.1.1). Joint vertical and oblique transmission will be of special interest in what follows. We shall usually assume that $0 \leq f \leq 1$; if f is relatively large, the model constructed using n contacts, each with chance g (see 3.1.1) is a poor approximation. Note that if $f < 1$, u is always decreasing if there is *only* oblique transmission of this form. The combination of vertical and oblique transmission of type (3.1.2) is shown in Table 3.1.1.

In this way the simplest form of frequency dependent transmission is introduced, but other oblique transmission functions can be designed which can produce an increase by themselves. Two examples are

$$\varnothing(u_t) = u_t + fu_t(1 - u_t) \tag{3.1.3}$$

or

$$\varnothing(u_t) = \frac{fu_t}{gu_t + e} \quad (e > 0), \tag{3.1.4}$$

both of which arise in ecology as models of density-dependent population growth (Christiansen and Fenchel 1977). The role of adults other than parents in transmitting phenotypes has been largely neglected as an evolutionary force, and little empirical information exists on which functions $\varnothing(\cdot)$ might be the most appropriate for any specific trait. (3.1.2) has the advantages of providing the simplest scheme and a fairly direct comparison with a model of horizontal transmission introduced later. Moreover, it is worth noting that (3.1.3) results from the combination of a very simple type of vertical transmission, in which $u' = u_t$ (see Table 2.3.1, cases 1a and 2) followed by oblique transmission according to (3.1.2). This second step affects the $(1 - u')$ fraction of the offspring that did not inherit the trait from parents:

$$\begin{aligned} &\text{First state:} && u' = u_t \\ &\text{Second stage:} && u_{t+1} = u' + (1 - u')\varnothing(u') \\ &&& \phantom{u_{t+1}} = u_t + fu_t(1 - u_t) \end{aligned}$$

producing (3.1.3).

In summary, the combination of simple models of vertical and oblique transmission can be represented by (3.1.3) and (3.1.4). The contact-transfer rate f may not, in practice, be less than one Negative f would have the interpretation that individuals with trait value H are converted to h on contact with h individuals.

3.2 OBLIQUE AND VERTICAL TRANSMISSION WITH NATURAL SELECTION

Oblique transmission can now be incorporated into our general vertical transmission model for a two-phenotype trait. The transmission rules are those in Table 3.1.1. Selection is assumed to operate before mating so that after selection in generation t we have

$$u_t^* = \frac{u_t(1 + s)}{1 + su_t}$$

(c.f. 2.5.1). As in section 2.6, asterisks refer to frequencies after selection (in "mature" individuals) and the absence of stars to frequencies before selection ("juveniles"). Now, mating occurs according to Table 3.1.1 and the new frequency of juveniles before selection in generation $t + 1$ is

$$u_{t+1} = 1 - [1 - \varnothing(u_t^*)][1 - B(u_t^*)^2(1 - m) - (C + Bm)u_t^* - b_0], \quad (3.2.1)$$

where B and C are as in (2.2.4) and (2.2.5). Since u_t^* is a function of u_t, this is a one-step recursion connecting successive juvenile H frequencies before selection. To construct frequencies after selection the adjustment is made

$$u_{t+1}^* = (1 + s)u_{t+1}/(1 + su_{t+1}), \quad (3.2.2)$$

with u_{t+1} in (3.2.1). Then (3.2.2) is a recursion in adult frequencies after selection. Either of (3.2.1) or (3.2.2) can be used to describe the evolution. The difference is merely the point in time at which the census is taken.

In the construction of (3.2.1) the transmission parameters from Table 3.1.1 are frequency dependent. There are two other equivalent ways of deriving (3.2.1). Suppose the parental transmission occurs first, after which

$$u_{t+1} = \Sigma p_i^{(t)} b_i.$$

Now of those offspring who did not receive phenotype H, a fraction $(1 - \Sigma p_i^{(t)} b_i)$ contact the H members of the parental generation, and are converted to H according to the function $\varnothing(u_t)$. After this the final value of u_{t+1} would be

$$u_{t+1} = \Sigma p_i^{(t)} b_i + (1 - \Sigma p_i^{(t)} b_i) \varnothing (u_t), \quad (3.2.1a)$$

as in (3.2.1). Alternatively, suppose the u_t members of the parental generation give rise by oblique transmission to $\varnothing(u_t)\, H$ members of the offspring generation. The remaining $[1 - \varnothing(u_t)]$ offspring then contact the $\Sigma p_i^{(t)} b_i$ children who became H by vertical transmission. The final value of u_{t+1} is the same as in (3.2.1). In what follows we shall use the linear function $\varnothing(u_t) = fu_t$ in (3.2.1); that is, we shall assume that the oblique transmission obeys a mass action law between H (generation t) and h (generation $t + 1$).

The recursion (3.2.2) in u_t^* can now be examined at equilibrium. It should be noted that if $B = 0$ (as in the examples of Table 2.3.1), then the dependence on m here disappears. The equation for equilibrium is obtained by setting $u_{t+1}^* = u_t^*$ in (3.2.2). The equation is of fourth degree:

$$\begin{aligned} &u^4 sf(1 - m)B + u^3[(C + mB)sf - (s + f + sf)(1 - m)B] \\ &\quad - u^2[sf - (1 + s)(1 - m)B + (f + s + sf)(C + mB) - b_0 sf] \\ &\quad\quad - u[1 - sf - f - (1 + s)(C + mB) + b_0(s + f + sf)] \\ &\quad\quad\quad + b_0(1 + s) = 0. \quad (3.2.3) \end{aligned}$$

For $B = 0$, (3.2.3) reduces to a cubic. If in addition $b_0 = 0$ or $s = -1$, then $\hat{u} = 0$ is an equilibrium. Otherwise it is impossible for H to disappear from the population. When $b_0 = 0$ it is of interest to consider the conditions under which the population is "protected" from losing the trait; that is, the equilibrium $\hat{u} = 0$ is unstable. For this we compute

$$\lim_{u_t^* \to 0} \frac{u_{t+1}^*}{u_t^*} = (1 + s)(f + C + mB). \quad (3.2.4)$$

Hence, if

$$(1 + s)(f + C + mB) > 1, \quad (3.2.5)$$

$\hat{u} = 0$ is unstable; the converse inequality allows $\hat{u} = 0$ to be stable so that, from some initial conditions at least, the trait H will be lost. The role of f in (3.2.5) is such that it enhances the possibility of keeping H in the population, that is, (3.2.5) is more easily satisfied with $f > 0$ than with $f = 0$ for a given set of the other parameter values.

When $b_0 = 0$ and $B = 0$ (see Table 2.3.1), apart from $\hat{u} = 0$ the equilibria are obtained from the quadratic equation

$$u^2(Csf) - u[sf + C(s + f + sf)] - [1 - (f + C)(1 + s)] = 0. \quad (3.2.6)$$

We present here a summary of the analysis. If (3.2.5) holds (with $B = 0$), then it is quickly seen that (3.2.6) has a single valid root

$$\hat{u} = \frac{sf + C(s + f + sf) - \sqrt{[sf + C(s + f + sf)]^2 + 4sfC[1 - (f + C)(1 + s)]}}{2fsC} \quad (3.2.7)$$

When $B = b_0$ the transformation (3.2.2) can be written

$$\hat{u}^*_{t+1} = \frac{(1 + s)[fu_t^* + (1 - fu_t^*)Cu_t^*]}{1 + s[fu_t^* + (1 - fu_t^*)Cu_t^*]}. \quad (3.2.8)$$

The righthand side is a monotonically increasing function of u_t^*. When (3.2.5) holds, this allows us to deduce the stability of the polymorphic equilibrium (3.2.7). When (3.2.5) fails and $s > 0$, neither root of (3.2.6) can be valid. In this case, from all starting conditions H is lost from the population. When (3.2.5) fails and $s < 0$ it is possible for $\hat{u} = 0$ to be stable and for both roots of (3.2.6) to be valid (i.e., both positive and less than unity). In this case the larger of these is stable and the smaller unstable. Consider, for example, the following cases from Table (2.3.1):

		b_3	b_2	b_1	b_0
case 1	$(b = 1, c = 0)$	1	1	0	0
case 2		1	1/2	1/2	0
case 3		1	$1 - b$	b	0.

In all of these, in addition to $b_0 = 0$, we have $C = 1$. The roots of (3.2.3) other than $\hat{u} = 0$ are $\hat{u} = 1$ and $\hat{u} = 1 + 1/f + 1/s$. If $f > 0$ the latter is valid only if $s < 0$. Then applying (3.2.5), $\hat{u} = 0$ is stable if

$$s < \frac{-f}{1+f}, \tag{3.2.9}$$

while $\hat{u} = 1$ is stable if

$$s > -f. \tag{3.2.10}$$

If

$$\frac{-f}{1+f} > s > -f, \tag{3.2.11}$$

then both $\hat{u} = 0$ and $\hat{u} = 1$ are stable and the sets of initial values u_0 which converge to them are separated by the unstable equilibrium $\hat{u} = 1 + 1/f + 1/s$. That is $u_0 < 1 + 1/f + 1/s$ entails $u_t \rightarrow 0$, while $u_0 > 1 + 1/f + 1/s$ entails $u_t \rightarrow 1$. The parameter ranges for stability of the two fixations are shown in Figure 3.2.1 while the dynamic behavior of (3.2.8) in some special cases is illustrated in Figure 3.2.2.

In case 4 of Table 2.3.1, with $b_0 = 0$ we have $b_3 = \alpha_p + \alpha_m$ and, since $b_3 < 1$, $\alpha < \frac{1}{2}$. If

$$0 < s < \frac{1}{f + \alpha_p + \alpha_m} - 1 \tag{3.2.12}$$

then $\hat{u} = 0$ is stable and is the only valid equilibrium. If $s > 1/f + \alpha_p + \alpha_m - 1$, then only (3.2.7) is stable. If

$$s < \min\left(0, \frac{1}{f + \alpha_p + \alpha_m} - 1\right),$$

then $\hat{u} = 0$ is stable and, in addition, both roots of (3.2.6) can be valid with the larger one stable. In general, $f > 0$ enhances the chance of preservation of H in the population over the situation

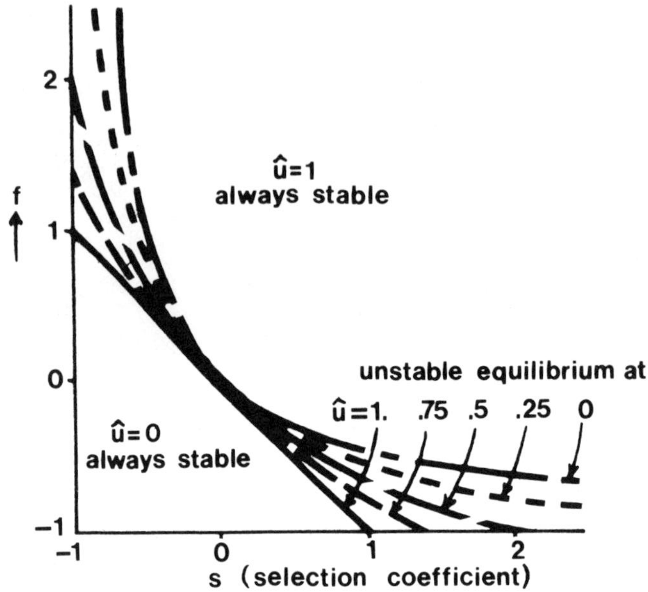

FIGURE 3.2.1. Conflict between cultural transmission (oblique) of strength f and natural selection of strength s, in the presence of vertical transmission leading to a neutral equilibrium. Curves represent the pairs of f, s values for a given equilibrium $\hat{u}$. The equilibrium is unstable, so that with a given pair f, s, if u_0 is less than $\hat{u}$, then $u \rightarrow 0$, and if u_0 is greater than $\hat{u}$ then $\hat{u} \rightarrow 1$. Note the similarity with Figure 2.6.1.

with only vertical transmission. Examples of trajectories of u_t are given in Figure 3.2.2.

If $b_0 \neq 0$, selection is absent, and $B = 0$, then (3.2.3) reduces to a quadratic:

$$u^2fC + u[1 - f - C + b_0f] - b_0 = 0 \qquad (3.2.13)$$

which always has a single valid root:

$$\hat{u} = \frac{f + C - 1 - b_0f + \sqrt{(f + C - 1 - b_0f)^2 + 4b_0fC}}{2fC}, \qquad (3.2.14)$$

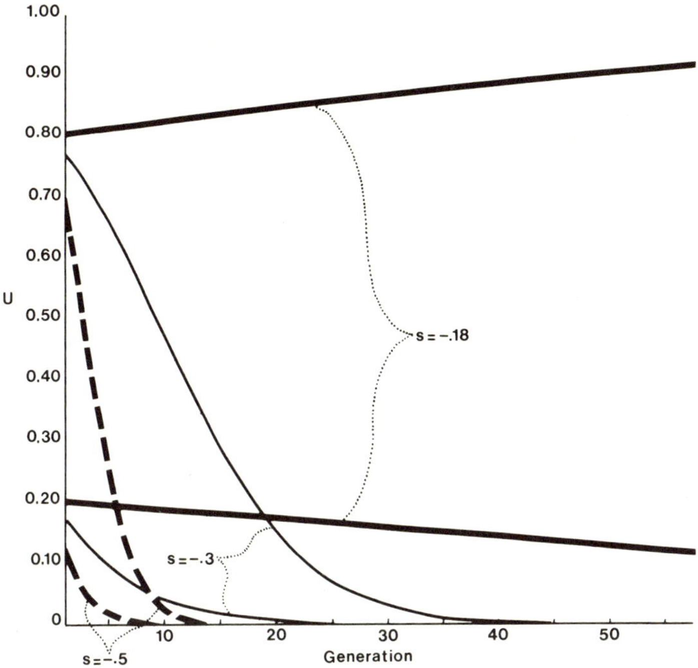

FIGURE 3.2.2. Trajectories for u_{t+1} as a function of u_t from (3.2.8) when $B = 0$, $C = 1$, $b_0 = 0$ (e.g. special cases 1, 1*a*, 1*b* of Table 2.3.1 for vertical transmission), oblique transmission determined by f, and selection s. f is set at 0.2. With $s = -0.3$ only $\hat{u} = 0$ is stable. With $s = -0.18$ the starting conditions u_0 determines which of $\hat{u} = 1$ or $\hat{u} = 0$ is reached (see (3.2.11)).

which is stable. Again in these cases without selection, $f > 0$ can prevent the loss of H when, with $f = 0$, it might have been eliminated. If $B = C = b_0 = 0$, then the interaction between selection and oblique transmission is sufficient to maintain the equilibrium $\hat{u} = (sf + f - 1)/sf$ provided $(1 + s)f > 1$.

The kinetics of increase of the trait H near an equilibrium are governed by the derivative of the recursion in the neighborhood of the equilibrium. In particular, if $\hat{u} = 0$ is an equilibrium, this derivative at $\hat{u} = 0$, (3.2.4), gives the initial rate of increase. If the left side of (3.2.5) is much larger than unity, an innovation (occurring as a "mutation") in a single individual would increase quickly in frequency. On the other hand, if s is small and f not large, initial increase will be slow. However, it is possible that a new social custom be imposed, by an act of force, or by successful persuasion, on most or all of a social group. Therefore, the trait may start at a high frequency—higher than biological evolutionists would concede for a genetic mutation. The equilibrium study above can then be invoked to determine whether the trait will move to equilibrium, fixation, or eventual loss. It may also be that rules of oblique transmission other than (3.1.1) hold. In particular, (3.1.2) and (3.1.3) both entail faster evolution of H than (3.1.1). In Table 3.2.1 some numerical examples are presented that allow the comparison of the kinetics of cultural change under various oblique transmission regimes.

TABLE 3.2.1. Vertical + oblique transmission. Vertical transmission for $\Sigma p_i b_i = u$, as in cases 1a ($b_3 = 1$), 2, 3, ($b_3 = 1$), 4 ($\alpha = 1/2$); always $b_0 = 0$, no selection, random mating. Oblique transmission: $\phi(u) = fu$ in model at left, discrete logistic (2.3.3) in model at right. t counts generations. Initial value of trait frequency $u_0 = 0.01$ in all cases.

	Oblique: $\phi(u) = fu$			Oblique: discrete logistic		
t	$f = 1$	$f = 0.1$	$f = 0.01$	$f = 1$	$f = 0.1$	$f = 0.01$
0	0.01	0.01	0.01	0.01	0.01	0.01
1	0.020	0.011	0.010	0.030	0.021	0.021
2	0.039	0.012	0.010	0.238	0.048	0.040
3	0.077	0.013		0.557	0.089	0.078
4	0.149	0.015		0.913	0.177	0.151
5	0.275	0.016	0.011	0.999	0.334	0.281
6	0.474	0.018		1.000	0.572	0.484
7	0.724	0.019			0.827	0.735
8	0.924	0.021			0.973	0.930
9	0.994	0.023			1.000	0.995
10	1.000	0.026	0.011			1.000
30		.152	.013			
100		.996	.027			

The models described above might apply equally well to the transmission of discrete valued cultural traits and infectious diseases. It is instructive to compare the expectations emerging from the foregoing analysis with data available concerning the transmission of hepatitis B virus. This infection, which is rare among Caucasians, reaches quite high frequencies among Orientals. In Chinese of Taiwan, for example, it is estimated that 15% of the population is positive for the virus, with relatively little geographic variation in the island. The presence of the virus, which is shown by laboratory tests, ordinarily has no effects on the health of the carrier, although it is believed to be responsible in some cases for clinical hepatitis and, in the long term, confers a high risk for liver carcinoma. About 2% of all deaths in Taiwan are attributable to liver carcinoma and although this is much higher than in other parts of the world it occurs mainly at middle age and older and probably has little effect on Darwinian fitness. Thus, in terms of the model (3.2.8), s is likely to be negative but small. It should be added that the proportion of individuals affected by the virus is higher than indicated by the 15% of people who carry the virus since, in addition to them, another 15% have antibodies to the virus. The latter must have, at some time, been infected but have no demonstrable virus and hence are not infectious.

It is known that about 45% of infectious expecting mothers transmit the virus transplacentally to the fetus. Transmission from the father is believed to be negligible. Thus we can estimate $b_1 = b_0 = 0$, and $b_3 = b_2 = C = 0.45$. Under these conditions, $C < 1/(1 + s)$ (see 2.6.14) and the infection could not be maintained in the population since $\hat{u} = 0$ would be the only stable equilibrium. Thus some other type of transmission would appear to be necessary for the maintenance of the infection in the population. Consider for example the equilibria under vertical *plus oblique* transmission obtained from equation (3.2.6). Here $C = b_2 = 0.45$, but f and s are unknown with only one other known data point, namely the frequency $\hat{u} = 0.15$. Of course, this assumes that the frequency of infection is at an equilibrium.

From what was said above we might ignore s, in which case (3.2.6) produces

$$\hat{u} = (f + C - 1)/fC = 0.15,$$

so that we have an estimate for f:

$$f = (1 - C)/(1 - C\hat{u}) = 0.59.$$

If s is slightly negative, f will be greater than 0.59.

Now from (3.2.5), we see that with s negligible the intermediate equilibrium $\hat{u} = 0.15$ is stable if $f = 0.59$ and $C = 0.45$. With small s we can expect the estimate of f not to change much, so that it is reasonable to infer that the same conclusion, maintenance of the virus, would occur as long as the Darwinian selection against carriers did not exceed more than 2–4%.

If improved hygiene were to stop the oblique transmission, leaving only vertical (i.e. transplacental) transmission, how soon would the disease disappear? This simply entails the removal of f in (3.2.8) and a starting value of $0.15 = u_0$. The dynamics are as follows:

time in generations	for $s = 0$	for $s = -.04$
0	$u_0 = 0.15$	0.15
1	$u_1 = 0.068$	0.065
2	$u_2 = 0.030$	0.028
3	$u_3 = 0.014$	0.012
4	$u_4 = 0.006$	0.005
5	$u_5 = 0.003$	0.002

In Chinese born in the United States, a lower frequency of virus carriers is found, suggesting that the process of decrease under improved hygiene has been occurring for one or two generations.

In summary, oblique transmission (from members of the parental generation other than parents), according to the models we have introduced, enhances the probability that u will tend eventually to unity. The oblique transmission chosen here would not be sufficient to maintain a trait in a population if it were acting entirely alone. In the presence of selection, the final fate of a trait depends on all these forces—vertical, oblique transmission, and selection. Internal, stable equilibria are not usual and demand a delicate adjustment of the parameters involved.

3.3 SEX-INFLUENCED TRANSMISSION

Males are more frequently affected by hepatitis and hepatoma than females. Another viral disease, Kuru, which affects the central nervous system, exhibits an even more pronounced sex difference to the extent that there are from four to six times more deaths in females from Kuru than occur in males. The disease is practically always lethal within one or at most a few years of the manifestation of its first symptoms. It is found exclusively among natives indigenous to a small region of the New Guinea highlands.

The disease has a familial pattern of occurrence that initially led geneticists (Bennett, et al., 1959), to suggest that Kuru is due to a gene recessive in males (almost all of whom die early) and dominant in females (where it determines early death in homozygotes and late death in heterozygotes). Later research (see Gajdusek, 1977), showed that the disease is due to a slow, latent virus that kills within a few years after the initial infection. Therefore the hypothesis of a single Mendelian gene must be ruled out, although it is not excluded that genetic differences in predisposition to the virus exist.

Infection takes place during ceremonies for the dead which, if the deceased had been affected, expose females and children who are close relatives of the dead to contagion. The ceremonies accompany the preparation of the body with manipulation of the brain, ceremonial sniffing, and culminate in necrophagy (cannibalism of the dead). The brain of the deceased is very rich in infectious material; its manipulation, sniffing, and eating involve a high risk of infection. Thus, the familial pattern might, as far as we know, be entirely due to the ceremonial transmission of the disease to close relatives. Since the cessation of cannibalism between 1957 and 1962, there has been continuous decline of Kuru deaths, and they have now almost disappeared. It is interesting to note that, in the case of hepatitis, geneticists were also first led to think of "Australia antigen" as a genetically determined trait, until the discovery that this antigen is associated with the presence of the virus (Blumberg et al., 1966; Blumberg, 1977).

Kuru itself is a biological entity and, as such, subject to biological evolution. But its transmission occurs through a cultural procedure, so that its inclusion as an item of cultural evolution is justified. Several other examples of diseases transmitted vertically as well as obliquely and horizontally are now known. The last two are more typical modes of transmission for infectious diseases.

In Table 3.3.1 a vertical and oblique transmission scheme is presented that distinguishes the vertical transmission in the two sexes. Random mating is assumed. The frequencies of H in adult males and females at a given generation are u_m and u_f. Those in the next generation $u_m^{*\prime}$ and $u_f^{*\prime}$ are given by the recursion relations

$$u_m^{*\prime} = \frac{(1+s)[1-(1-fu^*)(1-\Sigma p_i c_i)]}{1+s[1-(1-fu^*)(1-\Sigma p_i c_i)]} \tag{3.3.1}$$

and

$$u_f^{*\prime} = \frac{(1+s)[1-(1-fu^*)(1-\Sigma p_i b_i)]}{1+s[1-(1-fu^*)(1-\Sigma p_i b_i)]}, \tag{3.3.2}$$

TABLE 3.3.1. A model of sex-influenced cultural transmission. The constant f refers, as in Table 3.1.1, to horizontal transmission from affected members of the parental generation. The probability of h progeny is simply the complement to 1 of the probability of H given in the table. Trait H frequencies are u_m and u_f in males and females, respectively, with $u = (u_m + u_f)/2$. Values in the first column assume random mating.

p_i Frequency of Mating	Mother	Father	Probability of Female H Progeny	Probability of Male H Progeny
♂ ♀				
$u_m \times u_f$	H	H	$1 - (1 - b_3)(1 - fu)$	$1 - (1 - c_3)(1 - fu)$
$v_m \times u_f$	H	h	$1 - (1 - b_2)(1 - fu)$	$1 - (1 - c_2)(1 - fu)$
$u_m \times v_f$	h	H	$1 - (1 - b_1)(1 - fu)$	$1 - (1 - c_1)(1 - fu)$
$v_m \times v_f$	h	h	$1 - (1 - b_0)(1 - fu)$	$1 - (1 - c_0)(1 - fu)$

where

$$\Sigma p_i b_i = u_m^*(u_f^* b_3 + v_f^* b_1) + v_m^*(b_2 u_f^* + b_0 v_f^*) \quad (3.3.3a)$$

$$\Sigma p_i c_i = u_m^*(u_f^* c_3 + v_f^* c_1) + v_m^*(c_2 u_f^* + c_0 v_f^*) \quad (3.3.3b)$$

and

$$u^* = \frac{u_m^* + u_f^*}{2}.$$

The values b_i and c_i are transmission probabilities for female and male progeny respectively, with f, the chance of infection from extrafamilial members, taken to be equal in the two sexes. At equilibrium we have

$$u_f^* - (1 + s - su_f^*)[fu^* + \Sigma p_i b_i(1 - fu^*)] = 0 \quad (3.3.4)$$

$$u_m^* - (1 + s - su_m^*)[fu^* + \Sigma p_i c_i(1 - fu^*)] = 0 \quad (3.3.5)$$

These are simultaneous equations of high order and are unlikely, for general parameters, to yield closed solutions.

We will examine simple situations, in particular with $b_0 = c_0 = 0$, so that an equilibrium with $\hat{u}_m = \hat{u}_f = 0$ is possible. If the trait value H corresponds to having Kuru, and the condition is highly deleterious, s is expected to be substantially negative. If all infected (h) individuals die before reproducing, then $s = -1$. In the data by Bennett et al. (1959), about 37% of female deaths (88% of these in adult and the rest in younger females) and 6% of male deaths (30% in adults, 70% in younger males) were due to Kuru. Using the age distribution given in some detail by Bennett et al. (1959) in order to evaluate fitnesses as indicated by Fisher (1930) (see also Cavalli-Sforza and Bodmer, Chap. 5), selection is stronger in males ($s = -0.9$) than in females ($s = -0.4$). Averaging over the sexes, the combined s is less than -0.5, and Kuru cannot be maintained by vertical transmission alone; some oblique or horizontal transmission is required. The model (3.3.1) – (3.3.3) can be used to determine how strong these need to be to prevent the loss of Kuru from the population.

To determine whether Kuru would disappear when it is rare, a local analysis is made in the neighborhood of $(u_m, u_f) = (0, 0)$. If the disease does not disappear when it is initially rare, we say it is *protected* from loss. To determine whether protection or loss occurs, the transformation (3.3.3) is linearized near (0, 0). If u_m and u_f are small enough, this linearized version of (3.3.3) is

$$\begin{pmatrix} u_m \\ u_f \end{pmatrix}_{t+1} = (1 + s) \begin{pmatrix} \dfrac{f}{2} + c_1 & \dfrac{f}{2} + c_2 \\[2ex] \dfrac{f}{2} + b_1 & \dfrac{f}{2} + b_2 \end{pmatrix} \begin{pmatrix} u_m \\ u_t \end{pmatrix}_t \tag{3.3.6}$$

The point (0.0) will be unstable if the matrix on the right of (3.3.6) has an eigenvalue greater in absolute value than unity. The largest eigenvalue is

$$\lambda_L = \frac{(1 + s)}{2} \Big\{ f + b_2 + c_1 + \sqrt{(f + b_2 + c_1)^2 - 2f(b_2 + c_1 - b_1 c_2) + 4b_1 c_2 - 4b_2 c_1} \Big\} \tag{3.3.7}$$

λ_L is an increasing function of f. That is, it is easier to maintain Kuru with $f > 0$ than with $f = 0$. The condition that $\lambda_L > 1$ is in reality a condition on the selection coefficient s, which places a bound on how deleterious the disease can be yet still be prevented from leaving the population. Clearly, if there were some extra-familial transmission ($f > 0$) and a very high efficiency of transplacental and ceremonial transmission, then selection values $s < -\frac{1}{2}$ could be compatible with the maintenance of Kuru.

The data available for Kuru are not ideal for a quantitative analysis, so the conclusions that follow must be accepted with caution. Let us assume in what follows that $b_0 = c_0 = 0$, $b = b_1 = b_2$, $b_3 = 1 - 2b$; $c = c_1 = c_2$, and $c_3 = 1 - 2c$.

There are very few affected adult males, so that most of the available information concerns the second and fourth mating types in Table 3.3.1. Bennett et al. state that 12 out of 66 sons of women who died of Kuru also died of Kuru. The husbands of these women were free from Kuru, so $12/66 = 18\%$ is an estimate of c_2. Only one of 114 sons of mothers without Kuru were affected. Among adult daughters of women who died from Kuru (again with unaffected husbands) $18/27 = 67\%$ died from Kuru compared to $45/95 = 47\%$ of adult daughters from non-Kuru mothers. There were $6/25 = 24\%$ young affected daughters from affected mothers and $3/14$ from non-Kuru mothers. Thus we estimate b_2 as $(6 + 18)/(25 + 27) = 0.46$ and $c_2 = 0.18$. The transmission probability to daughters, b_2, appears to comprise two parts, one corresponding to infection as a child, and is comparable in magnitude to the same probability for sons, c_2. The other component arises by infection of adult females and may be due to the special role that women play in the ceremonial treatment of the dead.

From the existing data it is difficult to estimate u_f and v_f. The estimates based on relative frequencies of the various causes of death are $u_f = 0.37$ and $u_m = 0.06$, (Bennett et al., 1959), but since age at death is not specified for all causes of death it is impossible to tell whether these numbers apply before or after selection. Presumably the population value of u_f should be much higher than 0.37 because many Kuru deaths are known to occur post-reproductively. It is reasonable to assume that 0.37 represents a value between u_f and u_f^*, thereby underestimating the true u_f which is bound to be less than u_f^*. If we accept the value 0.06 for u_m, then the following calculation provides an estimate of u_f. The proportion of women whose parents are free from Kuru and who contract the disease (nonvertically) is $(45 + 3)/(95 + 14) = 0.44$. This must be equal to fu, where $u = (u_m + u_f)/2$ is the fraction of the population that is infectious. Thus

$$fu = 0.44 = f(0.06 + u_f)/2. \tag{3.3.8}$$

Further, from 3.3.2,

$$u_f' = 1 - (1 - fu)(1 - \Sigma p_i b_i), \quad \Sigma p_i b_i = b(u_f + u_m). \tag{3.3.9}$$

Because sons from unaffected matings have an extremely low chance of contracting Kuru we neglect nonvertical transmission in males. Then at equilibrium $u_f' = u_f$, and, from (3.3.8) and (3.3.9), $u_f = 0.44 + 0.56[1 - b(u_f + u_m)] = 0.62$ so that $f = 0.88/(0.06 + 0.62) = 1.29$. In other words, there must be a major contribution to infection by non-vertical transmission. Presumably this can be due to participation in ritual cannibalism of dead relatives and tribal members other than parents. This calculation has been formal only, and does not take into account the possibility that $f > 1$ can radically change the equilibrium structure of the underlying model.

The correlations between parents and offspring provides some insight into the original attribution of Kuru to genetics. With sex-influenced transmission it is important to evaluate separately the four parent-offspring correlations involving fathers (F) and daughters (D), fathers and sons (S), mothers (M) and daughters, and mothers and sons. Under the models defined above, these correlations are given in Table 3.3.2. In the same table, the third

TABLE 3.3.2. Expected correlation coefficients between parent P (F for father, M for mother) and offspring O (S for son, D for daughter) under sex-influenced cultural transmission as in Table 3.3.1 and under X-linked transmission as in Table 3.3.3. u_f, v_f and u_m, v_m are H and h frequencies in females and males, respectively. With primes they are the same frequencies in daughters and sons. u is the average H frequency in the parental generation.

	Sex-influenced cultural transmission General* r_{PO}	X-Chromosome Linked** r_{PO}
r_{FS}	$\frac{u_m v_m}{u'_m v'_m}(1 - fu)[u_f B_c + c_1 - c_0]$	0
r_{FD}	$\frac{u_m v_m}{u'_f v'_f}(1 - fu)[u_f B_b + b_1 - b_0]$	$\sqrt{q/(1 + q)}$
r_{MD}	$\frac{u_f v_f}{u'_f v'_f}(1 - fu)[u_m B_b + b_2 - b_0]$	$q/(1 + q)$
r_{MS}	$\frac{u_f v_f}{u'_m v'_m}(1 - fu)[u_m B_c + c_2 - c_0]$	$\sqrt{q/(1 + q)}$

*$B_b = b_3 + b_0 - b_1 - b_2$ $\quad B_c = c_3 + c_0 - c_1 - c_1$.
**q is the frequency of the allele g which makes gg females and g males healthy. See Table 3.3.3 and text for further explanations.

column lists the same correlation coefficients, with genetic inheritance linked to the X-chromosome.

X-chromosome-linked inheritance is a well-known special case of ordinary Mendelian inheritance. Females have two X-chromosomes, while males have one X and one Y chromosome. The Y-chromosome can, for our present purposes, be considered inactive. For a gene on the X-chromosome existing in the two forms ("alleles") G and g there are three possible types of females: GG, Gg, and gg. There are only two types of males, G and g, since they have only one X-chromosome. Let us assume GG and Gg females and G males have trait H (Kuru, for example), while gg females and g males have trait h. G is said to be *dominant* to g in this genotypic arrangement (and g *recessive* to

TABLE 3.3.3. X-chromosome-linked inheritance. p, $q = 1 - p$ are frequencies of G, g respectively in males. P, $Q = 1 - P$ are those in females. For simplicity, ratios of children are given at equilibrium $p = P$. G is dominant to g, so GG and Gg females and G males are H. Mating is random.

Mother Father	Frequency of Mating	Daughters H	Daughters h	Sons H	Sons h
$H \times H$	$(P^2 + 2PQ)p$	1	0	$1/(1 + q)$	$q/(1 + q)$
$H \times h$	$(P^2 + 2PQ)q$	$1/(1 + q)$	$q/(1 + q)$	$1/(1 + q)$	$q/(1 + q)$
$h \times H$	Q^2p	1	0	0	1
$h \times$ h	Q^2q	0	1	0	1

G). If genes G and g have, respectively, frequencies p and $q = 1 - p$, then at equilibrium there are p males with H, and q with h. Under random mating among females there are p^2 GG, $2pq$ Gg and q^2 gg. The transmission rule for this type of genetic scheme is: 1) a male obtains one of his mother's two X's, each being transmitted with the same probability. Sons of GG mothers will be G, sons of gg mothers will be g, one half of sons of Gg mothers will be G, one-half g, irrespective of their fathers. 2) Daughters take the X of the father and a randomly chosen one of the mother's two X's. Thus in the mating $Gg \times g$, for instance, there are one-half female offspring Gg (phenotype H) and one-half gg. The mating itself occurs with frequency $2pq \times q = 2pq^2$. Table 3.3.3 lists the frequencies of all matings and their outcomes in terms of H and h. From Table 3.3.3, a 2×2 table of H, h types can be constructed for, say, fathers and daughters (r_{FD}). The four correlations in the last column of Table 3.3.2 were calculated in this manner.

In Table 3.3.2, the Kuru transmission scheme (3.3.5) and X-linked transmission scheme are both expected to produce small r_{FS} values, while the other correlations are positive. Sampling fluctuation may make it difficult to distinguish this type of cultural inheritance from various types of Mendelian inheritance.

3.4 HORIZONTAL TRANSMISSION

The group of individuals that influences the members of a given generation t may not be restricted to the $(t-1)^{\text{st}}$, or parental, generation. In this section the influence of individuals in the tth generation on the *same* generation, t, is studied. This transmission from age peers is termed *horizontal.* As before, we ignore the effect of age structure within the generation. Horizontal transmission of the type envisaged here has great similarity with epidemics of infectious diseases, certain aspects of which have been discussed in Chapter 1.

Vertical transmission is assumed to occur first, resulting in a frequency of $u_{t+1}^{(C)}$ of H among the children (C) of generation $t+1$:

$$u_{t+1}^{(C)} = \Sigma p_i^{(t)} b_i. \tag{2.2.1}$$

Then horizontal transmission occurs, increasing the frequency of H to

$$u_{t+1}^{(A)} = u_{t+1}^{(C)} + f u_{t+1}^{(C)} (1 - u_{t+1}^{(C)}), \tag{3.4.1}$$

among adults (A), where it is assumed that $0 < f < 1$. Now selection occurs, after which the adults in generation $t+1$ just prior to their mating have H frequency:

$$u_{t+1} = \frac{(1+s)u_{t+1}^{(A)}}{1 + s u_{t+1}^{(A)}}. \tag{3.4.2}$$

This is a recursion in u_t, once the mating frequencies $p_{i,t}$ are specified as functions of u_t. The equilibrium equation from (3.4.2) is of fifth degree. If $b_0 = 0$, $b_3 \neq 1$, there is at least one root of this polynomial in the admissible range of $0 < \hat{u} < 1$. In fact, there are either one, three or five valid equilibria. If one is valid and if $B > 0$, $C > 0$ (see 3.3.6), then the transformation (3.4.2) is monotonic, and the equilibrium is stable. If three or five roots are valid with B, $C > 0$, then the roots are stable, unstable, stable, unstable, stable in increasing numerical order of the $\hat{u}$ values.

If $b_0 = 0$ then $\hat{u} = 0$ is an equilibrium of (3.4.2). If the mating is random,

$$\lim_{u_t \to 0} \frac{u_{t+1}}{u_t} = (1 + s)(1 + f)(b_1 + b_2). \tag{3.4.3}$$

If the right side of (3.4.3) is greater than unity, H is protected from loss. It is of interest to compare (3.4.3) with the condition for protection under oblique with vertical transmission, where the oblique transmission function is $\phi(u) = fu$. From (3.2.5), the condition for protection from extinction in the oblique transmission model is, with $b_0 = 0$:

$$\textit{oblique:} \quad (1 + s)(f + mB + b_1 + b_2) > 1, \tag{3.4.4}$$

while the condition from (3.4.3) with assorting is:

$$\textit{horizontal:} \quad (1 + s)(1 + f)(mB + b_1 + b_2) > 1. \tag{3.4.5}$$

Which of these allows easier protection of H depends on the transmission parameters. For example, if m is negligibly small, *then* $b_1 + b_2 > 1$ *is the condition in which horizontal transmission protects better than oblique.* The interpretation here is that, when H is rare, it will enter the offspring generation mainly from heterotypic matings. If these produce a large number of H offspring, then horizontal transmission does better.

With $B = 0$, as in the special cases 1, 3, and 4 of Table 2.3.1, and $b_0 = 0$, we have $\Sigma p_i b_i = bu$, where $b = \alpha_p + \alpha_m$ in case 4 and $b = 1$ in the others. (3.3.6) then reduces to

$$u' = \frac{(1 + s)(1 + f - fbu)bu}{1 + sbu(1 + f - fbu)}. \tag{3.4.6}$$

If $\hat{u} = 0$ is unstable (i.e., $(1 + s)(1 + f)b > 1$), then there is a unique equilibrium which, since (3.4.6) is a monotone increasing transformation, is stable. If $\hat{u} = 0$ is stable and $s > 0$, then no polymorphic equilibria exist and H is lost. If $\hat{u} = 0$ is stable and $s < 0$, then depending on b, f, and s, it is possible that two polymorphic equilibria exist, the larger of which is stable. This would be the case if, for example, $s = 0.4$, $b = 0.9$, $f = 0.8$. When $b = 1$, (3.4.4) and (3.4.5) are the same, since B is assumed to be zero. But if $b < 1$, then protection under vertical with horizontal transmission is more difficult than under vertical with oblique. Other examples from Table 2.3.1. can be handled in the same way as (3.4.6).

The mechanism of non-vertical transmission in the case of the hepatitis B virus discussed in section 3.3 is not well understood although such vectors as mosquitos and bedbugs are implicated in some situations. In addition aerial transmission may occur and it is known that a high proportion of infected individuals carry the virus in their saliva. Thus horizontal *and* oblique transmission may be important.

In the context of the present chapter we can examine the role of horizontal transmission acting only in conjunction with vertical under the assumption that s is negligible. From (3.4.1), then

$$\hat{u} = (f + fC - 1)/fC$$

is a valid equilibrium if $f(1 + C) > 1$ and is stable if $(1 + f)C > 1$. With $\hat{u} = 0.15$, $C = 0.45$, this produces the estimate $f = 1.31$, much larger than the value $f = 0.59$ estimated under vertical and oblique transmission. Again, $f > 1$ can radically alter the stability properties of the dynamical system. With more information on the age distribution of carriers, familial behavior and ecology, it might eventually be possible to use this framework to estimate the relative importance of oblique and horizontal transmission of hepatitis. In general we have seen that the qualitative differences between the two also depend on their interactions with vertical transmission.

3.5 SIB-SIB INTERACTIONS

A special type of horizontal interaction is that between sibs. For quantitative traits like IQ, it is well documented that there are important birth order effects. (A decrease of IQ with increasing birth order has been repeatedly observed.) These are believed to be due, in part, to parental stimulation decreasing on average with increasing birth order. Another theory invokes the effects of teaching of younger sibs by older ones, teaching which is likely to be more beneficial to the teacher than to the pupil. Certainly in the past, birth order strongly influenced the professions and economic position to which children could belong (e.g., primogeniture). This effect is now disappearing, but there still is an effect of birth order on IQ (see Zajonc 1976).

A model of parent and sib transmission suggested by S. R. Wilson (1974) and outlined in Table 3.5.1 showed that with two-generation data (parent-offspring) it would not be easy to distinguish this model from one of Mendelian inheritance.

In its original form, the model of Table 3.5.1 requires the estimation from real data of six parameters, given that

$$\sum_{j=0}^{2} p_{ij} = 1.$$

From the point of view of predicting evolution, however, the six parameters are not necessary. In fact, our b_i values are enough. In terms of Table 3.5.1 they would be estimated from the data using

$$b_i = p_{i2} + p_{i1}.$$

In Wilson's formulation there are only three such equations and three b_i values ($b_3, b_1 = b_2, b_0$) because no distinction has been made between $H \times h$ and $h \times H$ matings. If these matings are kept distinct, the four parameters b_3, b_2, b_1, b_0 are all necessary, and four additional parameters would be required to give the 4×3 matrix with rows indexed by the matings $H \times H$, $H \times h$, $h \times$

TABLE 3.5.1. The interaction between sibs in S. Wilson's model of nongenetic transmission. There are two children per family, and the probabilities given in the table refer to a family made of 2 H children, 1 H and 1 h and 2 h children, respectively. The probabilities are determined by the type of mating.

Mating	Progeny 2H	1H:1h	2h
$H \times H$	p_{22}	$2p_{21}$	p_{20}
$H \times h$	p_{12}	$2p_{11}$	p_{10}
$h \times h$	p_{02}	$2p_{01}$	p_{00}

H and $h \times h$. The additional four parameters (or three if $H \times h$ and $h \times H$ are pooled) would be necessary only for specifying the correlation between sibs. This correlation would differ from that given earlier by us if the probabilities, say p_{22} and p_{20} differ from b_3^2, $(1 - b_3)^2$, respectively $(b_3 = p_{22} + p_{21})$, and so on. Such nonrandom assortment of sibs would produce a sib-sib correlation of

$$r_{ss} = \frac{1}{u_{t+1}v_{t+1}} \{u_t^4(p_{22}p_{20} - p_{21}^2) + v_t^4(p_{02}p_{00} - p_{01}^2)$$
$$+ 4u_t^2v_t^2(p_{12}p_{10} - p_{11}^2) + u_t^2v_t^2(p_{00}p_{22} + p_{02}p_{00} - 2p_{21}p_{01})$$
$$+ 2u_t^3v_t(p_{12}p_{20} + p_{22}p_{10} - 2p_{21}p_{11})$$
$$+ 2u_tv_t^3(p_{12}p_{00} + p_{02}p_{10} - 2p_{11}p_{01})\}.$$

This reduces to (2.4.5) if sib-assortment is random.

Another approach to the problem of sib-to-sib influences extends the linear model of parent-to-offspring transmission used earlier. To do this, we might construct a three-pronged transmission scheme in which a particular offspring is affected by his mother, father, and a sib. The sib could be an older or younger brother or sister. The transmitting "triplet" and recipient could then be depicted in the following Table 3.5.2.

From a sample of families with two offspring, the parameters

TABLE 3.5.2.

Transmitting			Probability of receiving sib's type	
Father	Mother	Sib	H	h
H	H	H	$b_0 + \alpha_p + \alpha_M + \alpha_{SIB}$	$1 - b_0 - \alpha_p - \alpha_M - \alpha_{SIB}$
H	H	h	$b_0 + \alpha_p + \alpha_M$	$1 - b_0 - \alpha_p - \alpha_M$
H	h	H	$b_0 + \alpha_p + \alpha_{SIB}$	$1 - b_0 - \alpha_p - \alpha_{SIB}$
H	h	h	$b_0 + \alpha_p$	$1 - b_0 - \alpha_p$
h	H	H	$b_0 + \alpha_M + \alpha_{SIB}$	$1 - b_0 - \alpha_M - \alpha_{SIB}$
h	H	h	$b_0 + \alpha_M$	$1 - b_0 - \alpha_M$
h	h	H	$b_0 + \alpha_{SIB}$	$1 - b_0 - \alpha_{SIB}$
h	h	h	b_0	$1 - b_0$

could be estimated by maximum likelihood and the fit of the above model to data examined. This was done, for example, in the analysis of data collected by us (Chen, Feldman and Cavalli-Sforza, in preparation) on a sample of Taiwanese families. The particular question concerned Buddhist affiliation. The data are summarized in Table 3.5.3. Compared to parental effects, the older brother and older sister seem to strongly influence the younger sister and younger brother, respectively.

Yet another approach to intrafamilial problems involves a proposed time sequence for sib-to-sib influences. Consider, for

TABLE 3.5.3. Familial transmission of Buddhist religion (from Chen et al., unpublished).

Transmitters	Receiver	b_0	α_p	α_M	α_{SIB}	$\chi^2_{(4)}$
Father, Mother, Older Brother	Younger Brother	.17 ± .05	.17 ± .12	.11 ± .10	.46 ± .11	2.68
Father, Mother, Older Sister	Younger Sister	.11 ± .02	.28 ± .10	.25 ± .09	.32 ± .07	6.32
Father, Mother, Older Brother	Younger Sister	.10 ± .03	.26 ± .12	.06 ± .10	.43 ± .10	7.31
Father, Mother, Older Sister	Younger Brother	.10 ± .03	.10 ± .10	.20 ± .10	.50 ± .09	2.15

example, a family in which mother and father are H. The first offspring is H with probability $b_0 + \alpha_p + \alpha_M$. The second has a similar probability of being H, but if parents can dedicate less time to second children than to first ones (see, e.g. Zajonc, 1976), then for the second child the probability will be $b_0' + \alpha_p' + \alpha_M'$. If the second sib is h after parental effect, an event which has probability $1 - b_0' - \alpha_p' - \alpha_M'$, it might then be converted to H, at rate k, by contact with the first sib if this is H. The total probability that it is H at the time of measurement is then

$$b^* = b_p' + \alpha_p' + \alpha_M' + k(b_0 + \alpha_p + \alpha_M)(1 - b_0' - \alpha_p' - \alpha_M'),$$

assumed to be less than unity. Similar probabilities can be constructed for sibs from the other marriages. It might then be possible to estimate b_0, α_p, α_M, b_0', α_p', α_M' and k as before. Intrafamilial interactions have not been subject to a great deal of modeling and these must be regarded as preliminary suggestions.

3.6 MIGRATION BETWEEN POPULATIONS

The arrival of an immigrant may have the same effect as the birth of a mutant individual. Similarly, news of an invention made elsewhere may have the same effect as if it were an independent local discovery. In this sense immigration may be difficult to distinguish from mutation both in genetic and in cultural change. But genetic mutation is normally a rare event, whereas migration in both biological and cultural contexts can occur at high frequencies. Certainly the colonization of new continents and islands involved migration, and may have taken place repeatedly. Matrimony has historically been an important impetus to migration. Indeed, almost no human group is fully endogamous; at least a few spouses enter a given group from other populations and, especially if social groups are small, one of the two spouses is usually from another group (exogamy). When marriage is *virilocal* it is the prospective wife who moves to the husband's location, while in *uxorilocal* marriage the opposite happens.

Movement of people has consequences for genetic evolution. But when people move as adults, and carry their culture with them, migration has consequences also for cultural change. Today culture can move even without physical movement of individuals, without *demic migration,* through a more subtle process of information flow. But even in the past, when no telecommunications were available, and information was spread by contacts between neighbors, or at greater distance by travelers, traders, armies, and the like, permanent movement of people was not necessary for "*cultural migration*" to occur: the flow of information needs only some kind of social and cultural interaction without final replacement of the people concerned. The "borrowing" of words is a typical example (see, for example, Sapir, 1921). Such cultural migration is most noticeable when it takes place across the boundaries of different societies. Foreign words often enter into a language together with an innovation which they describe or with which they are associated. Clearly, cultural migration must be considered as a powerful force in the process of cultural change.

The process of exchange between populations, and the evolution deriving from it, can be modeled most simply as follows. Consider two populations P_1 and P_2 and a trait with values H and h. In population P_1 the frequency of H is u_1, in P_2 it is u_2. After a unit time (which would be a generation in genetic models or under vertical transmission) there has been a change to u_1' and u_2', according to

$$\begin{aligned} u_1' &= (1 - m_1)u_1 + m_1 u_2 \\ u_2' &= (1 - m_2)u_2 + m_2 u_1. \end{aligned} \tag{3.6.1}$$

The model (3.6.1) describes the situation where, in the absence of migration, trait frequencies do not change, that is, the equilibrium is neutral. This happens, for instance, in cases 1a and 2 of

Table 2.3.1, and in case 3 when $b_0 = 0$, $b_3 = 1$. Here m_1 measures the rate at which people from P_2 enter P_1, with $(1 - m_1)$ of P_1 staying at home. m_2 measures the rate at which individuals from P_1 enter P_2. If there is no actual shift of people, just flow of information, m_1 and m_2 might express the proportions of cultural information transported to the respective groups P_1 and P_2 from P_2 and P_1.

If this process continues for a sufficient time with constant m coefficients, the two populations will become identical. To see this, consider the difference $x = p_1 - p_2$. Then

$$x' = u_1' - u_2' = (1 - m_1 - m_2)x. \tag{3.6.2}$$

On further iteration,

$$x_t = (1 - m_1 - m_2)^t x_0. \tag{3.6.3}$$

As $m_1 + m_2 > 0$, then $\lim_{t\to\infty} x_t = 0$. The equilibrium values of the populations are

$$\hat{u}_1 = \hat{u}_2 = (m_2 u_1^{(0)} + m_1 u_2^{(0)})/(m_1 + m_2). \tag{3.6.4}$$

The proof of this is postponed until we discuss a more general model below.

It is also possible to construct simple migration models in cases where the transmission is such that it drives trait frequencies toward an equilibrium point. The point may be different for the different populations (say $\bar{u}_i$ for the ith population). Let $\alpha \leq 1$ be a measure of the force pulling each population toward its equilibrium, that is, $u_i' = \alpha(\bar{u}_i - u_i)$ if the populations were isolated. Then write

$$\begin{aligned} u_1' &= (1 - \alpha)[(1 - m_1)u_1 + m_1 u_2] + \alpha\bar{u}_1 \\ u_2' &= (1 - \alpha)[(1 - m_2)u_2 + m_2 u_1] + \alpha\bar{u}_2 \end{aligned} \tag{3.6.5}$$

Here α could be the rate of immigration from a population external to both populations 1 and 2, whose composition does not change in the process. It could also be the result of transmission with $b_0 \neq 0$, $b_3 \neq 1$, or the result of some stabilizing selection. Again using the variable $x = u_1 - u_2$ we have

$$x_{t+1} = (1 - \alpha)(1 - m_1 - m_2)x_t + \alpha(\bar{u}_1 - \bar{u}_2) \tag{3.6.6}$$

$$= (1 - \alpha)^{t+1}(1 - m_1 - m_2)^{t+1}x_0 + \alpha(\bar{u}_1 - \bar{u}_2)\sum_{i}^{t}(1 - \alpha)^i(1 - m_1 - m_2)^i. \tag{3.6.7}$$

In the limit as $t \longrightarrow \infty$ we have

$$\lim_{t\to\infty} x^{(t)} = \frac{\alpha(\bar{u}_1 - \bar{u}_2)}{1 - (1 - \alpha)(1 - m_1 - m_2)} = \hat{u}_1 - \hat{u}_2. \tag{3.6.8}$$

Substituting (3.6.8) into (3.6.5), the equilibrium values are obtained

$$\hat{u}_1 = [\bar{u}_1\alpha + (1 - \alpha)(\bar{u}_1 m_2 + \bar{u}_2 m_1)]/[\alpha + (1 - \alpha)(m_1 + m_2)]$$
$$\hat{u}_2 = [\bar{u}_2\alpha + (1 - \alpha)(\bar{u}_1 m_2 + \bar{u}_2 m_1)]/[\alpha + (1 - \alpha)(m_1 + m_2)]. \tag{3.6.9}$$

Notice that if $\alpha = 0$, and $\bar{u}_1$ and $\bar{u}_2$ are interpreted as the initial values of u_1 and u_2, namely $u_1^{(0)}$ and $u_2^{(0)}$, then (3.6.9) reduces to (3.6.4).

The above argument is easily generalized to n populations. Suppose m_{ij} is the ij element of a migration matrix **M**, such that m_{ij} is the fraction of individuals of population j which join population i to form the next generation and m_{ii} is the fraction of residents who remain. These elements are constant over time. (In terms of cultural influence the m_{ij} values measure the influence of population j on population i; we shall return to this structure later.) Suppose that at time $t + 1$ the frequency of the trait in the ith population is $u_{i,t+1}$. Then

$$u_{i,t+1} = \sum_{j=1}^{n} m_{ij}u_{j,t} \quad (i = 1, 2, \ldots n). \tag{3.6.10}$$

Or, in vector form

$$\mathbf{u}_{t+1} = \mathbf{M}\mathbf{u}_t = \mathbf{M}^{t+1}\mathbf{u}_0. \tag{3.6.11}$$

There is no outside stabilizing force analogous to α in (3.6.5), so that $\Sigma_j m_{ij} = 1$, for each i, that is, $\mathbf{M}$ is a stochastic matrix. The largest eigenvalue of $\mathbf{M}$ is therefore 1. If $\mathbf{r}_1, \mathbf{r}_2, \ldots \mathbf{r}_n$ are the right eigenvectors, and $\mathbf{l}_1, \mathbf{l}_2, \ldots \mathbf{l}_n$ are the left eigenvectors of $\mathbf{M}$, and $\mathbf{\Lambda}$ is the diagonal matrix of eigenvalues taken in the same order as the sequence of eigenvectors, then we can write

$$\mathbf{M}^t = \mathbf{R}\mathbf{\Lambda}^t\mathbf{L} \tag{3.6.12}$$

where the columns of $\mathbf{R}$ are $\mathbf{r}_i$ and the rows of $\mathbf{L}$ are $\mathbf{l}_i$. As the powers of Λ increase, the eigenvalues less than unity make an ever decreasing contribution to the product in (3.6.11). In the limit as $t \longrightarrow \infty$ only those terms involving the unit eigenvalue remain. Another way of writing (3.6.12) is

$$m_{ij}^{(t)} = \sum_{k=1}^{n} \lambda_k^t r_k(i) l_k(j) \tag{3.6.13}$$

where $r_k(i)$ is the ith component of the kth right eigenvector and $l_k(j)$ is the jth component of the kth left eigenvector; λ_k is the kth eigenvalue. Either of the forms (3.6.12) or (3.6.13) is convenient for use in (3.6.11) where the limiting frequencies can be obtained as functions of the initial vector $\mathbf{u}_0$. For example, if $\mathbf{M}$ is the migration matrix for (3.6.1)

$$\mathbf{M} = \begin{pmatrix} 1 - m_1 & m_1 \\ m_2 & 1 - m_2 \end{pmatrix}$$

$$\lim_{t\to\infty} \mathbf{M}^t = \begin{pmatrix} m_2/(m_1 + m_2) & m_1/(m_1 + m_2) \\ m_2/(m_1 + m_2) & m_1/(m_1 + m_2) \end{pmatrix}$$

$$= \mathbf{M}^*, \text{ say.} \tag{3.6.14}$$

The limiting frequencies $\mathbf{M}^*\mathbf{u}_0$ are therefore those in (3.6.4).

In the same way, extensions of (3.6.5) to n populations can be made using vector terminology.

$$\begin{aligned}\mathbf{u}_{t+1} &= (1-\alpha)\mathbf{M}\mathbf{u}_t + \alpha\bar{\mathbf{u}}\\ &= (1-\alpha)^{t+1}\mathbf{M}^{t+1}\mathbf{u}_0 + \{[1-(1-\alpha)^{t+1}\mathbf{M}^{t+1}]\\ &\quad\cdot [\mathbf{I}-(1-\alpha)\mathbf{M}]^{-1}\alpha\bar{\mathbf{u}}\},\end{aligned} \tag{3.6.15}$$

and at equilibrium

$$\lim_{t\to\infty} \mathbf{u}_t = \alpha[\mathbf{I}-(1-\alpha)\mathbf{M}]^{-1}\bar{\mathbf{u}}. \tag{3.6.16}$$

With this stabilizing force (measured by α) the equilibrium (3.6.16) must be independent of the initial frequencies.

The two classes of models discussed above are only the simplest representatives of a large class that can be proposed. It is important, however, to recognize the qualitative differences between them. The first class includes only forces intrinsic to the two populations and produces equilibrium values that depend strongly on the initial frequencies in each group. The second involves a linear form of selection, not precisely defined in terms of who lives and who dies, but rather as a stabilizing influence. Such a situation would be the case if small splinter societies exchange members with a large unchanging culture. The continued influx of these immigrants ensures that the major long-run force on the splinter groups is the culture from which the immigrants come, rather than the history of the small groups themselves.

Biological migration can be measured from demographic data on relative numbers of immigrant individuals, whereas cultural migration is difficult to evaluate directly. In certain cases it might be assumed that cultural migration is *proportional* to biological migration so that, roughly speaking, the more immigrants from a foreign source, the greater their cultural effect is likely to be. Biological migration data might then be used to evaluate cultural migration. In so doing, however, it should be kept in mind that the assumption of proportionality ignores any effects of cultural selection, which may make cultural migration more effective than expected on the basis of proportionality, either because of the desirability of the innovations offered by the immigrants, or indirectly by way of the authority and prestige of the immigrants (especially if these are conquerors, or are well known).

Occasionally, biological migration data are available in the matrix form of (3.6.14–3.6.15). With n populations, this requires $n(n-1)$ migration coefficients, two for each pair of populations (from i to j and from j to i). Such data are not easily available in practice and recourse is often made to a simple substitute, namely the prediction of migration on the basis of geographic distance. Much research has shown that intermigration between two populations is approximately proportional to a simple function of the distance between them (see, for instance, Hägerstrand, 1967; Sutter, 1962; Cavalli-Sforza and Bodmer, 1971). One such function currently employed is d^{-c}, where d is distance, and c is close to 2. For a slightly more complicated and more realistic prediction, with application to phonetic similarities in a microgeographic study of linguistic variation, see Trudgill (1974).

The existence of a close relationship between migration and genetic distance partly explains the success of simple models of genetic variation due to geographic isolation, first proposed by Malécot (see e.g. Malécot, 1966), and by Kimura and Weiss (1964). One of these models (usually called the linear stepping-stone model) predicts that the genetic similarity between two populations i and j, measured as a kinship coefficient ϕ_{ij}, will tend to decrease exponentially with geographic distance between i and j (d_{ij}):

$$\phi_{ij} = \phi_0 e^{-kd_{ij}} \tag{3.6.17}$$

where ϕ_0 and k are two coefficients whose interpretation will be given later.

It is of interest that in many cases this relationship has been

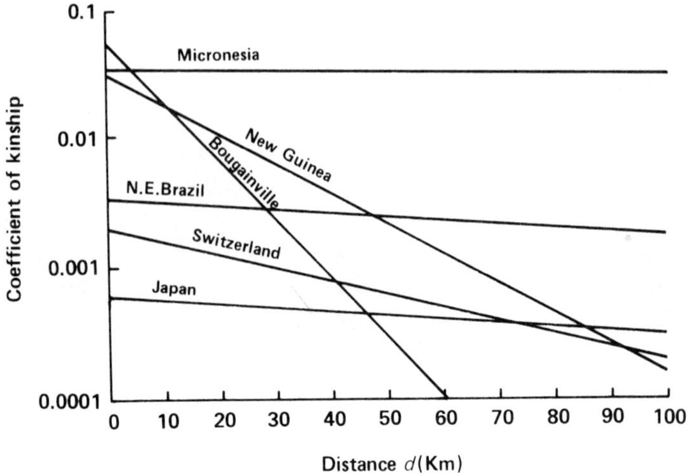

FIGURE 3.6.1. The abcissa is the geographic distance between pairs of Mendelian populations and the ordinate (on logarithmic scale) is a measure of their genetic similarity (kinship coefficient). The various straight lines correspond to studies in different areas of the world. Figure from Friedlaender, 1975.

found to be valid, with ϕ_0 and k varying considerably from place to place (see Fig. 3.6.1). The theory of the linear stepping-stone model assigns meanings to the two coefficients, ϕ_0 and k, that are of interest for our purposes. The geographic counterpart of the linear stepping stone is shown in Figure 3.6.2. It corresponds to a series of islands, each with N individuals, and each of which contributes at every generation a fraction m of its individuals to its

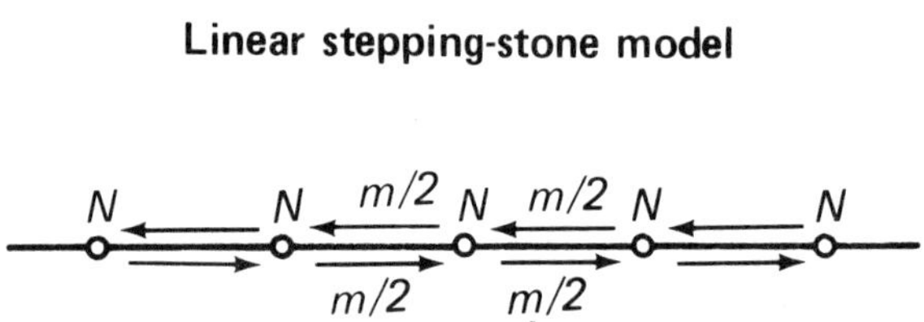

FIGURE 3.6.2. One-dimensional linear stepping stone migration model.

nearest neighbors, half to the left and half to the right. The rigidity of the model may seem unnatural. A less rigid approach, using a full migration matrix analogous to (3.6.15), with different N (number of individuals) for each population, shows that the variation in size of the population of each island, the variation in distance between islands, and the fact that migration is usually but not exclusively, to nearest neighbors, do not seem to create major deviations from the simple negative exponential given in (3.6.17). Even if the geographic pattern of the "stepping stones" (islands, villages, etc.) is not exactly a straight line, there is no serious deviation from the simple exponential, at least empirically. It should also be remembered that (3.6.17) is an *equilibrium* relationship, obtained from the balance of genetic drift (which increases genetic differences between the islands) and intermigration between islands, or other linear stabilizing forces (which tend to reduce them). Equilibrium is reached in relatively few generations if migration is large.

The two coefficients of (3.6.17) have the following interpretation, at equilibrium (Malécot, 1966; Crow and Kimura, 1970):

$$\phi_0 = \frac{1}{1 + 4N(2m\mu)^{1/2}}$$

$$k = (2\mu/m)^{1/2} \qquad (3.6.18)$$

where N is the number of individuals reproducing every generation in the standard island or population, m is the proportion of individuals which emigrate every generation from each island, $m/2$ going to the left, and $m/2$ to the right neighbor (Fig. 3.6.2). This emigration is compensated exactly by the same amounts of immigration from the nearest neighbors. The quantity μ is an aggregate of forces which introduce variation, like immigration from a remote population equal for all islands, or recurrent mutation, which keeps introducing new variation lost by drift, or still other forces.

It is very likely that a similar model may be applicable in

linguistics, and in particular for the substitution of cognates. When measuring genetic similarity, ϕ_{ij} is applied for the alternative forms (alleles) of a specific gene, represented as: $A_1, A_2 \ldots A_n$. The quantity ϕ_{ij} is the probability that, taking one individual at random from population i, and one from j, and examining gene A in both, they are found to be identical, that is, both A_1, or both A_2, and so on. (For organisms like humans, which have genes in duplicate, one of paternal and one of maternal origin, after sampling randomly one individual, one must also randomly sample the paternal or the maternal gene.) In the application to linguistics, there may be two or more noncognate words used to indicate the same meaning in one population at a given time. Wherever more than one cognate exists, one or more may be favored and replace the other or others (Sankoff, 1971), because of cultural selection or chance, as happens for allelic substitution in genetic evolution.

In an analysis of a linguistic survey of 17 Micronesian islands by Quackenbush (1968), Wang, et al. (unpub.) found that 512 glosses (different words) had the following distribution of roots (noncognatic words with the same meaning) considering the group of islands as a whole:

No. of roots per gloss:	1	2	3	4	5	6	7	8	9	10	11	12	18	Total
No. of glosses:	131	91	99	63	50	30	19	12	10	1	3	2	1	512

A new root may be an invention, or an introduction from a foreign language. That its spread is likely to go through a transitional period in which it will coexist with the other, as postulated above, is confirmed by some direct evidence to be found in the Micronesian data. Some informants were familiar with more than one root (rarely more than two), and the roots were stated to be almost equally common or, more rarely, one of the two was stated to be definitely more common than the other. The simplest interpretation is that for most of these glosses the process of substitution of one root with the other was in an intermediate state. New roots that have recently entered the language of an island are probably rare and therefore not known to every informant, and the same is true of roots that are close to being totally eliminated. Clearly, the analysis could gain by using carefully chosen samples of several informants per language or dialect, but this is time consuming and not commonly done in linguistic surveys.

As a numerical example, 44 glosses out of the 91 two-rooted glosses of Table 3.6.1 were studied in all 17 islands, and, for 13 of these, some islands had both roots in use. For instance, for the fourth gloss in Table 3.6.1 there existed the roots Fatur and

TABLE 3.6.1. Micronesian languages, data by Quackenbush (1968). Distribution of islands for thirteen glosses for which two different roots were found, and in at least one of the seventeen islands both roots were in use.

	Number of islands in which:		
English Equivalent of Gloss	Only More Common Root Found	Both Roots Found	Only Less Common Root Found
Bitter	15	2	0
Difficult	16	1	0
Hot	4	5	8
To walk	13	3	1
To fly	13	1	3
To fear	13	1	3
Swell	16	1	0
Want	7	4	6
Saw	11	1	5
Sex	14	1	2
A grave	15	2	0
Bowl	16	1	0
Turtle	13	1	3
Total	166	24	31

Mwet. The first was found in all islands except the eighth, where Mwet is the only root, but in the ninth, tenth and eleventh islands, both Fatur and Mwet were in use. For these 13 two-rooted glosses, transition is in process in some of the 17 islands. Transition may also be going on in many other islands, in which the heterogeneity of the gloss was not detected because of the very limited amount of information for each island. The distribution of islands for these 13 glosses is shown in Table 3.6.1.

To study the effect of geographic distance on linguistic isolation and evolution we will investigate more closely the qualitative observation that neighboring islands have more similar languages. One simple way to compute the coefficient of similarity (which corresponds very closely to the kinship coefficient for genes) is to estimate the proportion of glosses for which two islands, i and j, employ the same root. In order to take account of the complication that in some islands more than one root is used, we would like to know what proportion of people know (or use) a given root. If this proportion is p_{ik} for root k and island i, and p_{jk} for island j, then, summing over all n roots of a gloss, the similarity between the two islands is:

$$\sum_{k=1}^{n} p_{ik} p_{jk}. \tag{3.6.19}$$

In Micronesia the information on p_{ik}'s is limited to such statements that it is near 100% where only one root is claimed known, near 0% if unknown, and intermediate in a few cases. We simplify the situation by setting $p_{ik} = 1$ if a root is the only one or the predominant one, and $p_{ik} = 0$ otherwise. Then expression (3.6.19) is equal to unity where both islands have the same root and zero otherwise. The proportion of shared cognates (common roots) over all glosses is then the overall estimate of linguistic similarity of the islands on the basis of this criterion. The proportion observed for one pair of islands can be plotted against their geographic distance to test if (3.6.17) is satisfied.

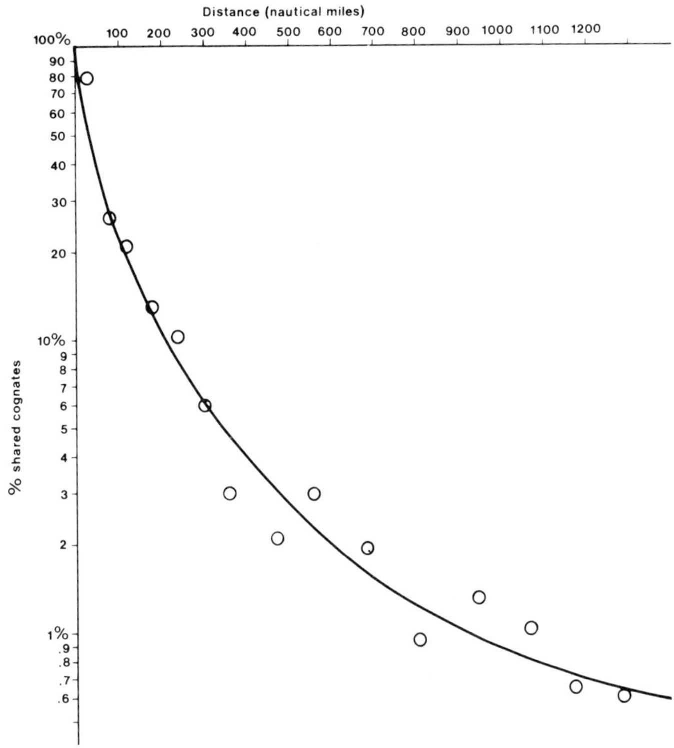

FIGURE 3.6.3. An unpublished analysis (Wang et al., unpub.) of the proportion y of shared cognates between pairs of languages spoken in different Micronesian islands and its dependence on the distance between islands, x, based on a selection of data by Quackenbush (1968).

Figure 3.6.3 shows that (3.6.17) is not satisfied. The logarithmic ordinate used in the figure should give a linear relationship, which is not observed. What went wrong? We can dispose, as we have already stated, of the objection that the model is oversimplified with respect to reality, by carrying out simulations with models closer to reality in terms of geographic distribution and demographic data. This appears to cause only minor departures from linearity. Another possibility is that equilibrium has not been reached. However, these islands have been inhabited for several thousands of years, long enough for the establishment of equilibrium, unless the process is very slow. There remain other possibilities. A major one is that individual words do *not* have the same rate of change. This has already been pointed out for historical (diachronic) change and is likely to be true for geographical (synchronic) variation also, because variation in space and in time are obviously closely related phenomena. The variation of the cultural substitution rate could be due to one of two different causes: the mutation rate may vary from one word to the other (and will then change the number of roots per gloss), or the cultural migration rate may vary so that different words differ in the ease with which they are picked up and adopted.

Both sources of variation would affect k, making it different from word to word. The result would be qualitatively in agreement with observation, in the sense that it would generate a concavity upwards, as in Fig. 3.6.3. A quantitative test requires knowledge of the variation of k. For instance, k may vary between zero and infinity with a "gamma" type distribution:

$$d\Phi(k) = \frac{1}{c\Gamma(n)} (k/c)^{n-1} e^{-k/c} dk, \qquad (3.6.20)$$

where c is a scaling coefficient, $\Gamma(n) = (n - 1)!$, and n is a parameter on which depends the relative variation of this distribution. Then, integrating over all possible values of k, the relationship between similarity ϕ and distance x would become

$$\phi = \int_0^\infty e^{-kx} d\Phi(k) = \frac{1}{(1 + cx)^n}, \qquad (3.6.21)$$

where c depends on the scale of distances. The shape of the curve relating the proportion of shared cognates to distance depends on the variation of k. If this is high, n is low and the curve of

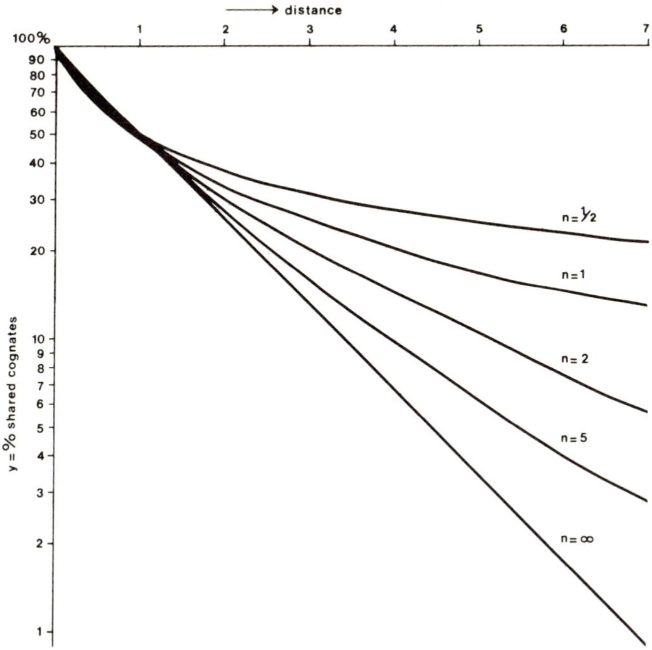

FIGURE 3.6.4. Effect of the variation of evolutionary substitution rate k of different glosses on the global proportion of shared cognates. The variation is given by the quantity n in the gamma distribution assumed for the variation of k in formulas (3.6.20) and (3.6.21), where n is equal to $1/(CV)^{1/2}$, and CV is the coefficient of variation of k (standard deviation divided by the mean). For ease of comparison, the distances of the various curves are scaled so that all curves cross at $y = 50\%$.

proportion of shared cognates versus distance is far from exponential. Figure 3.6.4 shows the effect of n on (3.6.21).

The curve drawn in Fig. 3.6.3 is obtained by fitting (3.6.21) to the data, and seems to give a very satisfactory representation of the observations. In (3.6.20), the mean and variance of k depend on c and n, which can be estimated from the observed curve of Fig. 3.6.3. There are several ways in which this analysis can be sharpened. First, we can test if it is true that a single gloss does indeed follow the negative exponential relationship (3.6.16), and if so we can evaluate its k value. This produces a distribution of k values by repeating the procedure for every gloss. Finally, if we can estimate the mean, $\bar{k}$, and variance, σ^2, of k, an estimate of n might be derived from

$$\sigma/\bar{k} = 1/\sqrt{n}, \tag{3.6.22}$$

after which the shape of the observed curve of linguistic similarity versus geographic distance can be predicted. A major value of this approach is that it would test the validity of the basic exponential for each word and thus give more confidence in the overall credibility of the interpretation. Another possible approach is to consider words with different numbers of roots separately, and test separately for the negative exponential of each of them. Words with only one root produce the same similarity for all pairs of islands (and here $k = 0$). Words with two roots should give an exponential with k smaller than that of three-rooted ones, and so on. The value of n would then be estimated, and the validity of the exponential tested for each number of roots. In conjunction with information on the number of roots per island, the process of substitution of roots might then be reconstructed.

The formal similarity of this process to the substitution of amino acids in proteins and of nucleotides in DNA, or in general of one allele by another, is of considerable interest because this similarity makes it possible to adapt to the process a substantial fraction of the theoretical machinery developed for allelic gene substitution. Naturally, caution should be exercised, because the two processes, biological and linguistic, have important differences. Thus, allelic substitutions do obey at least approximately the expectation of a negative exponential for the decrease of similarity with time (or with geographic distance). As we have seen, this is not so for linguistic root substitutions, which show upward concavity both in time and space. The heterogeneity of evolutionary rates noted for different words exists also in biology, but is much less marked and not sufficient to determine a detectable deviation from exponentiality. Another difference is that there is no strict equivalent of the genetic phenomena of homo- and heterozygosity in language.

Other interesting possibilities are offered by comparisons of linguistic with biological data. In Figure 3.6.1., an enormous difference is evident between the slopes of the relationship of genetic similarity with geographic distance in Micronesia, and the same slopes in New Guinea or Bougainville. One would expect, if cultural and biological migration are proportional, that there should be accordingly much more linguistic variation per mile in New Guinea than in Micronesia. Qualitatively, this conclusion seems correct, because in a relatively small area in New Guinea, over 500 different languages are known to be spoken.

Summarizing, migration, like mutation, can introduce cultural change, and if infrequent may be difficult to distinguish from mutation. If massive, it will usually leave other important marks. In contrast to biological migration, cultural migration does not necessarily involve physical exchange of individuals. It can be regarded as causing oblique cultural transmission. Cultural migration parameters should therefore measure flow of information irrespective of whether there has been permanent or temporary physical movement of individuals. The migration models we suggest are similar to those used for biological evolution, except for changes in the meaning of coefficients, and differences in evolutionary rates.

3.7 MIGRATION DEPENDENT ON EXTENT OF SEPARATION

Speciation in biology is the process by which barriers to interbreeding between groups of individuals are formed. Current evolutionary thought regards this as ordinarily due to the slow and almost continuous accumulation of differences between two or more populations of the same species, resulting finally in their reproductive incompatibility. There is argument over whether most of these differences are accidental or adaptive, an argument that has been at the center of population genetics for the past ten years. Isolation can be geographic but may also derive from utilization of different resources in the same geographic area, or from behaviors that separate one group of animals from another. Once the barriers are established to mating or mating success, the gene pools of the two (or more) species may diverge. The groups may interact for instance through competition for resources, but exchange of genes no longer occurs. (A recent survey of some ideas about speciation has been made by M. J. D. White, 1978.)

Processes analogous to biological speciation also exist in cultural evolution. The formation of two or more different languages from a single predecessor is a case in point. Two languages that are mutually incomprehensible are analogous to two different species. As the exchange of genes is impossible between species, so the exchange of thoughts between groups speaking completely different languages is impossible. A language may, however, accept words from another. The process is mediated by an agent familiar with both who finds a more convenient expression in one language and uses it in another. Although the phenomenon is relatively rare and may provoke outcry from linguistic purists, usually in the recipient group, nevertheless it takes place. For example, Migliorini (1978) lists origins and approximate times of introduction of foreign words into Italian.

Exchanges between two mutually comprehensible languages, or dialects spoken in adjacent areas, are likely to continue as long as mutual understanding is conserved. If the two languages diverge so that mutual understanding is more difficult, then linguistic exchange will tend to be less frequent, although borrowing of foreign words may still occasionally take place (see Trudgill, 1974). In the same way, incipient biological species may exchange migrants at first, but with increasing differentiation migrants may be less acceptable.

To model this state of affairs, the assumption of constant migration rates is removed. Migration is assumed to be a decreasing function of divergence. Depending on the cause of divergence, different formulations are possible. For example, if divergence between the groups is measured by the difference in equilibrium trait frequencies, then it will be controlled by the different transmission rules or selection coefficients that produce the different equilibria. Chance alone may cause divergence, but we ignore chance effects here and restrict our attention to deterministic forces that result in different equilibrium values $\bar{u}_1$ and $\bar{u}_2$ in two populations. The amount of migration between the populations is not constant but is a function of the difference in trait frequencies between the groups. First, return to equations (3.6.5), with equal migration rates $m_1 = m_2 = m$:

$$\begin{aligned} u_1' &= (1 - \alpha)[u_1 - m(u_1 - u_2)] + \alpha\bar{u}_1 \\ u_2' &= (1 - \alpha)[u_2 + m(u_1 - u_2)] + \alpha\bar{u}_2. \end{aligned} \tag{3.7.1}$$

Now suppose that m is a function ϕ of $(u_1 - u_2)$:

$$m = m_0\phi(u_1 - u_2). \tag{3.7.2}$$

The function ϕ should have the property that $\phi(0) = 1$. In other words, when the trait frequencies are the same, a constant amount of migration, m_0, takes place, and does not perturb the frequencies. $\phi(\cdot)$ should also take positive values such that $m \leq 1$. Functions such as

$$m_t = m_0[1 + k(u_{1,t} - u_{2,t})^2]^{-1} \tag{3.7.3}$$

and

$$m_t = m_0[1 - k(u_{1,t} - u_{2,t})^2] \tag{3.7.4}$$

with $0 \leq k \leq 1$ satisfy these properties. The latter migration function (3.7.4) is slightly simpler to analyse, and substituting this in (3.7.1) we have

$$x_t = u_{1,t} - u_{2,t} = (1 - \alpha)[x_{t-1} - 2m_0(1 - kx_{t-1}^2)x_{t-1}] + \alpha(\bar{u}_1 - \bar{u}_2). \quad (3.7.5)$$

For simplicity, set $\bar{u}_1 = 1, \bar{u}_2 = 0$. From (3.7.5) the equilibria are given by the roots of a cubic equation

$$2(1 - \alpha)m_0kx^3 + [(1 - \alpha)(1 - 2m_0) - 1]x + \alpha = 0. \quad (3.7.6)$$

Under the conditions $0 \leq k \leq 1$, (3.7.6) is seen to have a single root in the range $-1 \leq x \leq 1$, allowing valid equilibria for u_1 and u_2. Further, since (3.7.5) is monotonic, this root is achieved from any starting conditions.

The above analysis demonstrates that, if the level of exchange with equal frequencies in the two groups (m_0) is large enough in comparison with the stabilizing force α, then an intermediate level of divergence will result. If the migration were at the *constant rate* m_0, with $\bar{u}_1 = 1$, $\bar{u}_2 = 0$ then, from (3.6.9) the equilibrium attained would be

$$\hat{x} = \alpha/[\alpha + 2m_0(1 - \alpha)]. \quad (3.7.7)$$

This value is smaller than the valid root of (3.7.6), suggesting that trait frequency-dependent migration of the sort considered here produces an increase in the ultimate divergence of the groups.

in conclusion, the rate of migration, both biological and cultural, can be affected by the accumulation of differences between populations, which will cause the migration to decrease over time with increasing divergence. Thus a negative feedback of evolutionary divergence will result, accelerating its last phases. When the initial migration is relatively large compared to the forces carrying initial divergence between the two populations, then the difference may eventually stabilize at an intermediate level; otherwise it will proceed towards complete divergence. The latter process has been postulated for some instances of biological speciation.

3.8 POPULATION STRATIFICATION

As we have already discussed in Chapter 1, the degree of communication between all pairs of individuals chosen at random from a population is hardly the same. One cause of this asymmetry can be stratification of the whole population into subgroups within which a high level of communication occurs, but between which a different, reduced level exists. The strata can, for example, represent different socioeconomic classes, religions or castes. Now, the spread of new practices often has different dynamics among different groups in the population. The use of abortion (performed by medically trained personnel) was not very long ago restricted to those who could afford to travel to where it was legal, or who had a private and understanding physician. In very recent times it became more widely available. Ideas of conservation and preservation of nature were for many years regarded as characteristic of an eccentric and even crazy minority. In a short time the ideas have infiltrated into school classrooms. There are many examples of this general form, where among one set of individuals, be they academics, artists, or truck drivers, a practice has become widespread. The practice then moves into other strata of the population. The following is a simple model of such a phenomenon. Suppose in the first stratum, where the practice originates, the frequency of individuals who follow the practice is u_1, while that in the second stratum is u_2. In the first stratum the practice spreads according to the usual logistic equation

$$u_{1,t+1} = u_{1,t} + \alpha u_{1,t}(1 - u_{1,t}). \tag{3.8.1}$$

In the second stratum, at every time point, the practice spreads first because of its emigration from the first group at the rate γ. After this migration denote the frequency in the second stratum by $u_{2,t+1}^{(m)}$ with

$$u_{2,t+1}^{(m)} = (1 - \gamma)u_{2,t} + \gamma u_{1,t} \tag{3.8.2}$$

The second phase of spread in the second stratum is by contact at the rate β between those who have the practice and those who do not. The result is $u_{2,t+1}$:

$$u_{2,t+1} = u^{(m)}_{2,t+1} + \beta u^{(m)}_{2,t+1}(1 - u^{(m)}_{2,t+1}). \qquad (3.8.3)$$

Substitution of (3.8.2) in (3.8.3) produces a recursion which, with (3.8.1), describes the system. Note that we are modeling "cultural" migration here, rather than "demic" migration (which is migration of individuals). In fact, the u's represent proportions of populations, and the occurrence of migration from layer 1 to layer 2 is shown by the appearance of the term $\gamma u_{1,t}$ in (3.8.2). If the migration were demic there would be a simultaneous decrease in $u_{1,t+1}$ by the same amount. There are two equilibria of this system with $\hat{u}_1 = 0$. One has $\hat{u}_2 = 0$ and the other has $\hat{u}_2 = [\beta(1 - \gamma) - \gamma]/\beta(1 - \gamma)^2$. Both are unstable. The stable equilibrium is $\hat{u}_1 = \hat{u}_2 = 1$. The contact rates α and β are intrinsically positive so that this last equilibrium is stable. The ultimate rate of spread in the second group is $\beta(1 - \gamma) + \gamma$ per unit time. Thus for the spread in the second group to eventually outpace that in the first, $\gamma + \beta(1 - \gamma) > \alpha$ is required, or $\beta > (\alpha - \gamma)/(1 - \gamma)$. If $\gamma > \alpha$, this is automatic; if $\alpha > \gamma$, then the original group will keep the lead.

The kinetics of the system (3.8.1), (3.8.2), and (3.8.3) are depicted in Figure 3.8.1. Stratum 1 clearly has a logistic curve of increase in percent acceptance, independent of events in stratum 2. The latter has a more complex curve of increase but the dynamics here are practically indistinguishable from the logistic. Obviously if $u_{2,0} = 0$, then stratum 2 has a later start than population 1. The two groups have the same contagion rate ($\alpha = \beta$), and if the transfer from group 1 to group 2 is small, then the two curves of increase will be almost parallel. The amount by which u_2 lags behind u_1 is larger the smaller is γ. If, on the other hand, β and α are very different or γ is large, then the curves of increase depart from parallel.

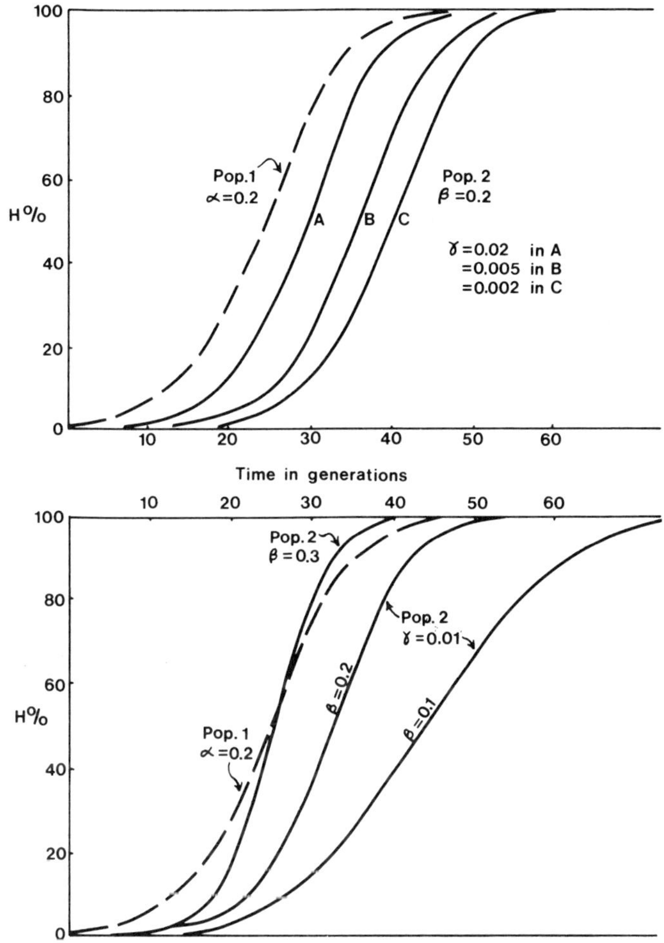

FIGURE 3.8.1. Cultural migration from population 1 to population 2. Population 1 obeys the logistic curve of increase with $\alpha = 0.2$. The cultural migration rate is γ and the conversion rate in population 2 is β. In the upper figure $\beta = \alpha$ and γ varies. In the lower figure γ is fixed and β varies.

In principle the same formulation can be applied to a system with more strata. If the amount of transfer across strata is small, then we expect each group to have a logistic dynamic with rate determined largely by the internal rate of contract. Such a cascade of strata with roughly parallel logistic curves separated by regular lags was observed for the acceptance of Malthusian practices in different social groups distinguished by degree of education. The data pertaining to this will be discussed in the next section. Here it is enough to say that similar situations are well known to sociologists and can be expected to arise in a great variety of circumstances. In some cases this can give rise to oscillations, as when a behavior that first passes through a social hierarchy later, in its original or modified form, catches on again with the original stratum, as in the semisatirical essay, "Noblesse oblige" by Ross et al.

The stratified population model is an important example of exceptions to the rule implicit in the classical mathematical treatment of epidemics, namely homogeneous mixing. This hypothesis, that all possible pairs of individuals in the population have equal transfer rates, is clearly likely to be violated in the physical world. N.T.J. Bailey (1975) summarizes some of the methods of overcoming this restriction, and in section 1.9 some other classical approaches are indicated.

3.9 THE RECENT DEMOGRAPHIC TRANSITION AS AN EXAMPLE OF STRATIFIED, VERTICAL AND OBLIQUE OR HORIZONTAL TRANSMISSION IN CULTURAL CHANGE

Recent demographic changes in Western societies have followed a common pattern which is characterized by at least three types of phenomena. (Fig. 3.9.1) Two of these phenomena—strong oscillations in mortality rates due to epidemics and famines have become less frequent and less important, and the overall mortality rate began a steady decline from rates of over

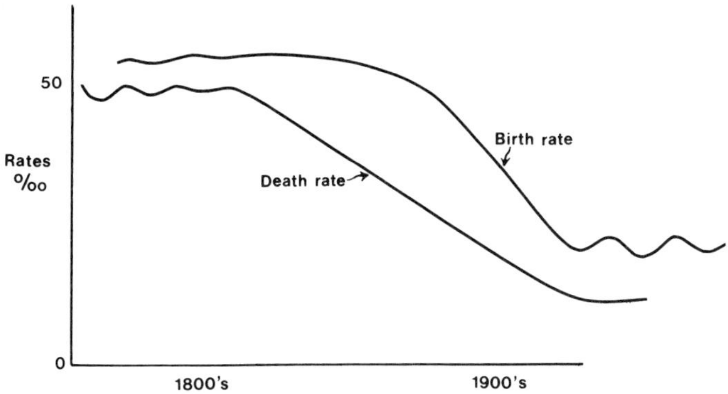

FIGURE 3.9.1. The demographic transition in Europe, idealized. Mortality rates stop oscillating and descend regularly towards a lower asymptote. After a lag, birth rates also descend and settle, with oscillations, at a lower asymptote. After A. Coale, (1974).

forty to fifty deaths per thousand inhabitants per year, to values four or five times smaller and still slowly decreasing—commenced in some parts of Europe at the end of the eighteenth century and spread to most other European countries during the nineteenth. They probably resulted from improved economic conditions, which favored improved transportation and distribution of food. Improved hygiene may also have played a role. The rate of decrease of death rates has slowed down considerably in recent times in the most advanced countries, but the downward trend has probably not ceased. The age classes most affected by the change have been the youngest ones. The mortality pattern was initially heavily weighted towards the youngest ages, with the first year of life ("infantile mortality") making the highest contribution. In one hundred years, the infantile mortality decreased by ten fold and sometimes more. Other age groups experienced a smaller fall of mortality, but no age group seems to have been totally unaffected (except for centenaries or near centenaries, for which adequate data are difficult to obtain). Although the change in mortality patterns (both the disappearance of oscillations and the downward trend) has been synchronous with improvement in nutrition, hygiene, and social conditions, progress in medicine has had an impact only in the last hundred years, and especially in the last thirty or forty years with the increase of the control of infectious diseases. The number of factors that have contributed to the evolution of mortality rates is large. The effect of each is difficult to assess because of the accompanying evolution in patterns of nutrition and public utilities. A major contribution has come from the introduction of vaccines, and, more recently, antibiotic and chemotherapeutic drugs effective against infectious diseases. The mortality due to the latter has thus now almost disappeared. Clearly, this change in mortality patterns is the consequence of evolution in cultural and economic practices as well as in medicine.

In the third type of phenomenon, the downward trend of mortality has been followed by a downward trend of fertility. Both trends are very approximately (descending) logistic. That is, there are long, approximately linear intermediate portions. Bumps are seen but can usually be accounted for by epidemics, wars, and the like. The trends are sometimes approximately parallel, but sometimes the birth rate descends faster. In areas in which the process has approached completion the birth rates have almost leveled near the death rates (sometimes with partial reversions of trend). Most western countries have in recent years shown only modest positive growth rates, of order 0.5% per year or less. In most developing countries, on the other hand, the demographic transition has barely started. Here the death rate has gone down but the birth rate has, if reduced at all, only recently started to descend. Until the growth rate has decreased to the level of the death rate, the population is growing. When the birth rate decrease is parallel to that of the death rate, there is a constant natural growth rate (the difference between birth and death rate) and thus the population grows exponentially. The sustained near-exponential growth in Europe over almost a century has probably been a major stimulus for European emigration to the Americas and other parts of the world. The growth of the United States and Europe has now slowed down considerably and is perhaps near the end; by contrast, developing nations are experiencing growth rates of 3% a year and above. At this growth rate, the doubling time of population is only a little over twenty years. It is well known that it is very difficult to match a large population growth rate with an equivalent increase of food production, let alone to control the increase in pollution due to overcrowding and general mismanagement of the planet's resources.

The decrease in birth rate following that in death rate is an empirically obvious and hopefully predictable fact. But the full explanation of the phenomenon is lacking. The length of the lag is unexplained. In spite of its great international importance, research on the *behavioral* motivation of fertility is rather sparse. Birth-control techniques have been the focus of frantic activity and this is, of course, a necessary step. But the propaganda in favor of these techniques in developing countries, where the need is most acute, has hit many social and political obstacles, most of which remain unresolved. Further studies of the process that led to the spontaneous lowering of birth rates in the developed countries could help in our search for ways of speeding up the process in the developing ones.

The changes in death rates are the consequence of cultural, social, and economic changes. Change in birth rates is, even more so, a cultural change, in that control of births is even more directly the consequence of specific activity and customs. Changes in mortality must be matched by changes in fertility or else the population in question is soon in danger of extinction or explosion. Prior to the industrial era most populations in the world had achieved a fertility level permitting a modest increase in total population size. Outside Europe, there was high mortality, and the fertility was near its maximum with relatively early marriage.

In Europe mortality was probably somewhat lower and fertility correspondingly below the maximum. Thus the usual average number of children ever born to a married postreproductive woman—that is, about five—represents a value below the possible maximum achievable, for instance, with earlier marriage.

Whenever there has been a sharp decrease in mortality the problem of adjustment of the fertility level has presented itself. In Sweden, mortality decreased between 1860 and 1960 from about 25°/∞ to about 7°/∞ (i.e. about .18% per year) in standardized rates. In France the decrease has been at the rate of .25% per year in standardized rates, but less than .16% in crude rates. In England the trend was also around .16%, while in Italy it was .36%. France is anomalous in that the curves for death and birth rates are almost superimposed. It is possible that the downward trend in European fertility started in France in some social strata even before the nineteenth century. In most of the rest of Europe reduction in fertility starts about 1870. In Italy the population as a whole shows an initial decline of fertility in 1890, although the more educated classes had already completed the transition to low fertility by 1880, even in the least developed southern provinces (see Fig. 3.9.2).

In modern industrial societies, fertility changes under the influence of a variety of economic and social stimuli. The availability of modern techniques of birth control has simplified the problem at the individual level, although in some societies a substantial proportion of women were able to reduce the number of children ever born from five to little more than two with more primitive methods. At the level of the individual woman the question can be asked: was there a continuous decrease of the planned number of children that was more or less the same for all women? (This is a model of "gradual adjustment" of the individual woman.) Or should the phenomenon be viewed as the acceptance of a new attitude or ideal, the small Malthusian family, about two children on average, which is in sharp contrast to traditions of high fertility? In this second case the model is one of acceptance of an innovation (the small family) by an increasing fraction of women. From the standpoint of overall birth rates the lowering of fertility will seem to be gradual in both cases.

These two possibilities are not mutually exclusive, and the one that predominates may vary from place to place. When social class is considered, however, a strong heterogeneity is noticeable: it is as if the *ideal* of the small family spread first (possibly through most of Europe) to the educated classes, and from them,

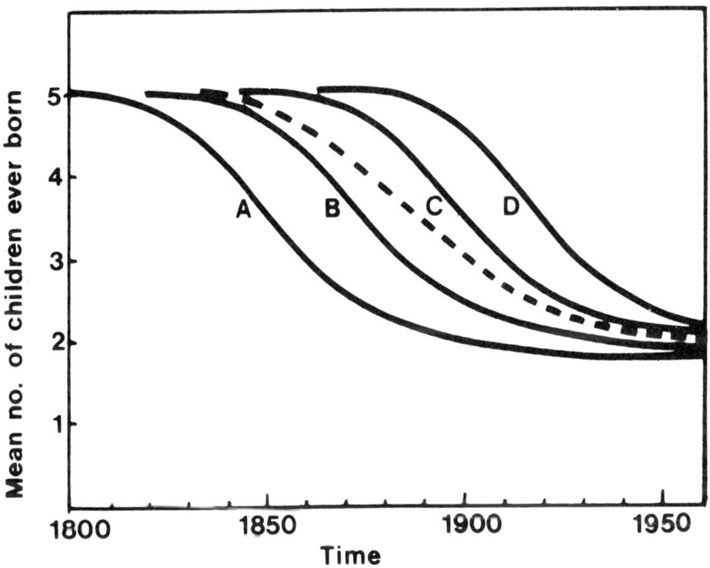

FIGURE 3.9.2. An idealized picture of the demographic transition in Italy. *A* is the most highly educated class; *B*, *C*, and *D* progressively less educated. The dotted curve is the population average assuming *A*, *B*, *C*, *D* are in proportion 1:5:5:1. The time scale is to be moved to the left for the southern provinces and to the right for the economically more prosperous provinces. From Zei and L Cavalli-Sforza (1977).

on a more local basis, to the less well educated (see Figure 3.9.2). For a more quantitative study of this evolution, means and variances of the number of children per woman were computed from available records. The results are compatible with the idea that the innovative ideal of the small family spread according to an approximately logistic curve, although the alternative model described above cannot be entirely ruled out (Zei and Cavalli-Sforza, 1977).

With this "spread of innovation" model in mind, the proportion of women who have adopted the small family ideal is a variable that might be studied dynamically according to the framework established in this book. It is reasonable to assume that women accept their mother's attitude in this respect so that there is vertical (uniparental) transmission (as in case 1 of Table 2.3.1). In addition, the example set by influential friends or public figures, both of the same and the older geneation, is likely to be important; this contributes both oblique and horizontal transmission. In addition to the transmission we must consider the role of natural selection *against* the small family ideal, since, in the absence of other factors of change, this is transmitted to about two children born as against five born to women maintaining traditional attitudes. The new custom, therefore, has a fitness $(1 + s) = 2/5$ relative to the old, so that $s = -0.6$. Thus with purely vertical uniparental transmission ($b_3 = b_2 = 1$, $b_1 = b_2 \neq 0$), the custom would disappear unless fathers also contributed strongly ($b_1 > 0$) to passing the custom. It seems likely that the action of women other than mothers who have adopted the small family ideal would be a more persuasive force.

In the model of vertical transmission just described we would choose $b_0 = b_1 = 0$, $b_2 = b_3 = c$, so that $B = 0$. After the superposition of oblique transmission, (3.2.8) applies. The innovation will be protected from extinction (see [3.2.5]) if $(1 + s)(f + c + mB) > 1$. With $s = -0.6$ this reduces to $f >$

$2.5 - c$ as the condition for $\hat{u} = 0$ to be unstable, in which case the innovation will spread until a value $\hat{u} > 0$ is attained. If $c = 1$ then $\hat{u} = 1$. Horizontal transmission produces the requirement $f > (2.5/c) - 1$, which is more stringent than that for oblique transmission. In the present cases, if there remains a proportion of traditional families at the end of this process, it is very small. In Table 3.9.1 we compare the distributions of the number of children of women in two different educational classes in the province of Verona. The two classes are (i) relatively younger (age 56–60 at the 1961 census) women of the highest educational class (women who took a university degree or higher diploma) and (ii) an older group (age 75 or older) of illiterate women. Their means are, respectively, 2.6 and 5.5. Clearly, at most two or three percent of women in the first sample manifest the traditional fertility pattern. Perhaps the few remaining families with

TABLE 3.9.1. Distribution of the number of children ever born per woman of postreproductive age (province of Verona, Italy, 1961 census). Sample 1: women with highest secondary school diploma, age 56–60; Sample 2: illiterate women, age 75 or older.

Number of children	Number of women	
	Sample 1	Sample 2
0	73	149
1	44	99
2	88	99
3	45	145
4	24	175
5	5	158
6	3	175
7	2	146
8	0	119
9	2	75
≥10	3	125
Total	289	1,465
Mean	2.597	5.536
Variance	2.930	7.910

very high fertility may be refractory to being converted to the new ideal—if so this would constitute another example of individual variation in resistance to adoption. In the second sample the few women with a low number of children were probably of low fertility for medical reasons.

Another way to view the process is by comparison with the model for endemic diseases given in Chapter 1. The transition to differential equations confounds horizontal and oblique transmission. Moreover, in the endemic disease model there is no vertical transmission. Suppose the individuals with "natural" fertility number n and those with voluntarily reduced fertility number m. Then one way to describe the changes in n and m sets is

$$\frac{dn}{dt} = \alpha n[1 - (n + m)/N] - \beta nm$$

$$\frac{dm}{dt} = \gamma m[1 - (n + m)/M] + \beta nm. \tag{3.9.1}$$

Here the natural fertility type grows at the rate α while the other type grows at rate γ. Both are retarded by a saturation factor which depends on the size of the total population, $n + m$, but the retardation is different for the two types, as specified by N and M. In some sense the constant β which specifies the rate of transition from high to low fertility patterns is the analogue of f used previously.

There are four equilibria of the system (3.9.1) which will be recognized as of the Lotka-Volterra type usually studied in ecology. The equilibria are ($\hat{n} = 0$, $\hat{m} = 0$), ($\hat{n} = 0$, $\hat{m} = M$), ($\hat{n} = N$, $\hat{m} = 0$) and the fourth, where both types coexist is:

$$\hat{n} = \frac{\gamma}{\beta}\left\{\frac{\alpha/M - \alpha/N - \beta}{\beta + \alpha/N - \gamma/M}\right\}$$

$$\hat{m} = \frac{\alpha}{\beta}\left\{\frac{\beta - \gamma/M + \gamma/N}{\beta + \alpha/N - \gamma/M}\right\}. \tag{3.9.2}$$

In the present context $\beta > 0$, $\gamma > 0$ and $\alpha > 0$. The stability of these equilibria can be studied locally and the point $(0, M)$ corresponding to the "take-over" of the reduced fertility idea is stable if

$$\beta > \alpha \left[\frac{N - M}{MN}\right] \tag{3.9.3}$$

independently of γ. "Natural selection" in this case would be indicated by $\gamma < \alpha$. For the complete take-over by the innovative fertility practice the condition (3.9.3) is sufficient. Qualitatively it expresses the fact that the attractiveness of the idea (as measured by β) must be so great as to overcome natural selection against it.

The model (3.9.1) is formally equivalent to that used to describe competition between two species in ecological theory. Cooke and Freedman (cited by Cooke, 1979) have suggested that such a model could describe the spread of innovations from one population to another. It should be clear that in the present context the model constitutes a substantial idealization, since in reality α, γ, and β would be expected to vary in response to changing mortality rates. A more realistic and considerably more complicated treatment would include growth rates for each type that changes over time.

In summary, the process of reduction of fertility during the demographic transition can be viewed as the spread of the notion of a smaller "ideal" family size. During the spread, continued vertical transmission of the older "natural" fertility habit would result in natural selection against the new idea. If the latter is attractive enough it can overcome the selection disadvantage and eventually dominate the population.

3.10 RANDOM SAMPLING DRIFT: VERTICAL AND OBLIQUE TRANSMISSION

In sections 2.8 and 2.9 the effect of random sampling was studied under cultural vertical transmission with or without natural selection. In the absence of natural selection, cultural transmission produces results parallel to those for genetic transmission plus natural selection. It is natural to ask two questions at this point, first, what are the effects of oblique and horizontal transmission on the drift models, and second, what happens when sampling schemes other than the positive binomial are studied.

The transmission scheme used will combine all those that have been used so far. The successive steps are the following.

(1) Mating, with correlation m between adults who have frequency u_t^* of H.

(2) Vertical transmission from parent to offspring producing the first-stage juveniles:

$$u'_{t+1} = B(u_t^*)^2 + CU_t^* + b_0 + mBu_t^*(1 - u_t^*). \quad (3.10.1)$$

(3) Then horizontal and oblique transmission occur (simultaneously) with contact rates α_1 and α_2 such that $\alpha_1 + \alpha_2 < 1$, and produce second-stage juveniles:

$$u''_{t+1} = u'_{t+1} + \alpha_1 u'_{t+1}(1 - u'_{t+1}) + \alpha_2(1 - u'_{t+1})u_t^*. \quad (3.10.2)$$

(4) Frequency independent change can occur with frequency μ for $H \rightarrow h$ and ν for $h \rightarrow H$, giving the frequency of third-stage juveniles:

$$\hat{u}_{t+1} = (1 - \mu)u''_{t+1} + \nu(1 - u''_{t+1}). \quad (3.10.3)$$

(5) Finally, natural selection for H occurs producing adults:

$$\begin{aligned} u^*_{t+1} &= (1 + s)\hat{u}_{t+1}/(1 + s\hat{u}_{t+1}) \\ &= (1 + s)\phi(u_t^*)/[1 + s\phi(u_t^*)] \end{aligned} \quad (3.10.4)$$

when $\phi(u_t^*)$ is the fourth degree polynominal:

$$\phi(u) = A_4u^4 + A_3u^3 + A_2u^2 + A_1u + A_0, \quad (3.10.5)$$

with coefficients:

$$A_0 = \nu + (1 - \mu - v)b_0[(1 - b_0)\alpha_1 + 1]$$

$$A_1 = (1 - \mu - \nu)[(Bm + C)\{(1 - 2b_0)\alpha_1 + 1\} + (1 - b_0)\alpha_2]$$

$$A_2 = (1 - \mu - \nu)[(1 - m)B\{(1 - 2b_0)\alpha_1 + 1\} - (Bm + C)\alpha_2 - (Bm + C)^2\alpha_1]$$

$$A_3 = -(1 - m)B(1 - \mu - \nu)[\alpha_2 + 2(Bm + C)\alpha_1]$$

$$A_4 = -(1 - m)^2B^2(1 - \mu - \nu)\alpha_1.$$

For the diffusion approximation we assume that the parameters satisfy the following order of magnitude conditions:

$$\begin{aligned} \sigma &= sN \\ \eta_i &= NA_i,\ i = 0, 3, 4 \\ \eta_1 &= N(A_1 - 1 + s) \\ \eta_2 &= N(A_2 - s) \\ a_i &= N\alpha_i,\ i = 1, 2, \end{aligned} \tag{3.10.6}$$

where σ, η_i are constants as $N \to \infty$. Then

$$\begin{aligned} M_{\delta u} &= \eta_0 + \eta_1 u + \eta_2 u^2 + \eta_3 u^3 + \eta_4 u^4 \\ V_{\delta u} &= u(1 - u). \end{aligned} \tag{3.10.7}$$

If $\nu = b_0 = 0$, then fixation in h will certainly occur. Fixation in H will certainly occur if $\mu = 1 - b_3 = 0$. If $\nu = \mu = b_0 = 1 - b_3 = 0$, then fixation in either state can occur. Under the conditions (3.10.6), and if α_1, α_2 are also small, the probability of fixation in H is given by the classical formula (Crow and Kimura, 1970, p. 425)

$$F(u_0) = \frac{e^{-2Du_0} - 1}{e^{-2D} - 1}, \tag{3.10.8}$$

where u_0 is the initial frequency of H,

$$D = \sigma + a_1 + a_2 + (1 - m)\beta_{12}, \tag{3.10.9}$$

and $\beta_{12} = N(b_1 + b_2 - 1)$.

If $D = 0$, then effectively the results are those for "neutral" genes: $F(u_0) = u_0$. Note that

$$F(u_0) > u_0 \text{ if } D > 0.$$

$$F(u_0) < u_0 \text{ if } D < 0.$$

As defined in (3.10.9), D summarizes the behavior of the trait frequencies. Thus a deterministic population of H and h should tend to fix H if $D > 0$, lose H if $D < 0$ and remain at a neutral equilibrium if $D = 0$. The stochastic behavior is governed by the same quantity, so that natural selection, represented by σ, horizontal and oblique transmission, represented by a_1 and a_2, and vertical transmission, represented by $(b_1 + b_2 - 1)$, all act additively in determining the evolution of the trait. Note that assortative mating dilutes the effect of vertical transmission.

The quantity D has the same significance in the present case as the parameter of natural selection in the analysis of purely genetic transmission. In addition to the probability of fixation (3.10.8), D also determines the expected time to fixation of H according to another result from the theory of genetic evolution (Kimura and Ohta, 1969) introduced in (2.8.22), (2.8.23). It is worth stressing that this expected time to fixation is independent of the sign of D to the limits of the approximations in (3.10.7) (Maruyama, 1974). However, these expectations are subject to considerable variation. The conditional expected time to fixation is obtained from (2.8.22) with D substituted for β_{12}.

3.11 A COMPARISON OF SPECIAL SCHEMES OF TRANSMISSION WITH RANDOM SAMPLING DRIFT: PARENTS VERSUS TEACHERS

In this section we compare two transmission schemes that must have major practical relevance: transmission primarily by parents and transmission primarily by teachers. In both models oblique transmission is also assumed, since other individuals (apart from the parent or the teacher) influence the transmission probability. The models are set up so that in the stochastic change from generation to generation no change is expected in the frequency of H. In both models the two trait values H and h are equally infectious, in the sense that an individual carrying the trait spreads it, whether it is H or h. The laws of transmission are constructed as follows.

In the uniparental model, each parent has one offspring to whom it transmits the trait it carries, with weight τ. In addition, every other (nonparent) individual in the population also contributes its trait to the offspring but with weight $(1 - \tau)/(N - 1)$. In the teacher's model only one individual (the teacher) contributes its trait, with weight τ, to everybody, while everybody but the teacher contributes $(1 - \tau)/(N - 1)$.

In both models the vector of individuals in the population can be written as an N-vector of 1's and 0's with 1 indicating that an individual in that index position is H, while 0 stands for h. Thus at some generation t the vector of the population could be $s^{(t)} = (1, 0, 0, 1, 0 \ldots 0, 1, 0)$ with the first, fourth ... and $(N - 1)^{\text{st}}$ individuals all H and the others h. Now the laws of transmission described above can be expressed through a transmission matrix $\mathbf{T}$. For the parental transmission model

$$\mathbf{T} = \begin{pmatrix} \tau & \dfrac{1-\tau}{N-1} & \cdots & \dfrac{1-\tau}{N-1} \\ \dfrac{1-\tau}{N-1} & \tau & \cdots & \dfrac{1-\tau}{N-1} \\ \cdot & \cdot & & \cdot \\ \cdot & \cdot & & \cdot \\ \cdot & \cdot & & \cdot \\ \dfrac{1-\tau}{N-1} & \dfrac{1-\tau}{N-1} & \cdots & \tau \end{pmatrix} \qquad (3.11.1)$$

by which we mean a matrix whose diagonal elements are all τ and non-diagonal elements all $(1 - \tau)/N - 1)$. The matrix is doubly stochastic, that is, rows and columns all sum to one. The corresponding representation for teachers is

$$\mathbf{T} = \begin{pmatrix} \tau & & \\ \tau & & \\ \cdot & \dfrac{1-\tau}{N-1} & \\ \cdot & & \\ \cdot & & \\ \tau & & \end{pmatrix}, \tag{3.11.2}$$

where one column (corresponding to the teacher) is made entirely of τ values and all other elements are $(1 - \tau)/(N - 1)$. The matrix is stochastic but not doubly stochastic, that is, the columns do not sum to unity.

The product $\mathbf{Ts}^{(t)} = \mathbf{p}^{(t+1)}$ is a column vector of probabilities such that in the next generation the 1, 2, . . . , N individuals will be H. Each vector $\mathbf{s}^{(t)}$ produces a probability vector $\mathbf{p}^{(t+1)}$ in this way. The probabilities for the states $\mathbf{s}^{(t+1)}$ of the population in generation t+1 are then computed by taking products of the $p_i^{(t+1)}$ and their complements according to whether $s_i(t+1)$ is one or zero, for example, if $N = 5$ and

$$\mathbf{p}^{(t+1)} = \begin{pmatrix} p_1 \\ p_2 \\ p_3 \\ p_4 \\ p_5 \end{pmatrix},$$

the probability of the state

$$\mathbf{s}^{(t+1)} = \begin{pmatrix} 1 \\ 0 \\ 1 \\ 0 \\ 0 \end{pmatrix}$$

is $p_1(1 - p_2)p_3(1 - p_4)(1 - p_5)$.

A model incorporating random sampling into the *uniparental* transmission mode can be constructed as follows (Feldman and Cavalli-Sforza, 1977c). If a parent is H, the probability that its child is H is $\tau + (k - 1)(1 - \tau)/(N-1)$ if there are $k - 1$ other nonparental H individuals in the parental population. This probability can be written as $1 - m(N - k)$ where $m = (1 - \tau)/(N - 1)$. Similarly the probability that the offspring of an H is h is $m(N - k)$. These arguments validate the following parent-child transition matrix

		child	
		H	h
parent	H	$1 - m(N - k) = p_1$	$m(N - k) = q_1$
	h	$mk = p_2$	$1 - mk = q_2$

. (3.11.3)

The transition probabilities are seen to depend on the number of H individuals in the parental population. This number changes randomly from generation to generation so that the matrix (3.11.3) specifies a time inhomogeneous Markov chain. We may now study the behavior over time of the frequency of H in the population.

The probability that there will be individuals with the H trait in the next generation can be obtained as follows. From matrix (3.11.3), H parents (of whom there are k out of N) generate H children with probability p_1 and h children with probability q_1; for h parents (of whom there are $N - k$) the respective probabilities are p_2 and q_2. The distribution of H offspring among the k parents of type H is binomial with probability p_1. The distribution of H offspring among the $N - k$ parents of type h is binomial with probability p_2. These two distributions are independent, so that the probability generating function $G(z|k)$ for the total number K' of H offspring in the population of size N, given that there are k parents of type H, is the product of the two binomial generating functions

$$G_P(z|k) = \sum_{k'=0}^{N} z^{k'} p[K' = k'|k] = (q_1 + p_1 z)^k (q_2 + p_2 z)^{N-k}$$
$$= B_{p_1,k}(z) B_{p_2,N-k}(z). \tag{3.11.4}$$

The analysis of the properties of $G(z|k)$ will not be presented in complete detail; only those results of immediate interest are reported. The first fact to note is that

$$E(K'|k) = \frac{\partial}{\partial z} G_P(z,k)\big|_{z=1} = k, \tag{3.11.5}$$

so that there is no *expected* change in H numbers. It is often useful to use $u = k/N$, the frequency of H in the population, instead of the absolute numbers, that is, $E[u_{t+1}|u_t] = u_t$, or, referring back to section 2.8,

$$M_{\delta u} = NE[u_{t+1} - u_t|u_t] = 0. \tag{3.11.6}$$

Again from (3.11.4)

$$E[(K')^2|k] = k^2[1 - 2m + m^2 N] + k[2mN - m^2 N], \tag{3.11.7}$$

so that

$$E[(u_{t+1} - u_t)^2|u_t] = mu_t(1 - u_t)(2 - mN), \tag{3.11.8}$$

or, in the same way as in section 2.8

$$V_{\delta u} = u(1 - u)[\alpha(2 - \alpha)], \tag{3.11.9}$$

where

$$\alpha = mN.$$

Thus, although there is no change in H frequency expected (3.11.6) under the influence of random sampling the process ends with the fixation of H (with probability u_0) or with the fixation of h (with probability $1 - u_0$). The rate at which fixation occurs is asymptotically equal to

$$\lambda_f^P = 1 - \frac{\alpha(2 - \alpha)}{N} \tag{3.11.10}$$

When the underlying matrix is (3.11.2), so that one individual is a teacher, then whether the teacher is H or h is important in determining whether a member of the offspring generation is H or h. If the teacher is H, then the probability that any individual in the offspring generation is H is $\tau + (k - 1)(1 - \tau)/(N - 1) = 1 - m(N - k)$. If the teacher is h, then the probability that any offspring is H is mk. Now the prior probability that the teacher is H, when the frequency of H in the parental generation is k/N is just k/N. Hence the overall probability distribution for the number of H offspring when each parent has exactly one child is binomial with parameters N and p_1 with probability k/N, and binomial with parameters N and p_2 with probability $1 - k/N$. The overall probability generating function for the number K' of H in the offspring generation is then, with the symbols used in (3.11.4):

$$G_T(z \mid k) = (k/N)B_{p_1,N}(z) + (1 - k/N)B_{p_2,N}(z). \tag{3.11.11}$$

Again, in this case

$$E(K' \mid k) = k \quad \text{or} \quad E(u_{t+1} \mid u_t) = u_t, \tag{3.11.12}$$

so that $M_{\delta u} = 0$.

A further calculation gives

$$E[(u_{t+1} - u_t)^2 \mid u_t] = \frac{u_t(1 - u_t)}{N}[1 + (N - 1)(1 - \alpha)^2]$$

with α as in (3.11.9). Hence

$$V_{\delta u} = u(1 - u)[\alpha(2 - \alpha) + N(1 - \alpha)^2], \quad (3.11.13)$$

which is clearly greater than (3.11.9). The rate of approach to fixation in this case is

$$\lambda_f^T = (1 - 1/N)(2\alpha - \alpha^2). \quad (3.11.14)$$

It is easy to see that

$$\lambda_f^P > \lambda_f^T$$

so that *fixation happens faster with the teacher type of transmission.* An important point to notice is that, in the case of parental transmission, the infinitesimal variance $V_{\delta u}$ from (3.11.9) is

$$\begin{aligned} V_{\delta u} = u(1 - u)[\alpha(2 - \alpha)] &= u(1 - u)\frac{N(1 - \tau)}{N - 1)}\left\{2 - \frac{N(1 - \tau)}{N - 1}\right\} \\ &= u(1 - u)\frac{1 - \tau}{1 - 1/N}\left\{2 - \frac{(1 - \tau)}{1 - 1/N}\right\} \\ &\rightarrow u(1 - u)(1 - \tau^2) \quad (3.11.15) \end{aligned}$$

as $N \rightarrow \infty$. The diffusion theory can be applied to $M_{\delta u}$ and $V_{\delta u}$ as they stand. However, in (3.11.13) we observe the added term $N(1 - \alpha)^2$. For the validity of the diffusion approximation in this case we must have $N(1 - \alpha)^2 \rightarrow \eta$ as $N \rightarrow \infty$ where η is the absolute constant. This entails that τ should asymptotically equal to the square root of (η/N) as $N \rightarrow \infty$. If this is satisfied, then the diffusion results of section 2.8 can be applied. If τ fails to satisfy this restriction then the qualitative fact that there is a faster approach to fixation in the teacher model becomes even stronger.

These arguments have been based on the assumption that every parent produces exactly one offspring. This assumption can be removed in a number of ways. One is to assume that the population size stays constant, N, but that the number of offspring per parent has some probability distribution. This gives rise to a process studied by Moran and Watterson (1959), Karlin and McGregor (1964) and Feldman (1966) which is called the direct product branching process. The details of construction of the process can be found in these original works. Here we present only the result that, if the probability generating functions for the number of offspring produced by H and h are $f(z)$ and $g(z)$ respectively, and if the matrices (3.11.3) and (3.11.5) describe the type of offspring (H or h), then the probability generating function for the number of H offspring in the offspring generation, given that there were k in the parental generation is

$$H^P(z) = \frac{\text{coefficient of } s^N \text{ in } f^k\{s[q_1 + p_1 z]\}g^{N-K}\{s[q_2 + p_2 z]\}}{\text{coefficient of } s^N \text{ in } f^k(s)g^{N-k}(s)}. \tag{3.11.16}$$

For teacher type transmission a generalization of (3.11.11) analogous to (3.11.16) is

$$H^T(z) = \frac{k}{N}\frac{\text{coefficient of } s^N \text{ in } f^N\{s[q_1 + p_1 z]\}}{\text{coefficient of } s^N \text{ in } f^N(s)} + (1 - k/N)\frac{\text{coefficient of } s^N \text{ in } g^N\{s[q_2 + p_2 z]\}}{\text{coefficient } s^N \text{ in } g^N(s)}. \tag{3.11.17}$$

When $f(z) = g(z)$, no selective differences exist between H and h; this assumption will be made in what follows. In the previous cases $H^P(z)$ and $H^T(z)$ were identically equal to z. No matter what the form of $f(\cdot) = g(\cdot)$, both (3.11.16) and (3.11.17) produce no change in expected numbers of H:

$$E(u_{t+1} \mid u_t) = u_t; \; M_{\delta u} = 0 \tag{3.11.18}$$

It is in the variances and rates of approach to fixation that there are differences among the generating functions. When $f(z) = g(z) = e^{\lambda(z-1)}$, so that each parent produces a Poisson distributed number of offspring, we have, from the parent model (3.11.16),

$$E[(u_{t+1} - u_t)^2 \mid u_t] = \frac{u_t(1 - u_t)}{N} \tag{3.11.19}$$

and

$$\lambda^P = (1 - 1/N). \tag{3.11.20}$$

When $f(z) = g(z) = (1 - \delta + \delta z)^\gamma$, $(\gamma \geq 2)$ the number of offspring per parent has a binomial distribution. In this case the parent model (3.11.16) produces

$$V_{\delta u} = u(1 - u)\left[1 - \frac{(1 - \alpha)^2}{\gamma}\right] \tag{3.11.21}$$

while the rate of approach to fixation is

$$\lambda_f^P = (1 - 1/N)\left[1 + \frac{(1 - \alpha)^2}{N\gamma - 1}\right]. \tag{3.11.22}$$

The negative binomial distribution for offspring number per parent has $f(z) = g(z) = [(1 - \delta)/(1 - \delta z)]^\gamma$, $(\gamma > 0)$. This produces

$$V_{\delta u} = u(1 - u)\left[1 + \frac{(1 - \alpha)^2}{\gamma}\right] \tag{3.11.23}$$

and

$$\lambda_f^P = (1 - 1/N)\left[1 - \frac{(1 - \alpha)^2}{N\gamma + 1}\right]. \tag{3.11.24}$$

These examples for $f(\cdot)$ in (3.11.16) and (3.11.17) are easy to calculate. The results can be presented more generally but the fact we wish to observe is that in all cases (except the Poisson) $m = \alpha/N$, from (3.11.3), enters the dynamics of the process in an important way. It should be observed that the negative binomial produces the fastest fixation with the Poisson intermediate between it and the binomial. The slowest rate of fixation, however, occurs when there is no fluctuation in offspring number per parent, that is, from the model (3.11.4).

The teacher model, represented by (3.11.17), has an entirely different response to random drift in the offspring number

TABLE 3.11.1. Results from random drift with oblique transmission. $\alpha = m$N, see text for assumptions allowing validity of diffusion. From top to bottom of the table the rates are slowest to fastest; corresponding expected fixation times are longest to shortest.

Model	Offspring Number Distribution	Rate of Approach to Fixation	$V_{\delta u} = \kappa u(1 - u)$ κ
Parental transmission	one per parent $f(s) = s$	$(1 - 1/N) + (1 - \alpha)^2/N$	$\alpha(2 - \alpha)$
	Binomial $f(s) = (1 - \delta + \delta s)^\gamma$	$(1 - 1/N)[1 + (1 - \alpha)^2/(N\gamma + 1)]$	$1 - (1 - \alpha)^2/\gamma$
	Poisson $f(s) = e^{-\lambda(s-1)}$	$1 - 1/N$	1
	Negative Binomial $f(s) = [(1 - \delta)/(1 - \delta s)]^\gamma$ $\gamma > 0$	$(1 - 1/N)[1 - (1 - \alpha)^2/(N\gamma + 1)]$	$1 + (1 - \alpha)^2/\gamma$
Teacher transmission (all distributions)		$(1 - 1/N)[1 - (1 - \alpha)^2]$	$\alpha(2 - \alpha) + N(1 - \alpha)^2$

distribution. In fact, no matter what the distribution, the same infinitesimal variance (3.11.13) and rate of approach to fixation (3.11.14) result. As one might have suspected, the form of the teaching process described by the matrix (3.11.2) removes parental effects from consideration. Among the distributions considered, the fastest rate of approach to fixation, including all the parental examples, is obtained with teacher type transmission (3.11.14). The expected number of generations to fixation (conditioned on not losing H from the population) can be calculated from the diffusion theory, when it applies. In all cases we have

$$V_{\delta u} = \kappa u(1 - u), \tag{3.11.25}$$

where κ varies from example to example. Such a function for $V_{\delta u}$ produces an average time to fixation (measured in actual generations), starting from an initial H frequency u_0, which we write as $\bar{t}_1(u_0)$ and

$$\bar{t}_1(u_0) = \frac{2N}{\kappa} \frac{(1 - u_0)}{u_0} \cdot \log_e (1 - u_0). \tag{3.11.26}$$

The constant factor κ can be obtained as follows:

$$\kappa = \lim_{N \to \infty} N(1 - \lambda_f^P) \text{ or } \lim_{N \to \infty} N(1 - \lambda_f^T), \tag{3.11.27}$$

whichever is relevant. The faster the rate of approach to fixation (i.e. the smaller is λ_f) the shorter is $\bar{t}_1(p)$, which can be explicitly written using (3.11.26). The results of this section are summarized in Table 3.11.1.

3.12 KINETICS OF CULTURAL CHANGE WITH OBLIQUE AND HORIZONTAL TRANSMISSION

Vertical transmission, as we have seen in Chapter 2, can be formally similar to genetic transmission, and is easily confused with it. It is clearly important for some cultural traits. But the modes of transmission most characteristic of culture are those involving contacts between any people, not necessarily related, but able to communicate with or learn from one another. This transmission has been separated, somewhat artificially, into two modes: oblique and horizontal. The dichotomy is convenient in the context of discrete generation models. The results indicate that the expected evolution is similar in the two situations but there are differences depending, for instance, on the intensity of vertical transmission. This stresses further the importance of considering age structure in greater detail.

As a general rule, when $b_0 = 0$, we have seen that oblique and horizontal transmission make it harder for H to be lost from the population; they enhance its protection. This was most easily seen when the frequency dependence came from the linear function fu, an especially simple form of transmission. Other functions could obviously produce more complicated dynamics. When $b_3 = 1$, however, the effect of oblique and horizontal transmission is to increase the rate of homogenization of the population in the infectious phenotype H. If $b_3 \neq 1$, $b_0 = 0$, then the ultimate equilibrium H frequency (if different from zero) is increased by nonvertical cultural transmission.

The general compounding of vertical, oblique, and horizontal transmission can be represented rather simply according to the formulations above. Suppose that after vertical transmission the H frequency u in the parents becomes

$$\begin{aligned} u^* &= u^2B + uC + b_0 \\ &= V(u), \quad \text{say,} \end{aligned}$$

in the offspring (see (2.2.6)). Now the $1 - V(u)$ offspring who are h contact the u members of the parental generation and are converted to H according to the function $g(u)$. After this oblique transmission we have

$$u^{**} = V(u) + [1 - V(u)]g(u).$$

In section 3.2, for the most part we took $g(u) = f \cdot u$. Now, finally the $(1 - u^{**})$ offspring who are h contact the u^{**} who are H and are converted according to some function $K(u^{**})$. The final offspring generation H frequency is then

$$u' = u^{**} + (1 - u^{**})\, K(u^{**}).$$

For $\hat{u} = 0$ *to be an equilibrium it must be an equilibrium of all three functions* g(u), V(u), K(u). This is certainly the case when $b_0 = 0$ and $g(u) = f \cdot u$, $K(u) = k \cdot u$ as in most of this chapter. *For* $\hat{u} = 1$ *to be an equilibrium it is sufficient that* $\hat{u} = 1$ *be a root of either* V(u) *or* g(u), so that it is enough that $b_3 = 1$.

The larger is f, the faster is the oblique transmission. The emergence of social organization has seen the rise to prominence of specialized teachers and of political chiefs who can address a large number of people, potentially even a whole population. This must have been a rare event before the origin of agriculture created the chance for village and then urban growth. In fact the social rules of hunters/gatherers would not easily permit this kind of transmission, but with the increase in community size it has become a necessity.

The teacher/leader type of transmission can determine an almost instantaneous change of a whole population. Here we have seen that drift is stronger with transmission through teachers than through parents. Thus the fragmentation of a population into smaller groups, each with its own teacher, can be expected to produce greater variation (taken over the whole population) than if each followed a parental type of transmission rule. We delay a more detailed discussion of this until our analysis of quantitative traits.

3.13 VARIATION AMONG POPULATIONS

Members of the human species exhibit profound biological and cultural differences. The idea of "races" stems from the perception of these differences and the desire to capture them in a typology. In practice, many different typologies have been proposed, defining from three to over 30 races. In comparing biological and cultural variation between populations, the latter appears, at first sight, to be rather fine-grained, so that even the first racial subdivision would leave us with groups that are culturally very heterogeneous. We therefore must consider finer biological subdivisions and start at the lower end of the taxonomic scale, where we find the "Mendelian population." A full story of its definition would be out of place, but it is of interest to consider that it is based on the idea that individuals of a Mendelian population mate randomly with respect to the genes being studied. This is a directly testable hypothesis: geneticists, however, usually test it indirectly by studying its consequences at the level of the gene pool. In fact, if mating is random, that is, to be socially more precise, if marriage can occur with equal probability between any members of the group (a condition sometimes called *panmixia*), the gene pool will obey some precise expectations. In particular, considering a gene for which at least two alternative types ("alleles") exist: A_1, A_2, A_3, . . . and so on, one testable expectation is that the frequency distribution of these alleles is homogeneous in subsamples of the population, and another, which is even easier to test, is that the contributions of the two parents show a random association. The two expectations are formally different and at first sight unrelated to random mating. To test the first, one asks if frequencies of alleles among different strata of the population differ no more than could be due to sampling fluctuations. The second is tested by the validity of the "Hardy-Weinberg" distribution in a single specific population. The two tests are closely related to each other and to the randomness of mating, as described in textbooks of population genetics (see, e.g., Cavalli-Sforza and Bodmer, 1971). It may be worth adding that panmixia may apply for some traits or genes, and not for others.

With a strict application of this definition, even tribes of aboriginal populations, which may number only a few hundreds or thousands of individuals, will not pass for homogeneous, unitary, Mendelian populations. For example, the Bedik of Senegal show some biological heterogeneity (Gomila, 1971), and the same is true of the Yanomamo Indians (Neel, 1973). Similarly, for nontribal societies in sparsely populated areas (e.g., mountain villages of Northern Italy or Switzerland), there is perceptible village-to-village variation generated by isolation due to geographic distance (tendency to marry preferentially in the same village or, if outside, with probability decreasing with increasing distance; see Cavalli-Sforza and Bodmer, 1971). However, when larger villages or towns of a restricted area are considered, little variation is perceived between them. In general, the greater the population density involved, the greater is the probability of admixture by migratory exchanges.

Nevertheless, most tribal societies approximately satisfy the criteria for definition of a Mendelian population. In nontribal societies that have not had recent immigration or clear differentiation of non-intermarrying groups, geographic districts corresponding to smallest or the next smallest administrative units might also satisfy this definition for most practical purposes. But if there is social stratification that prevents marriage ("castes," as in India), or if there is differentiation into clearly distinct ethnic groups of recent immigrants, then clearly these distinctions must be taken into account when defining strata expected to behave like Mendelian populations.

Our interest here is that criteria useful for the purpose of defining "Mendelian populations," which thus are considered as population units from the point of view of genetic sampling, can also be useful for a cultural study. In fact, the societies used as standard units for ethnographic descriptions, which are self-named groups whose members are obviously willing to share a common denomination, most often have the same language or dialect and very similar customs, and also have fewer intermarriage within the group. In general, differences in language or customs tend to prevent cultural exchange and intermarriage. The frequency of intermarriage can, in fact, be considered with few understandable exceptions as a very sensitive measure of cultural similarity, so that the genetic continuity established by free intermarriage is more likely to be present when there is also cultural continuity. While there are good operational criteria for testing if a population can be considered Mendelian, there are none for evaluating cultural homogeneity. In spite of this practical problem, or perhaps because of it, it might be suggested that the same sampling unit be used for genes and for culture. If unity of genetics and culture for defining an elementary population is accepted, we are next confronted with a dramatic hiatus between genetics and culture that proceeds from the comparison of the genetic and cultural variations between these populations.

Let us consider the genetic variation first and examine specific genes, e.g. the genes determining the ABO and the Rh blood groups. For the first we will consider three alleles, A, B, and O, and four for the latter, R_0, R_1, R_2, r. The sum of the last four will be less than 100%, the residual being formed by many rarer alleles which we will ignore. Table 3.13.1 indicates the frequencies of these alleles in some of the major aboriginal population groups of the world. Although in some cases an allele present in one group may be absent in another, for a *bona fide* gene we never

TABLE 3.13.1. Approximate frequencies of alleles of the *ABO* and some *Rh* genes in some of the major aboriginal ethnic groups of the world.

	ABO Gene			*Rh* Gene			
	A	B	O	R_0	R_1	R_2	r
Africans	18%	11%	71%	74%	3%	4%	12%
Europeans	28%	6%	66%	2%	40%	17%	38%
Orientals	19%	17%	64%	4%	76%	20%	0

have a case in which a human ethnic group has 100% of an allele, while other groups have 0% of it but have 100% of another. The R_0 allele is among those that come closest to this amount of differentiation. If we plot the distribution of gene frequency of one specific allele in Mendelian populations sampled from one of these ethnic groups, we find an even smaller variation. Figures 3.13.1 and 3.13.2 show some distributions in Africa where the range of variation is quite limited. It would be higher in the whole world, but would never span 0–100%.

If similar populations are sampled for cultural traits, the variation between populations is, comparatively, much higher, and a 0–100% range is common rather than being the exception. In fact, for some traits most societies have either 0% or 100%, and intermediate ("polymorphic") societies are rare, perhaps representing transitional stages. It is not easy to find data comparable

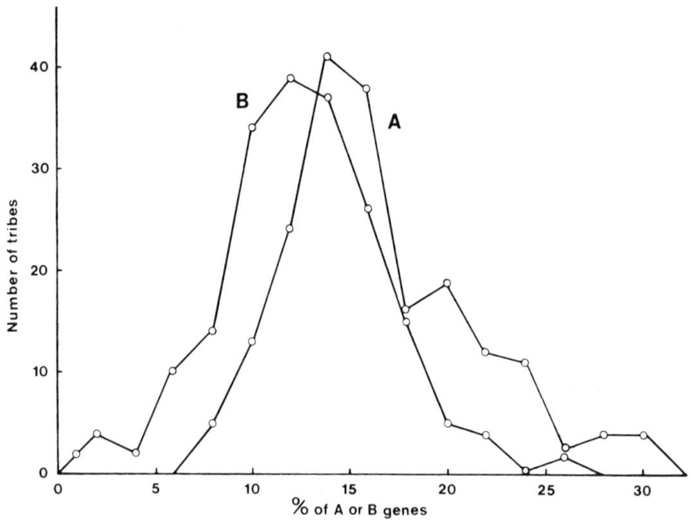

FIGURE 3.13.1. Distributions of the gene frequencies of alleles A, B, and O in sub-Saharan Africa. The ordinate is the proportion of tribes showing the gene frequency given on the abscissa.

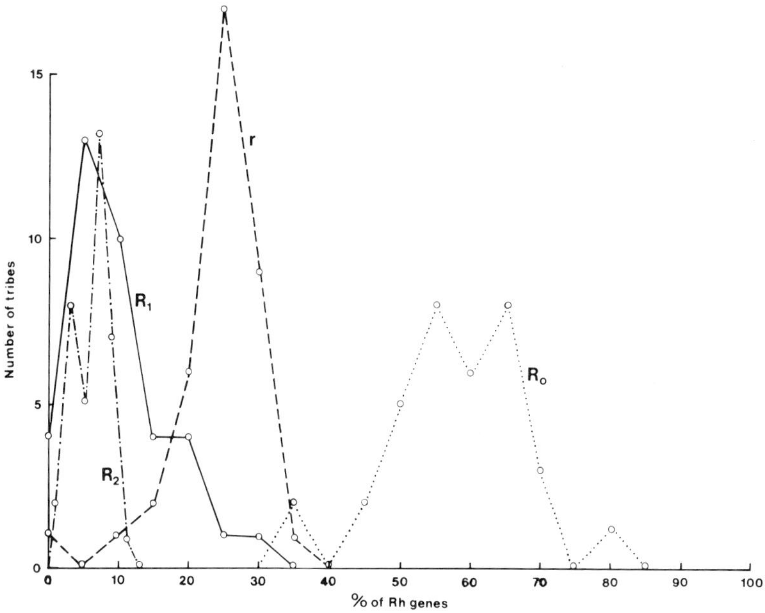

FIGURE 3.13.2. Distributions of the gene frequencies of alleles R_0, R_1, R_2 and r of the Rh system in sub-Saharan Africa. Abscissa and ordinate as in 3.13.1.

to those we have seen for blood group genes for the construction of empirical distributions of the frequencies of cultural traits in different populations. However, a simple consideration of the linguistic variation shows that the mutual intelligibility of languages spoken even in two neighboring countries can be close to zero, and the number of different languages in existence is extremely large. In other words, cultural variation is expected to be potentially enormous, compared to biological variation.

The perusal of ethnographic surveys gives qualitative confirmation of this extreme cultural variation. One would like to be able to formalize this impression, either by measuring a ratio of variation between populations to that within populations, or by building distributions of frequencies of a trait among various populations. We will be content here with qualitative or semi-quantitative examples. Figures 3.13.3 and 3.13.4 are derived from Murdock's ethnographic atlas. Figure 3.13.3 shows the distribution per African tribe of the plan of the house which exists in various forms, the circular and rectangular ones being most common. Although the graph does not give information on the variation within societies, the common experience of a traveller

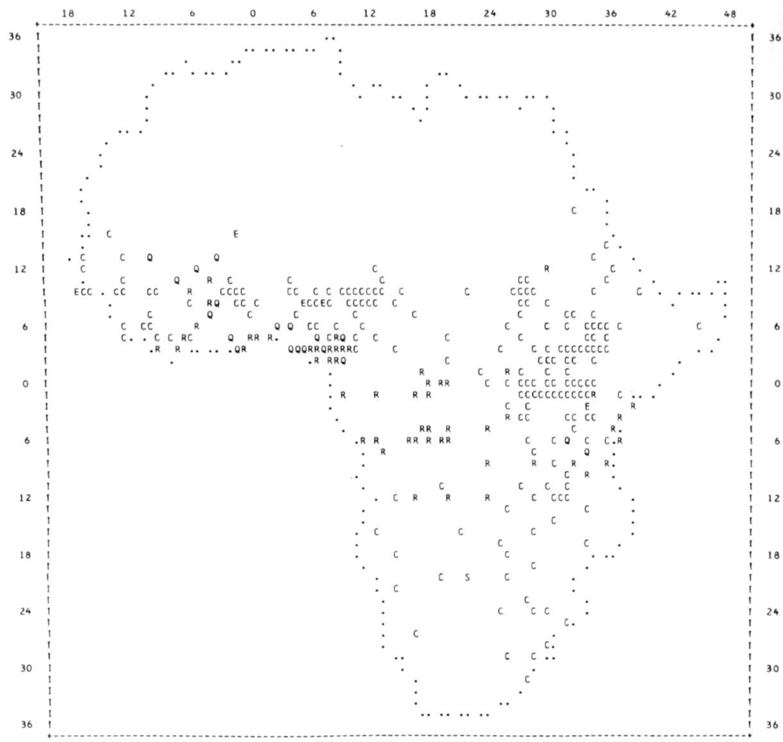

FIGURE 3.13.3. Geographic distribution of ground plan dwellings of sub-Saharan African societies by shape. C = circular, E = elliptical, P = polygonal, Q = quadrangular around (or partially around) an interior court, R = rectangular or square, S = semicircular, — = absence of information (source: see Fig. 3.13.4).

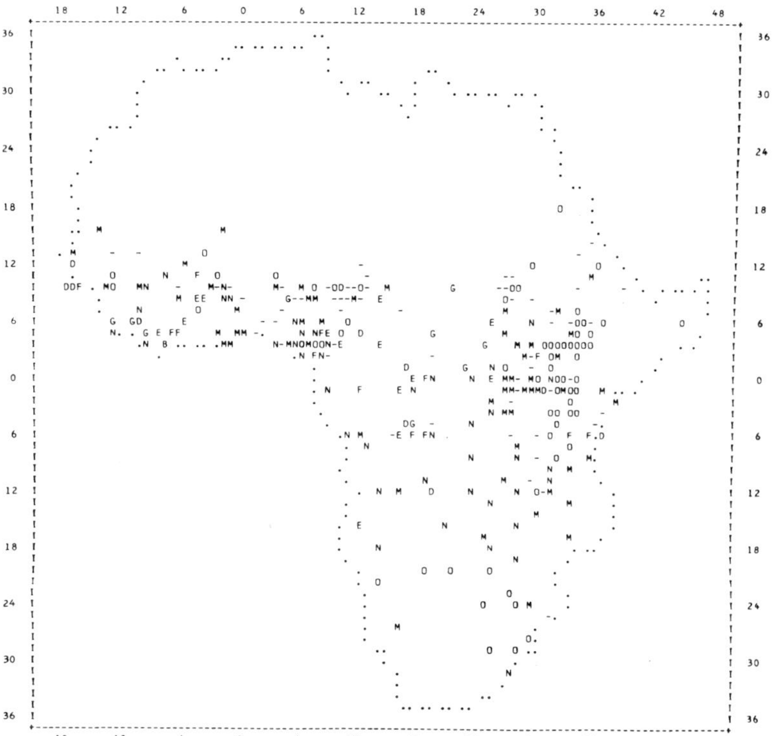

FIGURE 3.13.4. Geographic distribution of participation in fishing by sex in sub-Saharan African societies. D = differentiation of specific tasks by sex but approximately equal participation by both sexes in the total activity; E = equal participation by both sexes without marked or reported differentiation in specific tasks; F = females alone perform the activity, male participation being negligible; G = both sexes participate, but females do appreciably more than males; I = sex participation irrelevant, especially where production is industrialized; M = males alone perform the activity, female participation being negligible; N = both sexes participate, but males do appreciably more than females; O = the activity is absent or unimportant in the particular society; P = the activity is present, but sex participation is unspecified in the sources consulted; — = absence of information. Information from Murdock's Ethnographic Atlas (1967); graphic rendition courtesy of C. Viganotti.

through Africa suggests that most societies show almost no variation of the house ground plan within the tribe.

Figure 3.13.4 shows one of the traits with the smallest between/within tribe variation among those given by Murdock. The trait is participation in fishing by sex. In this trait (as in almost all others), the variation is evaluated qualitatively, by indicating tribes in which all males, almost all males, about equal numbers of males and females, almost all females, or only females participate in fishing. Counting societies of the various types from this figure, in most societies males are the exclusive or almost exclusive performers. In slightly fewer societies, there is only some limited female participation. Equal participation by both sexes is infrequent, and only a few societies have a larger participation by females. But there are more societies with exclusive participation of females. This distribution seems to have two maxima, one near 0% and one near 100%, with a minimum for intermediate fractions, that is, it is U-shaped. The distribution of frequency of Micronesian islands having a certain percentage of individuals using one out of two roots, using data such as two-rooted glosses of Table 3.6.1, would, if enough data were available, most probably show a U-shape.

Families of theoretical distributions of binary trait frequencies observed in different populations were generated originally by S. Wright (1931b) in predicting the effects of random genetic drift. Examples are shown in Figure 3.13.5. Four theoretical distributions are shown for a trait which has a mean frequency of 50% (arranged over all populations). The abcissa is the frequency of the trait in a population; it can vary between 0% and 100%. The parameter that determines the shape of the distributions is Nm, where N is the number of individuals reproducing per generation in a specific population, and m is the coefficient of migration (the proportion each population exchanges per generation). The shape of the distribution is that of a narrow bell if Nm is large, that is, each population is large and there is much

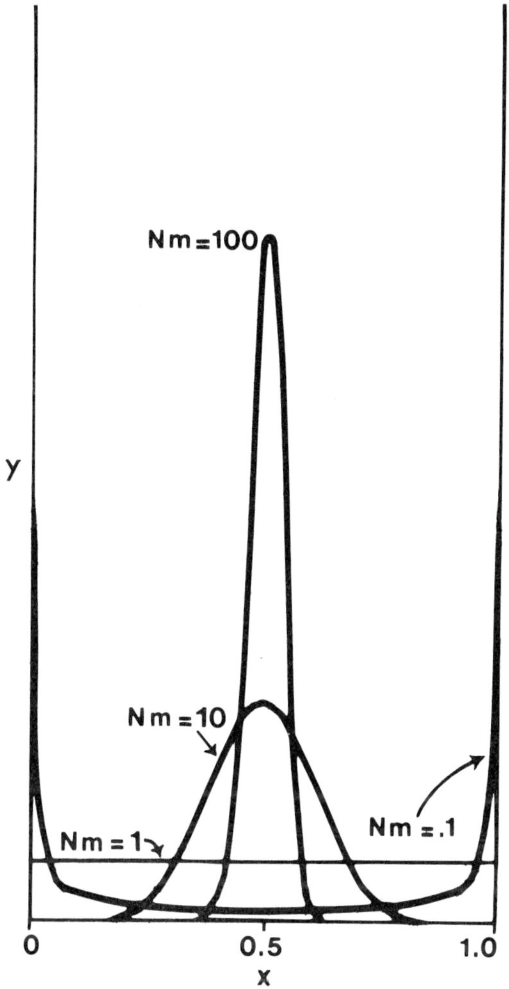

FIGURE 3.13.5. Theoretical frequency distribution of gene frequencies in independent populations under genetic drift and migration (or mutation). The abcissa is the gene frequency in each population and averages 50% (over populations). Each population is made of N individuals, and exchanges them with a large external source, receiving Nm and losing Nm individuals per generation by emigration. Thus, m is the proportion of immigrants and also of emigrants in each population per generation. If Nm is large, the distribution of population has small variation and peaks highly. As Nm decreases the variation increases; for $Nm = 1$ the distribution is flat, and for $Nm < 1$ it takes a U-form (after S. Wright, 1940). As explained in the text, biological variation is usually bell-shaped and cultural variation is usually U-shaped.

migration. A reduction of migration (i.e., an increase of isolation) and/or a decrease of population size will increase the variation between populations, and in extreme cases the distribution becomes U-shaped, as seems to be the case for many cultural traits. For the genetic case this is true only for extreme isolates; the observed distributions of Fig. 3.13.1 and Fig. 3.13.2 are clearly bell-shaped.

The equivalent to Nm in cultural drift is D, which can be expressed as follows:

$$D = (\sigma + \tau + \mu_m)N, \tag{3.13.1}$$

where τ represents the sum of all contributions of oblique, vertical and horizontal transmission, including, as usual, cultural selection, while σ represents natural selection ($N\sigma = s$) and μ_m represents cultural migration ($N\mu_m = m$). For cultural variation among populations to be U-distributed, we expect D to be small. In biology, N is the number of reproducing individuals. In cultural transmission, N can be equated to the number of teachers in the population. When transmission is vertical and biparental, N is the number of parents, and is therefore the same number as in biological transmission. In teacher-leader transmission, N is the number of teachers or leaders, and it may be as small as $N = 1$. Thus, one can think in terms of a *cultural equivalent* N. For types of transmission other than biparental and teacher/leader, one would expect $N/2$ for uniparental, while oblique or horizontal may be expected to be usually intermediate between vertical and teacher-leader. Thus, the cultural equivalent N can be very small under conditions of transmission in which cultural evolution inside populations tends to be very rapid, as in a teacher-leader model. In this case, homogeneity within each population is generated quickly, as are maximal differences between populations, which is characteristic of a U-shaped distribution.

There are other factors that contribute to increasing the cultural variation between populations, and that are subsumed under σ and τ. The usefulness or acceptability of a cultural trait is a function of the environment, both physical (ecological) and cultural. Individuals will use leaves, mud, bricks, stone, or other stuff for building housewalls, depending on climate, the availability of raw material, and degree of technology. In the inheritance of property, succession may be in the female or in the male line, and there will be the strong influence of whether the society is matrilineal or patrilineal in other respects.

Isolation by distance is of importance also in the spread of cultural traits from one population to the other. An analysis of the geographic distribution of cultural traits in African societies shows that many traits are geographically highly clustered (Guglielmino et al., 1981), as can be seen in Figure 3.13.3 and Figure 3.13.4. The clustering may be the outcome of one or more of many different factors. Gross ecological diversity may favor the concentration of certain traits in specific ecological areas; historical events may generate similarity between tribes that have a common origin, or may have been and perhaps still are under comon neighborhood rules; good roads may not exist to favor cultural contacts; and so on. The outcome is a highly patchy geographic distribution that will accentuate the tendencies to a U-shape in the frequency distributions.

The extreme variation between populations seen for many linguistic and other cultural traits does not exist for *every* cultural trait. On the contrary, there are cultural traits that are very constant. Interest in such constant traits has promoted the generation of a list of "universals" (for language: Greenberg, 1966). What is characteristic of cultural variation is the extreme breadth of the range of variations, from total conservation to extreme differences. This range in turn reflects extreme variation of evolutionary rates, from near zero to extremely high. It is obviously with the latter that one is more likely to observe the greatest degree of spatial variation. However, synchronic variation may also be increased by factors other than a fast rate of cultural evolution, for instance, the necessity to adapt to a great variety of environmental conditions.

It is known that on average biological traits vary little between populations as compared with the variation within; (for a more extensive analysis of this point, see Nei and Roychoudhury, 1974; Lewontin, 1972). In summary, the variation between populations may be much larger than that within, there being a strong contrast between the tendency to homogeneity for many cultural traits within populations and the existence of sharp differences between populations. But different cultural traits vary considerably in this respect, from extremes with exceptionally high variation between populations (and almost no variation within) to extremes with no variation at all. This situation can be explained, at least in part, by considering features of cultural transmission, and it invites further quantitative speculation.

3.14 CORRELATION OF CULTURAL AND BIOLOGICAL VARIATION

It is almost unavoidable in studying the extent of biological and genetic variation to ask the further question: is there a correlation between genetic and cultural variation? This is a relatively simple question to ask. It is also easy to give a formal answer in terms of a correlation coefficient. But spatial distributions can easily generate spurious correlations, and the same can be said for correlations between two quantities that vary over time. The difficulty lies in the interpretation of these correlations.

Two such correlations are of special interest. We have discussed in section 3.13 the correlation of the pattern of marriage with that of genetic differentiation. The latter is so strongly dependent on the former that marriage patterns, which would otherwise be of interest only to sociologists, are a topic of study also in population genetics.

Another related field of interest is that of the correlation of linguistic and genetic variation between populations. With standard methods of multidimensional analysis, it is possible to correlate genetic differences between two population pairs (estimated on the basis of many genes) and their linguistic differences, estimated for instance on the basis of proportions of shared cognates (see Section 1.8). This analysis (examples: Spielman et al., 1974, on the Yanomamo; Friedlaender, 1971, in Melanesia; Spuhler, 1972, on several American Indian tribes) has given results varying from zero to high correlations between languages and genes. How do such contradictory results arise?

The contradiction is not entirely unexpected. Many of the demographic and historical events that affect genes also affect languages in the same direction: isolation causes drift of genes and of languages, and migratory exchanges cause populations to become more homogenous, both for genes and for language. If these were the only two factors of evolution there might be considerable parallelism in the linguistic and genetic differentiation. Languages, however, evolve much faster than genes: a large percentage of words are replaced in a thousand years, but it takes many millions of years to replace an equally large fraction of amino acids in a protein or of nucleotides in DNA. Thus, similarities between languages tend to be practically cancelled after, say 10,000 years of separation, while the same time span causes only minor differences in gene frequencies. Moreover, certain political and historical events may change languages very drastically during the lifetime of individuals, while having very little effect on genes. The Romans were successful in imposing their language on a number of countries. One of these, Romania, even though fairly far from Rome, has even today a large fraction of words in the current language (even the name of the country) that are clearly of Roman origin. Languages express the importance of past political domination, as well as other cultural influences, by the number of foreign words and roots borrowed. But while a handful of administrators and soldiers can sometimes change the language of a country, they can hardly have a profound effect on the gene pool.

CHAPTER FOUR

Multiple State Traits

4.1 MENDELIAN TRANSMISSION AS AN EXAMPLE OF A MULTIPLE STATE TRAIT

The models described in Chapter 2 were for traits that could take one of two values. Frequency changes over time can be studied in those cases with unidimensional recursion systems whose properties are relatively easy to unravel. As the number of states that a trait can assume increases, so the complexity of possible vertical transmission schemes also increases. In comparing nonbiological and biological transmission, however, it is necessary to consider three and higher state numbers, since these appear naturally in simple genetic schemes for diploid organisms. Such models are in use, for example, in medical genetics, where some authors advocate their use in determining risks that certain diseases will arise and in starting counseling on the basis of such risk estimates. These are generally not evolutionary treatments, but statistical. Here we pursue the role of different regular transmission schemes in determining the equilibrium frequency configurations.

It is easy to write down the relations describing the generalization of the two dimensional scheme to one involving n states. But the dynamics of the generalized system may be very difficult to study. Thus, if the frequencies of the n states at time t are the components of the vector

$$\mathbf{u}_t = (u_t^{(1)}, u_t^{(2)}, \ldots u_t^{(n)})^T \tag{4.1.1}$$

(where $\mathbf{u}_t$ is considered as a column vector and the superscript T denotes the transpose), then under random mating we can write the column vector of size n^2 whose components are the frequencies of the n^2 matings as

$$\mathbf{U}_t^T = \{\mathbf{u}^{(t)} \otimes \mathbf{u}^{(t)}\}^T = (u_1^{(t)} \cdot u_1^{(t)}, u_1^{(t)} \cdot u_2^{(t)}, \ldots, u_n^{(t)} \cdot u_n^{(t)}), \quad (4.1.2)$$

where $\otimes$ is the symbol for the Kronecker product of vectors or matrices. The offspring of these n^2 matings occur in frequencies described by the elements of $\mathbf{u}_{t+1}$. The correspondence between $\mathbf{u}_t$ and $\mathbf{u}_{t+1}$ is given in terms of an $n^2 \times n$ matrix $\mathbf{T}$ as

$$\mathbf{u}_{t+1} = \mathbf{U}_t^T \mathbf{T} \qquad (4.1.3)$$

For general transmission laws, $\mathbf{T}$ (4.1.3) is a quadratic recursion scheme, which for n large can be difficult to analyse in detail. A well-known, simple three-state example of (4.1.3) takes as the states the genotypes AA, Aa and aa at a single gene locus. According to Mendel's rules, the transmission matrix $\mathbf{T}_{[9\times 3]}$ is as follows in Table 4.1.1. If the frequencies of AA, Aa, aa are u_0, v_0, and w_0, respectively, in the initial generation, and we write $\mathbf{u}_0 = (u_0, v_0, w_0)$, then the vector of matings is (suppressing the subscript) $\mathbf{U}_0^T = (\mathbf{u}_0 \otimes \mathbf{u}_0)^T = (u^2, uv, uw, uv, v^2, vw, wu, wv, w^2)$. The three genotypes in the offspring have frequencies

$$\mathbf{u}_1 = \mathbf{U}_0^T \mathbf{T} = [(u_0 + v_0/2)^2, 2(u_0 + v_0/2)(w_0 + v_0/2), (w_0 + v_0/2)^2]. \qquad (4.1.4)$$

As is proved by the Hardy-Weinberg theorem, there is subsequently no change in the genotype frequencies:

$$\mathbf{u}_1 = \mathbf{u}_2 = \ldots = \mathbf{u}_t = \ldots. \qquad (4.1.5)$$

The quantities

$$p = u_0 + v_0/2, q = 1 - p \qquad (4.1.6)$$

are the allele frequencies of A and a, and (4.1.6) can be rewritten as

$$\mathbf{u} = (p^2, 2pq, q^2), \qquad (4.1.7)$$

TABLE 4.1.1. Mendelian genetic transmission.

Mating	Mating frequencies	Progeny *AA*	*Aa*	*aa*
1. *AA* × *AA*	u_0^2	1	0	0
2. *AA* × *Aa*	u_0v_0	1/2	1/2	0
3. *AA* × *aa*	u_0w_0	0	1	0
4. *Aa* × *AA*	u_0v_0	1/2	1/2	0
5. *Aa* × *Aa*	v_0^2	1/4	1/2	1/4
6. *Aa* × *aa*	v_0w_0	0	1/2	1/2
7. *aa* × *AA*	u_0w_0	0	1	0
8. *aa* × *Aa*	v_0w_0	0	1/2	1/2
9. *aa* × *aa*	w_0^2	0	0	1

where the time subscript is suppressed due to (4.1.5). The state (4.1.7) that obtains after the first generation, is the well known *Hardy-Weinberg equilibrium*, reached (as is seen from (4.1.5)) after one generation of random mating.

Since the outcomes of matings 2 and 4 are identical, as are those of 3 and 7, 6 and 8, the transmission matrix can be reduced to size 6 × 3 by ignoring parental sex. Mendel's "laws" are in fact the rows of the transmission matrix. When two of the three types are indistinguishable, as happens with dominance (described in Chapter 2 as an example of frequency-dependent transmission coefficients), then the transmission matrix is no longer time-homogeneous; only in the extended 9 × 3 form are the entries of **T** constant. If instead of the two alleles in the example above we have k, then there are $k(k + 1)/2$ genotypes or states. Although **T** is larger the process remains as simple as in the two-allele case.

We now turn to the details of some three-state models that are not genetic, so that a comparison can be made of the complexities inherent in genetic and non-genetic transmission. In many senses genetic transmission will be seen to represent one of the simplest of the quantitative models.

4.2 VERTICAL TRANSMISSION FOR THREE-STATE MODELS

The transmission matrix exemplified in Table 4.1.1 by the Mendelian matrix can be regarded as a special case of a general transmission system with twenty-seven coefficients, as in Table 4.2.1. The three states have, at time t, frequencies u, v, w. The column headed "Random mating frequencies" is computed under the assumption of random mating, from (4.1.2). Here a_{ij} represents the probability that phenotype j ($j = 1, 2, 3$) comes from mating i ($i = 1, 2, \ldots, 9$). Since every row in Table 4.2.1 adds to unity there are only eighteen free parameters.

It will be rare that the coefficients in Table 4.2.1 can all be estimated satisfactorily. Later we present a case of multistate uniparental transmission, for the transmission of profession from father to son in which the estimation is attempted. For biparental transmission, the transmission matrix itself is not sufficient and the mating system must be specified in addition. For the moment random mating is assumed, and in a later section assortative mating is introduced.

The recursion relations between generations for the frequen-

TABLE 4.2.1. General transmission matrix for a three-state trait.

Mating ♀ ♂	Random Mating Frequencies	Offspring H_1	H_2	H_3
$H_1 \times H_1$	u^2	a_{11}	a_{12}	a_{13}
$H_1 \times H_2$	uv	a_{21}	a_{22}	a_{23}
$H_1 \times H_3$	uw	.	.	.
$H_2 \times H_1$	.	.	.	.
$H_2 \times H_2$	.	.	.	.
$H_2 \times H_3$	.	.	.	.
$H_3 \times H_1$	.	.	.	.
$H_3 \times H_2$	.	.	.	.
$H_3 \times H_3$	w^2	a_{91}	a_{92}	a_{93}

cies of H_1, H_2, H_3, namely u, v, w, respectively, are quadratic and, in the general case, very complicated. We shall consider some special cases: additive, multiplicative, and symmetric transmission schemes, each of which can in turn be further simplified.

4.2a. Additive Transmission

We assume here that the eighteen independent entries can be written according to the following rule: Father H_i contributes to the transmission coefficient for progeny H_k an amount α_{ik}, and mother H_j contributes a moiety α'_{jk} to offspring H_k. The two are added to produce the complete transmission probability. Table 4.2.2 illustrates the resulting rule. In Table 4.2.2 all α's are less than unity and each of the nine rows must sum to unity. As can be easily seen the recursions in this case become linear:

$$\begin{aligned} u' &= (\alpha_{11} + \alpha'_{11})u + (\alpha_{12} + \alpha'_{12})v + (\alpha_{13} + \alpha'_{13})w \\ v' &= (\alpha_{21} + \alpha'_{21})u + (\alpha_{22} + \alpha'_{22})v + (\alpha_{23} + \alpha'_{23})w \\ w' &= (\alpha_{31} + \alpha'_{31})u + (\alpha_{32} + \alpha'_{32})v + (\alpha_{33} + \alpha'_{33})w. \end{aligned} \tag{4.2.1}$$

Write $\alpha_{ij} + \alpha'_{ij} = t_{ij}$. Then we have $(u', v', w')^T = \mathbf{T}(u, v, w)^T$ where $\mathbf{T} = \|t_{ij}\|$ is a stochastic matrix and has at least one eigenvalue equal to unity. There are a number of alternative equilibrium outcomes of the iteration of (4.2.1). For example, if all entries of $\mathbf{T}$ are non-zero, then an equilibrium is approached

TABLE 4.2.2. Additive transmission.

Mating ♂ ♀	Mating Frequencies	Offspring H_1	H_2	H_3
$H_1 \times H_1$	u^2	$\alpha_{11} + \alpha'_{11}$	$\alpha_{12} + \alpha'_{12}$	$\alpha_{13} + \alpha'_{13}$
$H_1 \times H_2$	uv	$\alpha_{11} + \alpha'_{21}$	$\alpha_{12} + \alpha'_{22}$	$\alpha_{13} + \alpha'_{23}$
$H_1 \times H_3$	uw	$\alpha_{11} + \alpha'_{31}$	$\alpha_{12} + \alpha'_{32}$	$\alpha_{13} + \alpha'_{33}$
$H_2 \times H_1$	uv	$\alpha_{21} + \alpha'_{11}$	$\alpha_{22} + \alpha'_{12}$	$\alpha_{23} + \alpha'_{13}$
⋮	⋮	⋮	⋮	⋮

at which all phenotypes are represented. Or, if $t_{12} = t_{13} = 0$ and all other entries are positive the population is eventually all H_1. The rate of approach to these states is the second largest eigenvalue of **T**, namely

$$\lambda_2 = \frac{t_{11} + t_{22} - t_{13} - t_{23}}{2} + \left[\frac{(t_{11} + t_{22} - t_{13} - t_{23})^2}{2} + (t_{12} - t_{13})(t_{21} - t_{23}) - (t_{11} - t_{13})(t_{22} - t_{23})\right]^{1/2}.$$

Note that, if $\alpha_{11} = \alpha_p + b_0/2$, $\alpha'_{11} = \alpha_m + b_0/2$, $\alpha_{21} = \alpha'_{21} = \alpha_{22} = \alpha'_{22} = b_0/2$, elimination of H_3 reduces Table 4.2.2 to the two state additive scheme of Table 2.3.1.

4.2b. Multiplicative Transmission

Parents mate at random and contribute independently to the offspring. The contributions of the two parents are not necessarily equal. The model is "almost" genetic in that it envisages two types of particles, + and −, with the three phenotypes H_1, H_2, H_3 representing $++$, $+-$, $--$. Then H_1, H_2, and H_3 transmit + with probabilities τ_1, τ_2, τ_3 respectively. Taking into account the independence of the maternal and paternal transmission, the resulting multiplicative transmission law is set out in Table 4.2.3. Here, u_i is the frequency of H_i.

TABLE 4.2.3. Multiplicative transmission.

		Offspring		
Mating	Frequency	H_1	H_2	H_3
$H_1 \times H_1$	u_1^2	τ_1^2	$2\tau_1(1-\tau_1)$	$(1-\tau_1)^2$
$H_1 \times H_2$	$2u_1u_2$	$\tau_1\tau_2$	$\tau_1(1-\tau_2) + \tau_2(1-\tau_1)$	$(1-\tau_1)(1-\tau_2)$
$H_1 \times H_3$	$2u_1u_3$	$\tau_1\tau_3$	$\tau_1(1-\tau_3) + \tau_3(1-\tau_1)$	$(1-\tau_1)(1-\tau_3)$
$H_2 \times H_2$	u_2^2	τ_2^2	$2\tau_2(1-\tau_2)$	$(1-\tau_2)^2$
$H_2 \times H_3$	$2u_2u_3$	$\tau_2\tau_3$	$\tau_2(1-\tau_3) + \tau_3(1-\tau_2)$	$(1-\tau_2)(1-\tau_3)$
$H_3 \times H_3$	u_3^2	τ_3^2	$2\tau_3(1-\tau_3)$	$(1-\tau_3)^2$

From Table 4.2.3 we compute

$$u_1' = (\tau_1 u_1 + \tau_2 u_2 + \tau_3 u_3)^2 \tag{4.2.2a}$$

$$u_2' = 2[\tau_1 u_1 + \tau_2 u_2 + \tau_3 u_3][(1 - \tau_1)u_1 + (1 - \tau_2)u_2 + (1 - \tau_3)u_3] \tag{4.2.2b}$$

$$u_3' = [(1 - \tau_1)u_1 + (1 - \tau_2)u_2 + (1 - \tau_3)u_3]^2. \tag{4.2.2c}$$

The reader will recognize that the case $\tau_1 = 1, \tau_2 = 1/2, \tau_3 = 0$ is the genetic case described in section 4.1. This transmission model was introduced for the purpose of fitting actual human pedigrees by Elston and Stewart (1971). The idea was to use this scheme to determine roughly how far the observed rule of transmission of a trait departs from Mendel's rule by comparison of the estimated τ_1, τ_2, τ_3 with the 1, 1/2, 0 expected. Small deviations of τ_1 from 1 or of τ_3 from 0 could of course be due to mutation, while departure of τ_2 from 1/2 could be explained under the rubric of gametic "selection"; for example, segregation distortion, meiotic drive, or gametic competition.

From (4.2.2), writing

$$z = \tau_1 u_1 + \tau_2 u_2 + \tau_3 u_3, \tag{4.2.3}$$

we have

$$z' = z^2(\tau_1 + \tau_3 - 2\tau_2) + 2z(\tau_2 - \tau_3) + \tau_3 \tag{4.2.4}$$

and $u_1' = z^2$, $u_2' = 2z(1 - z)$, $u_3' = (1 - z)^2$. This last shows that the three states are always in "Hardy-Weinberg" equilibrium except that the equivalent of the "gene frequency," $u_1 + u_2/2$, changes until equilibrium is reached. For most parameter sets the recursion (4.2.4) converges to an equilibrium

$$\hat{z} = \frac{1 + 2\tau_3 - 2\tau_2 - [1 + 4\tau_2^2 + 4\tau_3 - 4\tau_2 - 4\tau_1\tau_3]^{1/2}}{2(\tau_1 + \tau_3 - 2\tau_2)}. \tag{4.2.5}$$

But it is possible that (4.2.5) not be achieved, and that a two point cycle be approached. Certain special parameter sets are worth identifying primarily because of their relationship with the one-locus two-allele genetic case

Special case I. $\tau_3 = 0, \tau_1 \neq 1, 1 > 2\tau_2$. Here only $\hat{z} = 0$ is valid. (4.2.5) reduces to $\hat{z} = 0$ is this case.

Special case II. $\tau_3 = 0, \tau_1 \neq 1, 2\tau_2 > 1$. Here in addition to $\hat{z} = 0$ we have the root

$$\hat{z} = \frac{2\tau_2 - 1}{2\tau_2 - \tau_1}. \tag{4.2.6}$$

Parallel to cases I and II we have

Special case III. $\tau_1 = 1, \tau_3 \neq 0, 2\tau_2 > 1$. Only the equilibrium $\hat{z} = 1$ is valid.

Special case IV. $\tau_1 = 1, \tau_3 \neq 0, 2\tau_2 < 1$. Both $\hat{z} = 1$ and

$$\hat{z} = \tau_3/(1 + \tau_3 - 2\tau_2) \tag{4.2.7}$$

are valid.

Special case V. $\tau_1 + \tau_3 = 2\tau_2$. The only equilibrium is

$$\hat{z} = \tau_3/(1 + 2\tau_3 - 2\tau_2) \tag{4.2.8}$$

Special case VI. $\tau_1 = 1, \tau_3 = 0$. Both $\hat{z} = 0$ and $\hat{z} = 1$ are equilibria.

The stability analyses of the equilibria in these special cases, and indeed for the most general case (4.2.5) are very simple. This arises from the fact that the 2×2 local-stability matrix at any equilibrium of this system always has one eigenvalue of zero. The other, which of course governs the stability of any given equilibrium, is

$$\lambda = 2\hat{z}(\tau_1 - \tau_3) + 2(1 - 2\hat{z})(\tau_2 - \tau_3). \tag{4.2.9}$$

Reviewing some of the special cases (4.2.9) we have the conclusions:

Special case I. $\hat{z} = 0$ stable.
Special case II. $\hat{z} = 0$ unstable and (4.2.6) stable.

Special case III. $\hat{z} = 1$ stable.

Special case IV. $\hat{z} = 1$ unstable and (4.2.7) stable.

Qualitatively, these results follow from the fact that, for example, in special case IV, when $2\tau_2 < 1$, not enough H_1 offspring are produced from the mating $H_2 \times H_2$ to push H_1 to fixation, even though $\tau_1 = 1$. In fact, if τ_3 is small, the advance of H_1 can be stopped at quite a low frequency (4.2.7).

Special case V. The polymorphic equilibrium (4.2.8) is always stable.

Special case VI corresponds to a model of segregation distortion or litter selection studied by Haldane (1924) (see also Karlin, 1972a). Here $2\tau_2 > 1$ assures H_1 should fix, while $2\tau_2 < 1$ results in H_3 fixing.

In the general case, substitution of (4.2.5) into (4.2.9) produces

$$\lambda^* = 1 - [1 + 4\tau_2^2 + 4\tau_3 - 4\tau_2 - 4\tau_1\tau_3]^{1/2}. \quad (4.2.10)$$

Clearly, $\lambda^* < 1$. For most parameter values $\lambda^* > -1$. But with τ_3 large, τ_1 and τ_2 small, this last condition can be violated. For example, if $\tau_1 = 0.05$, $\tau_2 = 0.1$, $\tau_3 = 0.95$, $\lambda^* < -1.1$. In this case the unique equilibrium cannot be stable. The system quickly moves into a two-point cycle. Numerical studies we have made indicate that this is the only cycle involved, no matter how high τ_3 is. This should be compared to the two-point cycle observed in Chapter 2.2 for the two-state model. Again, in the present case, offspring must be unlike both parents before the cycle can be achieved.

Despite the departure from the genetic case (for which $\tau_1 = 1$, $\tau_2 = 1/2$, $\tau_3 = 0$), the Hardy-Weinberg structure still obtains at every generation, but unlike the genetic case, the "gene frequency" z changes until in the limit it reaches equilibrium.

The multiplicative model involves at most three parameters. It has formed the basis for studies by Elston and Stewart (1971) and others of the transmission of diseases in families. The idea is that the closer the estimates of τ_1, τ_2, and τ_3 are to 1, 1/2, 0, respectively, the more likely that the disease has a genetic etiology. We now introduce another model that under special conditions also reduces to the genetic model, but can reduce to a multiplicative system in which $\tau_1 = (1 - \tau_3)$, and $\tau_2 = 1/2$.

4.2c. *Symmetric Transmission*

Here H_1, H_2, H_3 are viewed as ordered measurements of some behavioral, phenotypic, or cultural variable, equally spaced along some axis so that H_2 is "equidistant" from H_1 and H_3. In any mating $H_i \times H_j$ the probability of producing H_k is the same when i, j, or $k = 1$ as when those of i, j, or k that are 1 are replaced by 3, and vice versa. This results in the mating Table 4.2.4 below.

Note that since the probabilities in each row of the three righthand columns must add to unity there are in fact six free parameters in the system. The recursion relations are

$$u_1' = \alpha_1 u_1^2 + 2\delta_1 u_1 u_2 + 2\eta_1 u_1 u_3 + 2\delta_3 u_2 u_3 + \rho_1 u_2^2 + \alpha_3 u_3^2 \quad (4.2.11)$$

$$u_2' = \alpha_2(u_1^2 + u_3^2) + 2\delta_2(u_1 u_2 + u_2 u_3) + 2\eta_2 u_2 u_3 + \rho_2 u_2^2 \quad (4.2.12)$$

$$u_3' = \alpha_3 u_1^2 + 2\delta_3 u_1 u_2 + 2\delta_1 u_2 u_3 + 2\eta_1 u_1 u_3 + \rho_1 u_2^2 + \alpha_1 u_3^2. \quad (4.2.13)$$

Subtracting (4.2.13) from (4.2.11) we have

$$u_1' - u_3' = [(\alpha_1 - \alpha_3)(u_1 + u_3) + 2(\delta_1 - \delta_3)u_2](u_1 - u_3), \quad (4.2.14)$$

so that $\hat{u}_1 = \hat{u}_3$ is an equilibrium, which we henceforth call the *symmetric equilibrium*. In addition the equilibrium

$$\hat{u}_2 = \frac{(1 - \alpha_1 + \alpha_3)}{\alpha_3 - \alpha_1 + 2(\delta_1 - \delta_3)} \quad (4.2.15)$$

TABLE 4.2.4. Symmetric transmission.

Mating	Frequency	Offspring H_1	H_2	H_3
$H_1 \times H_1$	u_1^2	α_1	α_2	α_3
$H_1 \times H_2$	$2u_1u_2$	δ_1	δ_2	δ_3
$H_1 \times H_3$	$2u_1u_3$	η_1	η_2	η_1
$H_2 \times H_2$	u_2^2	ρ_1	ρ_2	ρ_1
$H_2 \times H_3$	$2u_2u_3$	δ_3	δ_2	δ_1
$H_3 \times H_3$	u_3^2	α_3	α_2	α_1

may exist. The condition for existence of (4.2.15), that is, that $0 \le u_2 \le 1$, is

$$2\delta_1 > 1 + 2\delta_3. \tag{4.2.16}$$

If $\hat{u}_1 \neq \hat{u}_3$, then from (4.2.11) and (4.2.15) the equilibrium value of u_1 must satisfy

$$u_1^2[\alpha_1 + \alpha_3 - 2\eta_1] + u_1[2\eta_1 - \alpha_1 - \alpha_3](1 - \hat{u}_2) + \rho_1\hat{u}_2^2 + 2\delta_3\hat{u}_2(1 - \hat{u}_2) + \alpha_3(1 - \hat{u}_2)^2 = 0. \tag{4.2.17}$$

Equilibria with $\hat{u}_1$ satisfying (4.2.17) where $\hat{u}_2$ is given by (4.2.15) and $\hat{u}_3 = 1 - \hat{u}_1 - \hat{u}_2$ will be called *asymmetric*. Clearly, if $\alpha_1 + \alpha_3 - 2\eta_1 < 0$, these cannot exist.

The symmetric equilibria can be obtained by substituting $u_1 = u_3$ and $u_2 = 1 - 2u_1$ in (4.2.11). The result is the quadratic equation

$$u_1^2\{\alpha_1 + \alpha_3 + 2\eta_1 - 4(\delta_1 + \delta_3) + 4\rho_1\} + u_1\{2\delta_1 + 2\delta_3 - 4\rho_1 - 1\} + \rho_1 = 0, \tag{4.2.18}$$

with the admissibility criterion $0 \le u_1 \le 1/2$. Because there are a large number of parameters the number of special cases conceivable is correspondingly large. No attempt will be made here to be exhaustive in covering these. An obvious one defines H_1, H_2, and H_3 as the genotypes *AA*, *Aa* and *aa* with $\alpha_1 = 1$, $\delta_1 = \delta_2 = \rho_2 =$

$1/2$, $\rho_1 = 1/4$, $\alpha_2 = \alpha_3 = \delta_3 = \eta_1 = 0$, which produces Mendelian transmission.

One special case of interest sets $\rho_1 = 0$. In terms of the relationship between phenotypes and genotypes mentioned in the previous paragraph, this is a long way from the genetic scheme. If $\rho_1 > 0$, in fact, it can be shown that there is always a unique nontrivial symmetric equilibrium. When $\rho_1 = 0$, then $\hat{u}_2 = 1$ is clearly a symmetric equilibrium, but another, given by

$$\hat{u}_1 = \frac{1 - 2(\delta_1 + \delta_3)}{\alpha_1 + \alpha_3 + 2\eta_1 - 4(\delta_1 + \delta_3)} \quad (4.2.19)$$

exists, provided $2(\delta_1 + \delta_3) > 1$. If $2(\delta_1+\delta_3) < 1$, then with $\rho_1 = 0$, $\hat{u}_2 = 1$ is the only symmetric equilibrium. Three cases of the symmetric transmission model involving simple assumptions on the parameter set are worth discussing in more detail.

Special case I. $\rho_1 = 0$, $\alpha_1 = 1$. In this case the only asymmetric equilibria are $\hat{u}_1 = 1$ and $\hat{u}_3 = 1$. If $2(\delta_1 + \delta_3) < 1$, then $\hat{u}_2 = 1$ is the only symmetric equilibrium. The graphical representation of the stability configuration in Figure 4.2.1, obtained by local linear analysis near the respective equilibria, tells the story.

Special case II. $\rho_1 = 0$, $\alpha_1 \neq 1$. When $2\delta_1 > 1 + 2\delta_3$ and $\alpha_1 + \alpha_3 > 2\eta_1$, then from (4.2.15), (4.2.17), and (4.2.19) the symmetric equilibria (4.2.19) and $\hat{u}_2 = 1$ both exist while the asymmetric equilibria exist if

$$(1 - \hat{u}_2)^2 \left\{\frac{3\alpha_3 + 2\eta_1 - \alpha_1}{4}\right\} + 2\delta_3\hat{u}_2(1 - \hat{u}_2) < 0 \quad (4.2.20)$$

where $\hat{u}_2$ is defined by (4.2.15). If $\alpha_1 + \alpha_3 < 2\eta_1$, no asymmetric equilibria are possible. If $2\delta_1 < 1 - 2\delta_3$, no interior equilibria can exist, and $\hat{u}_2 = 1$ is stable. If (4.2.20) holds, then the symmetric equilibrium cannot be stable. In fact, when both asymmetric equilibria exist simultaneously with the symmetric point the latter cannot be stable, and the former are stable. Further, if the

FIGURE 4.2.1

(4.2.19) does not exist	(4.2.19) stable	(4.2.19) unstable
$\hat{u}_1 = \hat{u}_3 = 1$ unstable,	$\hat{u}_1 = 1, \hat{u}_2 = 1, \hat{u}_3 = 1$	$\hat{u}_1 = 1, \hat{u}_3 = 1$ both stable
$\hat{u}_2 = 1$ stable	all unstable	$\hat{u}_2 = 1$ unstable

$1 - 2\delta_3$ $\qquad$ $1 + 2\delta_3$

$2\delta_1$ ⟶

Legend: Special case $\rho_1 = 0$, $\alpha_1 = 1$; $2\delta_1$ is plotted horizontally. The key values $1 - 2\delta_3$ and $1 + 2\delta_3$ are obtained from the local linear analysis.

asymmetric equilibria exist for some range of parameters, then the only criterion for the stability of the symmetric equilibrium is the violation of (4.2.20), assuming, of course, that $2\delta_1 < 1 + 2\delta_3$. On the other hand, if the asymmetric equilibria do not exist, and the symmetric point is valid, it may become unstable due to cycling (that is, its smallest local eigenvalue is < -1). This happens only if α_3 is very large, δ_3 is very large, and η_1, α_1, δ_1, δ_2 are all sufficiently small.

Special case III (one mutation-like parameter). A class of symmetric transmission rules can be constructed as follows. If a parent is H_1, then with probability m its offspring is H_2, and with $1 - m$ its offspring is H_1. If a parent is H_3, then with probability m its offspring is H_2, and with $1 - m$ it is H_3. An H_2 parent has H_1, H_2, H_3 offspring in the proportions $m/2$, $1 - m$, $m/2$, respectively. Now for the $H_i \times H_j$ mating the probabilities of H_1, H_2, and H_3 offspring are assumed to be the arithmetic mean of the separate probabilities defined above for each parent separately. The result is the mating Table 4.2.5 with $0 \leq m \leq 1$. If m is small, then the matings that cause the departure from a genetic situation are $H_1 \times H_3$ and $H_2 \times H_2$. The equilibrium analysis of this system is extremely simple. No boundary states are equilibria, asymmetric equilibria cannot exist, and the only equilibrium is $\hat{u}_1 = \hat{u}_3 = 1/4$, $\hat{u}_2 = 1/2$. It is always stable for $m > 0$.

This model has in it an implicit "mutation-like" structure, in

TABLE 4.2.5. One-parameter symmetric transmission.

Mating	Frequency	Offspring H_1	H_2	H_3
$H_1 \times H_1$	u_1^2	$1 - m$	m	0
$H_1 \times H_2$	$2u_1u_2$	$1/2 - \frac{m}{4}$	$1/2$	$\frac{m}{4}$
$H_1 \times H_3$	$2u_1u_3$	$\frac{1-m}{2}$	m	$\frac{1-m}{2}$
$H_2 \times H_2$	u_2^2	$\frac{m}{2}$	$1 - m$	$\frac{m}{2}$
$H_2 \times H_3$	$2u_2u_3$	$\frac{m}{4}$	$1/2$	$1/2 - \frac{m}{4}$
$H_3 \times H_3$	u_3^2	0	m	$1 - m$

that an offspring's type cannot be shifted from the parental types by too great an amount. It is seen that the equilibrium is of the Hardy-Weinberg form, despite the nonmultiplicative, nongenetic transmission scheme.

The delicate dependence of the equilibrium structure on the transmission rule can be exemplified by changing the previous transmission rule very slightly. Suppose that an H_1 parent has an H_2 offspring with probability $m/2$ (instead of m above) with $1 - m/2$ for H_1 offspring. A symmetric assumption is made for H_3 parents, but everything else about the model is retained. This apparently small change causes the equilibrium now to be $\hat{u}_1 = \hat{u}_2 = \hat{u}_3 = 1/3$. For example, if $m = 0.1$, the first system, with an equilibrium of (1/4, 1/2, 1/4), has transmission scheme

$$\begin{array}{lll} 0.9 & 0.1 & 0.0 \\ 0.475 & 0.5 & 0.025 \\ 0.45 & 0.1 & 0.45 \\ 0.05 & 0.9 & 0.05 \\ 0.025 & 0.5 & 0.475 \\ 0.0 & 0.1 & 0.9, \end{array}$$

while the second, with an equilibrium of (1/3, 1/3, 1/3) has transmission scheme

0.95	0.05	0.0
0.5	0.475	0.025
0.475	0.05	0.475
0.05	0.9	0.05
0.025	0.475	0.5
0.0	0.05	0.95.

This class of "mutation-like" transmission schemes can clearly generalize in many ways to multiple states. The practical problem remains to break the parameter range into an easily described set of classes; as we have seen with three phenotypes this may be extremely difficult. The last class of special cases that are couched in terms of a single parameter allows some flexibility in this respect although the genetic case does not fit naturally into the framework.

When $\rho_1 \neq 0$ in the general symmetric model, intricate meshing of the parameters is necessary to describe the equilibrium structure. When the asymmetric equilibria exist, they are stable, while the symmetric point is not. Boundary points are not relevant in this more general case. Cycling may still occur, but only if ρ_1, η_1 are very small in addition to the natural "switching" requirements discussed in the $\rho_1 = 0$ case.

4.2d. Progeny Exclusively of Parental Type

Recall from Chapter 2 that when $b_3 = 1$, $b_0 = 0$, the population fixes either on $u = 1$ or $u = 0$ depending on whether $b_1 + b_2$ is greater or less than unity. A natural extension to three states of this model allows a mating between types i and j only to produce types i and j. In particular, a homotypic mating $H_i \times H_i$ only produces H_i offspring. The transmission matrix for this case is given in Table 4.2.6 where reciprocal matings have been pooled.

TABLE 4.2.6. Progeny exclusively of parental type.

Mating	Frequency	Offspring H_1	H_2	H_3
$H_1 \times H_1$	u^2	1	0	0
$H_1 \times H_2$	$2uv$	a	$1 - a$	0
$H_1 \times H_3$	$2uw$	$1 - b$	0	b
$H_2 \times H_2$	v^2	0	1	0
$H_2 \times H_3$	$2vw$	0	c	$1 - c$
$H_3 \times H_3$	w^2	0	0	1

It is not difficult to see that only equilibria with a single type present can be locally stable. In fact

$$\begin{aligned}\hat{u} = 1, \hat{v} = \hat{w} = 0 &\text{ is stable iff } a > 1/2, b < 1/2,\\ \hat{u} = 0, \hat{v} = 1, \hat{w} = 0 &\text{ is stable iff } c > 1/2, a < 1/2,\\ \hat{u} = \hat{v} = 0, \hat{w} = 1 &\text{ is stable iff } b > 1/2, c < 1/2.\end{aligned}$$

Thus, if any of a, b, c is above and another is below $1/2$, a state of fixation is locally stable. But if all of a, b, $c > 1/2$ or all $<1/2$, then an internal equilibrium exists and is unstable. The point

$$\hat{u} = (1 - 2c)/L, \hat{v} = (1 - 2b)/L, \hat{w} = (1 - 2a)/L \quad (4.2.21)$$

with $L = 3 - 2a - 2b - 2c$, is an equilibrium. When the equilibrium exists, then from any starting population frequency vector a cycle is entered that passes ever closer to the fixation boundaries, as illustrated in Figure 4.2.2.

Models analogous to that in Table 4.2.6 can be constructed for any number of states. Certain generalizations are worth stating. First, if there is an even number of possible phenotypes, the matrix of the linear equilibrium equation is singular. In such cases it appears that one of the states (which one depends on the transmission parameters) generally achieves fixation. On the other hand, with an odd number of states a polymorphic equilibrium is possible, but never stable. Instead the population cycles away from the equilibrium to the boundaries of the frequency

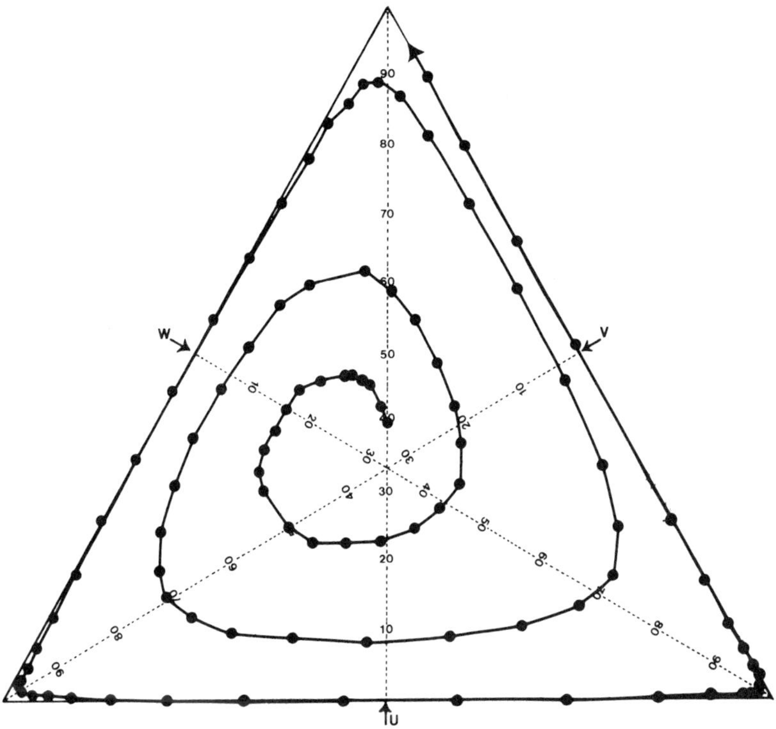

FIGURE 4.2.2 Baricentric coordinate representation of a 3-state discrete trait. Vertical transmission with progeny exclusively of parental type, with transmission matrix as in Table 4.2.6, with $a = 0.3$, $b = 0.2$, $c = 0.2$. Initial values $u_0 = .4$, $v_0 = w_0 = 0.3$.

domain, wherein it continues to cycle; first one state attains very high frequency, then another, and so on in a repeating sequence. These phenomena have parallels in certain competition systems (May 1973) studied in population ecology. In the present class of models, when the interior equilibrium does not exist, there is quick passage to a state of phenotype fixation. Only one such state can be stable at at time. More discussion of models of this kind follows in section 4.5.

4.2e. Parental Proximity Models

Other multistate systems can be constructed which might be called *parental proximity* models, in that an offspring of the mating $H_i \times H_j$ is most likely to be close to H_i and H_j in phenotype. A four-state example is given in Table 4.2.7 and extensions to more states follow naturally. Such models involve symmetric one-parameter forms of transmission. For small m, offspring are clustered close to the parents in trait value. For larger m, there is a more even distribution. These models are not constructed to represent genetic schemes, although there are many reasonably simple genetic structures, for example, one locus with three alleles, or two loci with two alleles each, which could conceivably fit into the schemes of Table 4.2.7. The models might represent transmission of a trait ordered on some scale. Of course the positivity of the entries in the two tables entails that m should not be too large. From the structure of the transmission scheme it might be conjectured that, for example in Table 4.2.7, the equilibrium would have $\hat{u}_1 = \hat{u}_4$ and $\hat{u}_2 = \hat{u}_3$. There is indeed a symmetric equilibrium of this form but it is stable only for larger values of m. For small m there are two "asymmetric" equilibria, with $\hat{u}_1 \neq \hat{u}_4$ and $\hat{u}_2 \neq \hat{u}_3$ to which convergence occurs. It is likely to be difficult to distinguish qualitatively between equilibrium structures emerging from models of the various classes studied in this chapter and certain biological transmission schemes, especially those employing the concepts of distribution of liability to disease and manifestation thresholds (e.g. Falconer 1960, Kidd and Cavalli-Sforza, 1973), which are receiving a certain amount of attention in medical and human behavorial genetics. (See also Feldman and Cavalli-Sforza, 1979a.)

To summarize, we have seen a variety of transmission schemes for three-state traits, some of which can be naturally extended to traits with higher numbers of states. The evolutionary dynamics depend on the transmission parameters and, in cases where there are many stable states, also on the initial frequencies. For three

TABLE 4.2.7. Four-state parental proximity model.

Mating	Frequency	Offspring probability H_1	H_2	H_3	H_4
$H_1 \times H_1$	u_1^2	$1 - m - m^2 - m^3$	m	m^2	m^3
$H_1 \times H_2$	$2u_1u_2$	$(1 - m - m^2)/2$	$(1 - m - m^2)/2$	m	m^2
$H_1 \times H_3$	$2u_1u_3$	m	$1 - 2m - m^2$	m	m^2
$H_1 \times H_4$	$2u_1u_4$	m	$(1 - 2m)/2$	$(1 - 2m)/2$	m
$H_2 \times H_2$	u_2^2	m	$1 - 2m - m^2$	m	m^2
$H_2 \times H_3$	$2u_2u_3$	m	$(1 - 2m)/2$	$(1 - 2m)/2$	m
$H_2 \times H_4$	$2u_2u_4$	m^2	m	$1 - 2m - m^2$	m
$H_3 \times H_3$	u_3^2	m^2	m	$1 - 2m - m^2$	m
$H_3 \times H_4$	$2u_3u_4$	m^2	m	$(1 - m - m^2)/2$	$(1 - m - m^2)/2$
$H_4 \times H_4$	u_4^2	m^3	m^2	m	$1 - m - m^2 - m^3$

states the number of parameters varies from nine to eighteen. This type of analysis is relevant to the study of diseases of complex etiology or behavioral traits for which equilibrium distributions are assumed and transmission modes estimated from observed data. The distribution of the traits at equilibrium is very sensitive to the transmission rule. Many rules admit a Hardy-Weinberg distribution of phenotypes, though not necessarily a stable one. All results in this section assume random mating.

4.3 NUMERICAL EXAMPLES OF MULTISTATE TRANSMISSION

An example of the transmission of a trait that can take on more than two values concerns the comparison of occupations in father and son, a problem referred to as "transgenerational social mobility." Several authors have used the representation of this transmission by a $k \times k$ Markov chain matrix $\mathbf{T}$, where k is the number of occupations. The elements of this matrix, t_{ij}, specify the probability that a child whose parent is of state i is itself of state j, with $\Sigma_j t_{ij} = 1$ (that is, the matrix is *stochastic*). Parents and children should, of course, be considered at the same age or

TABLE 4.3.1. Glass and Hall's data on social mobility in Britain.

Father's Social Class	No. of Pairs	Son's social class 1	2	3	4	5	6	7
1	129	38.8	14.6	20.2	6.2	14.0	4.7	1.5
2	150	10.7	26.7	22.7	12.0	20.6	5.3	2.0
3	345	3.5	10.1	18.8	19.1	35.7	6.7	6.1
4	518	2.1	3.9	11.2	21.2	43.0	12.4	6.2
5	1,510	0.9	2.4	7.5	12.3	47.3	17.1	12.5
6	458	0	1.3	4.1	8.8	39.1	31.2	15.5
7	387	0	.8	3.6	8.3	36.4	23.5	27.4
Total no. of pairs	3,497	103	159	330	459	1,429	593	424

developmental stage for the process to be valid as a description of evolution.

The representation of the transmission by a Markov chain matrix **T** is inevitably very approximate. Even so, in the classical example of social mobility data of 1949 from England and Wales (Prais's treatment (1955) of Glass and Hall's data, see Table 4.3.1) the distribution of social classes is very similar in parents and children, and very similar to the equilibrium distribution of the observed transmission matrix. This equilibrium is obtained from the well known theory of Markov chains, as the eigenvector associated with the greatest eigenvalue smaller than 1. The comparison is shown graphically in Figure 4.3.1. From the computation it might be inferred that change in the elements of the transmission matrix is sufficiently slow over the period studied that equilibrium is not a bad assumption, or that such a state is reached quickly because the leading non-unit eigenvalue of **T** is small.

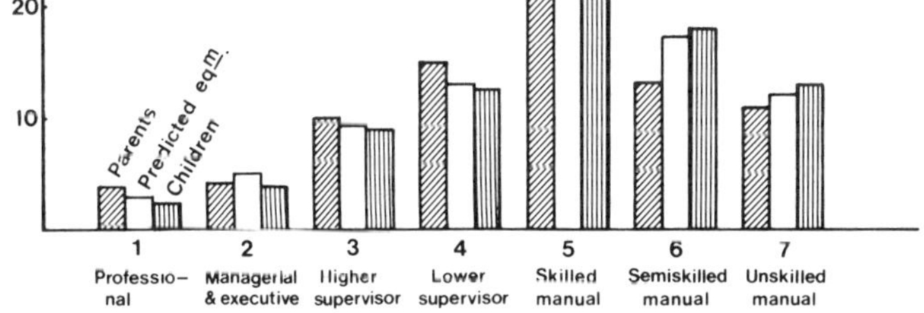

FIGURE 4.3.1. Distribution of social classes in England and Wales, 1949. Data from Glass and Hall. Central bar in each cluster of three bars: prediction of frequencies at the stationary state of the Markov chain according to Prais (1955). Left bars: parents' distribution. Right bars: distribution in children.

Several objections have been raised to the application of Markov chains to social processes. Among them, the fact that individual variation is usually ignored is an important one. Recent statistical treatments have pointed out the difficulty of estimating the entries in the matrix when the process actually represents one that is occurring continuously in time (Singer and Spilerman, 1974). Following its origin with the pioneering work of P. A. Sorokin (see e.g. the 1959 reprinting of his 1927 volume) the theory of social mobility has developed considerably. The interested reader is referred to the volumes by Bartholomew (1973) and Boudon (1973) for extensive reviews of the application of Markov chains and other matrix approaches to social mobility.

A simpler treatment of the example of social mobility makes recourse to a continuous approximation. The distribution of occupations can be reduced to normal by taking a scale with unequal class intervals, such that the observed frequencies of the various classes correspond exactly to those of a normal distribution of mean zero and variance one. For instance, from Table 4.3.1 for the distribution of fathers we have:

Original Class	1	2	3	4	5	6	7
Observed relative frequency*	.037	.043	.099	.148	.432	.131	.111
Cumulative frequency	.037	.080	.179	.327	.759	.890	1.000
Corresponding normal variate	−1.79	−1.41	−.92	−.45	+.70	1.23	
Mean class value	−2.17	−1.58	−1.13	−.69	.13	.94	1.71

*This now is computed from the values in the last column of Table 4.3.1, e.g., 129/3497 = 0.037, and so on.

The evaluation of the mean class value for each class interval is based on a simple, well-known property of the normal integral (see, e.g., Cavalli-Sforza and Bodmer, 1971, p. 327). Mean class values could be used to compute a Pearsonian correlation between parent and son ($r = +0.41$ in this case). The treatment of a continuous variate may seem less rigorous, but has the practical advantage of introducing a different conceptualization, which may be more satisfactory than that of a Markov chain. One may, in fact, suppose that the transmission from father to son is imperfect because of random noise, indicated by a variate ϵ with mean 0 and variance σ^2. Then X_{t+1} (occupation of son) is given by that of the father, X_t, as follows:

$$X_{t+1} = X_t + \epsilon. \tag{4.3.1}$$

Under these conditions, as we shall see in Chapter 5, the variance of X would be expected to grow indefinitely over time. In fact, it seems instead to reach an equilibrium. This may well be the consequence of two types of factors: (i) the contribution to the transmission by influences other than from the father, in particular the social environment, which may tend to restrict the choice of occupations; (ii) even more important, the limitations in the availability of jobs can stabilize the variance of the distribution of occupational statistics very rapidly. In principle, the introduction of more complex transmission and selection is possible also in multistate models, but it is simpler in the continuous models which we develop in the next chapter.

When more than one parent contributes to transmission, the transmission matrix is much more complicated. We have seen, in the models of the last section, the biparental transmission of a k-state trait is specified by a transmission matrix of size $k^2 \times k$ and the mating rule; when there is no effect of parental sex, the size of the matrix can be reduced to $k(k + 1)/2 \times k$. These matrices are big, and even with relatively few states many elements of observed matrices may be zero. This is, for instance, the case in the four-state transmission of religious preferences in Table 4.3.2. The religions are P (Protestant), C (Catholic), J (Jewish), O (Other). The number of zero entries is increased here by very high assortative mating (to be analyzed in the next section). It may seem somewhat discouraging that none of the

TABLE 4.3.2. Transmission of religious preferences among Stanford students. C = Catholic, P = Protestant, J = Jewish, O = Other.

	Students				
Father × Mother	C	J	P	O	Total
C × C	22	0	0	5	27
C × J	0	0	0	0	0
C × P	0	0	0	0	0
C × O	0	0	1	1	2
J × C	1	0	0	0	1
J × J	0	20	0	5	25
J × P	0	0	1	0	1
J × O	0	0	0	1	1
P × C	1	0	0	0	1
P × J	0	0	0	1	1
P × P	1	0	42	24	67
P × O	0	0	1	9	10
O × C	4	1	0	1	6
O × J	0	0	0	0	0
O × P	0	0	9	9	18
O × O	0	0	3	24	27
Totals	29	21	57	80	187

models developed in the earlier section seems at first sight to be useful in this example. However, we may well have here the superposition of vertical and horizontal transmission, the former determining a choice between the religions of father and mother when these are different (as in model 4.2d), and the latter introducing the chance of conversion to another religion, which would involve the relative frequencies of the religious affiliations of the proselitizers and of those to be proselitized, as well as the rate of proselitization.

Obviously the similarity of children to their parents is great, and as discussed in section 2.5 the role of assortative mating may be extraordinarily important here. In comparing these data to the models, we note the peculiar role played by the "other" category. Most departures from parental religions are "other." However,

TABLE 4.3.3. Transmission of religious preference in homotypic matings (from Table 4.3.2). Numbers in parentheses = expected numbers if no parental transmission.

	Students		
Matings	Like Parents	Unlike Parents	Total
C × C	22 (4.19)	5 (22.81)	27
J × J	20 (2.81)	5 (22.19)	25
P × P	42 (20.42)	25 (46.58)	67
O × O	24 (14.79)	3 (12.21)	27
Totals	108	38	146

For absence of transmission: $\chi^2(3) = 253.6$, $P < 0.1\%$.

the departure of offspring from parental religion seems highest when parents are Protestant.

Tables 4.3.3 and 4.3.4 illustrate ways in which difficulties for statistical analysis generated by the large number of zeros might be circumvented, while retaining some of the key features of the data. Thus, Table 4.3.3 shows the relative frequency of departures among offspring of homotypic matings from the parental type. The test for transmission is given by a comparison of observed with expected if there were no transmission, that is, if children of all matings were *C*, *P*, *J* or *O*, according to the relative

TABLE 4.3.4. Transmission of religious preference in heterotypic matings (from Table 4.3.2).

	Observed	Expected if no Vertical Transmission
Like father	11	14.57
Like mother	27	12.71
Unlike parents	3	13.72
Totals	41	41.00

Father vs. mother: $\chi^2(1) = 6.74$, $P < 5\%$
For absence of transmission: $\chi^2(3) = 25.32$, $P < 0.1\%$.

TABLE 4.3.5. Transmission of movie preferences in Stanford students. ADV = adventure; LT = light; HVY = heavy movies.

Father × Mother	Children ADV	LT	HVY	Total
ADV × ADV	2	5	1	8
ADV × LT	5	11	2	18
ADV × HVY	6	16	9	31
LT × ADV	1	1	3	5
LT × LT	3	20	7	30
LT × HVY	5	15	9	29
HVY × ADV	0	4	3	7
HVY × LT	2	15	1	18
HVY × HVY	8	23	19	50
Totals	32	110	54	196

TABLE 4.3.6. Homotypic matings for movie preferences (from Table 4.3.5). Numbers in parentheses equal numbers expected if there is no vertical transmission.

Father × Mother	Like parent	Unlike parent	Total
ADV × ADV	2 (1.31)	6 (6.69)	8
LT × LT	20 (16.84)	10 (13.16)	30
HVY × HVY	19 (13.78)	31 (36.22)	50
Totals	41	47	88

For absence of transmission: $\chi^2(3) = 4.52$, $P > 5\%$.

TABLE 4.3.7. Heterotypic matings for movie preferences (from Table 4.3.5). Observed and expected in the absence of vertical transmission.

	Observed	Expected
Like father	31	33.97
Like mother	45	38.69
Unlike parents	32	35.34
Totals	108	108.00

For absence of transmission: $\chi^2\ (2) = 0.32$, $P > 80\%$
Father vs. mother: $\chi^2\ (1) = 2.58$, $P > 5\%$.

frequencies among all children. The enormous value of the χ^2 testing this hypothesis confirms earlier conclusions in the binary analysis. Again, Table 4.3.4 summarizes heterotypic matings and illustrates the importance of vertical transmission, and also of the maternal effect, in agreement with the results obtained earlier using the binary approach.

Another example of multistate transmission is for movie preferences (Table 4.3.5). Here $k = 3$. ADV indicates movies (western, thrillers), LT indicates light (comedies, musicals), and HVY indicates heavy (dramas, serious documentaries, foreign films). Here again there may be a certain amount of assortative mating, to be examined in the next section. For a study of transmission independent of assortative mating, we can compare, as before, homotypic and heterotypic matings, obtaining Tables 4.3.6 and 4.3.7. There is no significant evidence in favor of transmission, in agreement with binary data.

4.4 ASSORTATIVE MATING

Assortative mating may cause important changes in the dynamics of traits under biparental transmission. It is inevitably more complex to study with multiple state traits, although some simple results can be obtained under special conditions. As with the binary analysis, the major problem is that the correlation between spouses may change over time if there is transmission between spouses, or if there are other causes of change (e.g., divorce might have a different chance of occurring as a function of dissimilarity between individuals). Longitudinal studies would be required for clarification of these issues. If similarity between parents changes during life, one would like to evaluate the correlations between spouses at the time they are likely to have an influence on the child for the trait in question. But this is probably not an instantaneous event, and effective periods of

TABLE 4.4.1. Assortative mating for religious preferences among parents of Stanford students. RM = random mating, AM = assortative mating, C = Catholic, P = Protestant, J = Jewish, O = Other.

Father × Mother	Total	RM	AM
C × C	27	5.48	23.65
C × J	0	4.62	1.46
C × P	0	14.12	4.45
C × O	2	7.79	2.46
J × C	1	4.62	1.46
J × J	25	3.90	19.73
J × P	1	11.91	3.78
J × O	1	6.57	2.07
P × C	1	14.12	4.45
P × J	1	11.91	3.76
P × P	67	36.40	67.98
P × O	10	20.08	6.33
O × C	6	7.79	2.46
O × J	0	6.57	2.07
O × P	18	20.08	6.33
O × O	27	11.07	34.66
Total	187	187.00	187.04
χ^2:		260.53	49.83
d.f.:		12	11
P:		0.001	0.001

influence are not known with much precision. Non-constancy of the spouse correlations makes the task of evaluating the effect on transmission of the similarity between parents more arduous.

From inspection of Table 4.3.2 there is very strong assortative mating for religion. Ways of testing and estimating it are indicated in Table 4.4.1. Here expected frequencies under random mating are computed on the basis of the relative frequencies, p_i, of the four denominations among spouses (C, J, P, O)

$$p_C = 64/374 = 0.1711$$
$$p_J = 54/374 = 0.1444$$
$$p_P = 165/374 = 0.4412$$
$$p_O = 91/374 = 0.2433.$$

The deviation from random mating expectation (labelled *RM* in Table 4.4.1) is extremely high. One can fit a simple hypothesis of assortment assuming that a proportion m of matings occurs assortatively, with $1 - m$ at random. The maximum likelihood estimate for m is obtained by equating to $1 - m$ the ratio of the observed frequency of heterotypic pairs ($41/187 = 0.2193$) to that expected under random mating, which is 0.6960. The estimate of m is thus 0.685. Alternatively, m could also have been estimated as the square root of χ^2 (computed in the classical way, divided by the total number of observations and by the number of states minus one; see, e.g., Li (1975)). To test this simple assortment hypothesis, the homotypic mating of the i-state has expected frequency $mp_i + (1 - m)p_i^2$, and that of the heterotypic mating of states i and j is $2(1 - m)p_ip_j$. These expected numbers are given in the last column of table 4.4.1. Although there is considerable improvement in the fit compared with that of the random mating hypothesis, the fit is still unsatisfactory. Even

TABLE 4.4.2. Movie preferences among parents of Stanford students (see also Table 4.3.5). ADV = adventure, LT = light, HVY = heavy, RM = random mating, AM = assortative mating.

	Child					
Father × Mother	ADV	LT	HVY	Total	RM	AM
ADV × ADV	2	5	1	8	5.82	10.61
ADV × LT	5	11	2	18	19.19	16.38
ADV × HVY	6	16	9	31	31.99	27.30
LT × ADV	1	1	3	5	6.53	5.57
LT × LT	3	20	7	30	21.55	27.92
LT × HVY	5	15	9	29	35.92	30.66
HVY × ADV	0	4	3	7	7.65	6.53
HVY × LT	2	15	1	18	25.26	21.56
HVY × HVY	8	23	19	50	42.09	49.48
Total	32	110	54	196	196.00	196.01
χ^2:					9.56	2.54
d.f.:					4	3
					P < 5%	P > 5%

taking account of differences in frequencies between the sexes, the fit does not improve greatly. The cause of the discrepancy is to be found in a different rate of assortment for different denominations. In particular, the observed homotypic matings are less numerous than expected, with less assortment for "other," and more for Jews and Catholics. Therefore more complex models of assortative mating have to be sought.

The other example used in the previous section, that of movie preferences, might be expected to show a substantial similarity between spouses, given the similarity found among friends, and there is ample time in over twenty years of married common life for convergence of interests. Table 4.4.2 shows the relevant data and expectations. Here, random mating frequencies have to be obtained with the use of different preference frequencies for the two sexes, because there is a substantial sex difference among parents.

		fathers	mothers
preference for			
adventure movies	(*a*)	29.08%	10.20%
light movies	(*l*)	32.65%	33.67%
heavy movies	(*h*)	38.27%	56.12%

This difference is significant ($\chi^2 = 24.44$ with 3 d.f., $P < 1\%$), and in the expected direction, given the traditional roles of men and women. Interestingly enough, the difference practically disappeared among the students.

The test of random mating indicates a significant departure at the 5% level, and thus we proceed to fit a hypothesis of assortative mating. The evaluation of the expected frequencies under assortative mating presents, however, a problem, because of the sex difference in frequencies. It is possible to adapt the assortment model to take account of sex differences by assuming that among the m assorted pairs, in a fraction f, the male partner influences the choice of the spouse, while in a fraction $(1 - f)$ the situation of the sexes is reversed. In the absence of this information we might assume that males and females choose equally frequently $(f = 1/2)$. Then the following model might be appropriate:

$$\begin{array}{ll} A \times A & (1 - m)aa' + m(a + a')/2 \\ H \times H & (1 - m)hh' + m(h + h')/2 \\ L \times L & (1 - m)ll' + m(l + l')/2 \end{array} \tag{4.4.1}$$

where a, a' h, h', l, l' are the frequencies of the three movie preferences among males and females respectively $(a + h + l = a' + h' + l' = 1)$ given above. Reciprocal heterotypic matings are handled separately:

$$\begin{array}{ll} \female \quad \male & \\ A \times H & (1 - m)ah' \\ H \times A & (1 - m)a'h \\ \text{and so on} & \end{array} \tag{4.4.2}$$

Using the estimates of a, h, l and a', h', l' obtained above, an estimate of m can be obtained from the total number of heterotypic matings observed in Table 4.4.2, namely 108. The expected total of these matings is $196(1 - m)(h'a + ah' + a'l + l'a + h'l + hl')$. Setting the total of the expected heterotypic matings equal to the observed total number, 108, and solving for m we obtain $m = 0.147$. The expectations using this m are given in the last column of Table 4.4.2. The resulting fit is good. m is significantly different from zero, as indicated by the improvemnt of χ^2. In fact, the difference between the χ^2 for random and that for assortative mating is itself approximately χ^2 (here $\chi^2 = 7.02$ with 1 d.f., $P < 1\%$).

As we have seen with two-valued traits, assortative mating affects the equilibrium to which the dynamic system is expected (theoretically) to tend. Consider again as an example the multiplicative model of transmission. As was shown earlier, the frequencies u_1, u_2, u_3, of H_1, H_2 and H_3 can be summarized by the single variable $z_i = \tau u_1 + \tau_2 u_2 + \tau_3 u_3$. With assortative mating such economy is no longer possible; a second variable

$$w = \tau_1^2 u_1 + \tau_2^2 u_2 + \tau_3^2 u_3 \tag{4.4.3}$$

must be introduced. Then we have

$$u' = (1 - m)z^2 + mw \tag{4.4.4}$$

for example. The recursion in z and w is

$$\begin{aligned} z' &= z^2(1 - m)(\tau_1 - 2_2 + \tau_3) + 2z(\tau_2 - \tau_3) + \tau_3 \\ &\qquad + mw(\tau_1 - 2\tau_2 + \tau_3) \\ &= Q_1(z) + mw(\tau_1 - 2\tau_2 + \tau_3), \end{aligned} \tag{4.4.5}$$

say, and

$$\begin{aligned} w' &= z^2(1 - m)(\tau_1^2 - 2\tau_2^2 + \tau_3^2) + 2z(\tau_2^2 - \tau_3^2) \\ &\qquad + \tau_3^2 + mw(\tau_1^2 - 2\tau_2^2 + \tau_3^2) \\ &= Q_2(z) + mw(\tau_1^2 - 2\tau_2^2 + \tau_3^2), \end{aligned} \tag{4.4.6}$$

say. Substituting from (4.4.6)

$$w = Q_2(z)/[1 - m(\tau_1^2 - 2\tau_2^2 + \tau_3^2)]$$

into (4.4.5) the equilibrium quadratic is

$$Q_1(z) - z + \frac{m(\tau_1 - 2\tau_2 + \tau_3)Q_2(z)}{1 - m(\tau_1^2 - 2\tau_2^2 + \tau_3^2)} = 0. \tag{4.4.7}$$

Again, for most parameter sets (m, τ_1, τ_2, τ_3) there is convergence to the single valid root of (4.4.7), although parameter ranges exist that allow limit cycles to be established.

In conclusion we note that, although the type of assortative mating considered here affects the equilibrium value, it does not affect the degree of the equilibrium polynomial in the models treated here. Assortative mating can be detected and estimated for some cultural traits but may be confounded with coincidental changes in the parents during married life. Only longitudinal studies would separate them. Models can be constructed that allow for assortative mating when the trait distribution differs by the sex of the parent. It might be expected that in many cases a single parameter will be insufficient to explain the observations. The mating pattern must be considered in conjunction with the vertical transmission process in comparing observations with transmission models.

4.5 HORIZONTAL AND OBLIQUE TRANSMISSION

As seen in the previous section religion has a strong vertical component in its transmission. But the history of religion abounds with examples of horizontal or oblique transmission that ordinarily have been called "conversion." Now, both horizontal and oblique transmission of multistate and dichotomous traits differ in that there is only a single parameter necessary to describe the latter. Thus if the frequencies of H and h are u and v ($u + v = 1$) and h is converted to H by oblique transmission only, with the effective contact rate f_1, while H is converted to h with effective contact rate f_2, then after conversion

$$u' = u + f_1uv - f_2uv. \qquad (4.5.1)$$

Obviously the simple coefficient $f_1 - f_2$ is sufficient to describe the process and, as mentioned in Chapter 3, $u_t \longrightarrow 1$ or 0 according to whether $f_1 > f_2$ or $f_1 < f_2$ respectively.

With k states for the trait the effective contact rates can be represented by the off diagonal elements of a $k \times k$ matrix that might not be symmetric, in general. Thus, for such a k-state system with oblique or horizontal transmission $k(k - 1)$ transmission parameters are required and might be represented as in the matrix **T** below

$$\mathbf{T} = \begin{pmatrix} 0 & f_{12} & f_{13} & \cdot & \cdot & \cdot & f_{1k} \\ f_{21} & 0 & f_{23} & \cdot & \cdot & \cdot & f_{2k} \\ \cdot & & & & & & \cdot \\ \cdot & & & & & & \cdot \\ \cdot & & & & & & \cdot \\ f_{k1} & & & & & f_{k,k-1} & 0 \end{pmatrix}. \tag{4.5.2}$$

Note that the rows of this matrix would not ordinarily sum to unity. The simplest resulting recursion system for oblique transmission using **T** is exemplified by

$$\begin{aligned} u_1' &= u_1 + f_{12}u_1u_2 + \cdot\cdot\cdot + f_{1k}u_1u_k - f_{21}u_1u_2 \cdot\cdot\cdot - f_{k1}u_1u_k \\ &= u_1\left[1 + \sum_{j=1}^{k} (f_{1j} - f_{j1})u_j\right]. \end{aligned} \tag{4.5.3}$$

In general, with the convention that $f_{ii} = 0$,

$$u_i' = u_i\left[1 + \sum_{j=1}^{k} (f_{ij} - f_{ji})u_j\right], \tag{4.5.4}$$

With the recursion system (4.5.4) the assumption is that vertical transmission leaves the trait frequencies unchanged; the only change is due to contact with members of another state. Systems like (4.5.4) have been studied in mathematical ecology and exhibit certain peculiarities. For example, any interior ($u_1 \neq 0$, all i) equilibria must satisfy the system of linear equations

$$\sum_{j=1}^{k} (f_{ij} - f_{ji})u_j = 0 \quad (i = 1, 2, \ldots, k). \tag{4.5.5}$$

Only if the number of states k is odd can these equations have an isolated solution. The reason is that, since $\Sigma_{j=1}^{k} u_j = 1$, (4.5.4) in fact represents $k - 1$ simultaneous linear equations the matrix of which is easily seen to be skew-symmetric. If k is odd then the dimension of the matrix is even and a solution to the linear equations is possible. But if k is even the skew-symmetric matrix is of odd dimension and is therefore singular, admitting no isolated solution.

Even when a solution exists it is not stable. To see this, consider a 3-state example with three-dimensional matrix **T** as in (4.5.2). The possible equilibria are as follows.

i) Pure fixation states, for example, $u_3 = 1$, $u_1 = u_2 = 0$ is stable if $f_{31} > f_{13}$ and $f_{32} > f_{23}$. In general, fixation in $u_k = 1$ requires $f_{kj} > f_{jk}$ for all j for stability. No equilibria are possible with two states present.

ii) An interior equilibrium

$$\hat{u}_1 = (f_{32} - f_{23})/N, \hat{u}_2 = (f_{31} - f_{13})/N, \hat{u}_3 = (f_{21} - f_{12})/N, \quad (4.5.6)$$

with $N = f_{32} + f_{31} + f_{21} - f_{23} - f_{13} - f_{12}$, exists if either

$$f_{21} > f_{12}, f_{31} > f_{13}, f_{32} > f_{23}$$

or

$$f_{21} < f_{12}, f_{31} < f_{13}, f_{32} < f_{23}. \quad (4.5.7)$$

If either condition of (4.5.7) holds, no pure fixation state is stable. In fact (4.5.6) is itself not stable. The frequencies cycle out from it as shown in Figure 4.5.1. Similar cyclic behavior is exhibited by all models of this kind with an odd number of states. With an even number of states, fixation in a pure state appears to be the rule.

The anomalies that arise in this class of model do not occur if, for example, linear vertical transmission occurs prior to horizontal:

$$u_i' \propto r_i u_i + \sum_{j=1}^{k} \alpha_{ij} r_i r_j u_i u_j, \quad (4.5.8)$$

with the added requirement that $\Sigma_i u_i' = 1$. If all α_{ij} were zero this

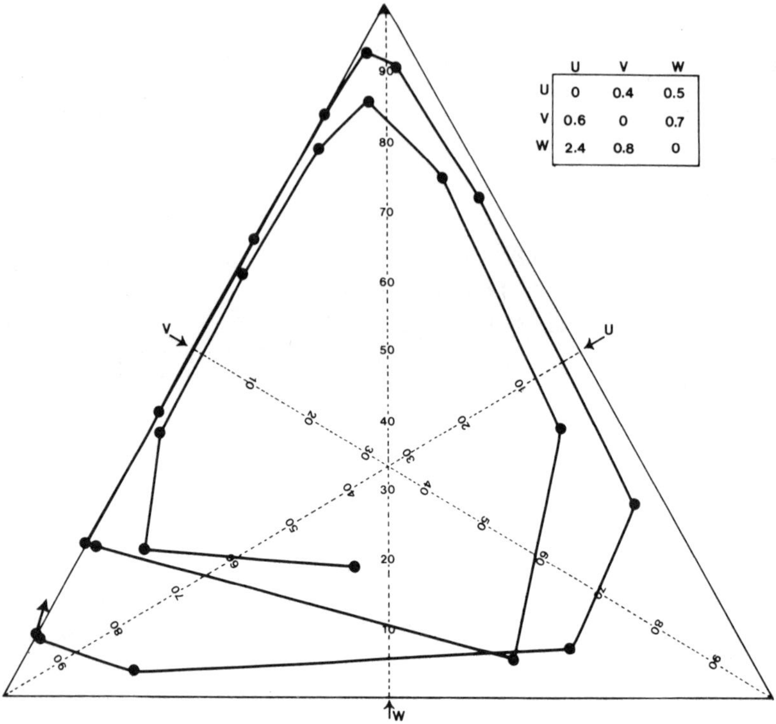

FIGURE 4.5.1. Baricentric coordinate representation of horizontal and oblique transmission for a 3-state trait (after vertical transmission, leaving trait frequencies unchanged). The transmission matrix is indicated at the right. Initial values: $u_0 = 0.45$, $v_0 = 0.36$, $w_0 = 0.19$. States of population represented only every ten generations.

would produce the linear fractional recursion

$$u_i' = \frac{r_i u_i}{\Sigma\, r_i u_i}, \tag{4.5.9}$$

which, while it does result in fixation of that state with the highest r value, is very well behaved. An appropriately normalized version of (4.5.8) would be expected to allow stable polymorphic equilibria without cycling for at least some classes of r_i and α_{ij}. In effect, the modification (4.5.8) introduces selection or competition between the states.

The sequence in which the various modes of transmission operate is clearly important in determining the dynamics of evolution. In section 5.10 it is pointed out that the combined effect of teachers and parents depends on who acts first.

4.6 THE EVOLUTION OF SURNAMES: AN EXAMPLE OF DRIFT IN MULTISTATE CULTURAL TRANSMISSION

Early in the development of modern statistics, workers were attracted to the numerical properties of surnames of populations. In particular, Galton and Watson originated the theory of branching processes in their probabilistic studies on the extinction properties of surnames. In most cultures surnames are transmitted only by fathers; women have the father's family name but cannot pass it on. Here quantitative modeling will be restricted to males. A surname may be carried by n individuals in a population, each of whom may have $0, 1, 2, \ldots, k, \ldots$ sons (according to some probability rule) who will inherit their father's name. If all n individuals at a given time have no sons then the surname will become extinct. Clearly, the probability of extinction and the time to extinction will depend on the probability distribution of families with $0, 1, 2, \ldots, k, \ldots$ sons. Watson and Galton (1874) proposed some general theory concerning extinction probabilities for surnames. A. J. Lotka (1931) applied the Galton-Watson theory to American fertility data to determine the probability of extinction of a male line of descent, and showed that they are reasonably well described by an "augmented geometric distribution":

$$P_k = bc^{k-1} \quad \text{with} \quad k = 1, 2, \ldots$$
$$p_0 = 1 - b/(1 - c), \tag{4.6.1}$$

where $b, c > 0$, $b + c < 1$. The following discussion is based on the treatment by Karlin and McGregor (1967). We begin by computing the probability generating function

$$\phi(s) = \sum_{k=0}^{\infty} p_k s^k \tag{4.6.2}$$

of the distribution (4.6.1). For this distribution we have

$$\phi(s) = \frac{1 - (b + c)}{1 - c} + \frac{bs}{1 - cs}. \tag{4.6.3}$$

Suppose that at generation zero there are X_0 individuals with a given surname, X_1 in generation one, X_2 in generation two, and so on. Then define

$$\phi_n(s) = \sum_{k=0}^{\infty} \{\text{Probability } X_n = k\} s^k. \tag{4.6.4}$$

If only one individual carries the surname in question in the original generation then $\phi_0(s) = s$, $\phi_1(s) = \phi(s)$ as in (4.6.3). In general the probability generating function (p.g.f) in the nth generation can be defined in terms of the previous by

$$\phi_n(s) = \phi(\phi_{n-1}(s)) = \phi_{n-1}(\phi(s)). \tag{4.6.5}$$

When $\phi(s)$ has the linear fractional form (4.6.3), $\phi_n(s)$ can be obtained explicitly:

$$\phi_n(s) = 1 - \frac{m^n(1 - s_0)}{m^n - s_0} + \frac{m^n[(1 - s_0)/(m^n - s_0)]^2 s}{1 - [(m^n - 1)/(m^n - s_0)]s}, \tag{4.6.6}$$

where $s_0 = [1 - (b + c)]/c(1 - c)$ and $m = b/(1 - c)^2$, provided $m \neq 1$. Here m is the mean of the distribution (4.6.3), that is, the mean number of sons per family. If the population remains constant in size there will be on average one son per family (and one daughter). If $m > 1$ we expect the population to increase while if $m < 1$ it should decrease.

The probability that a given surname will be extinct at the nth generation is

$$\text{Prob}\,\{X_n = 0\} = \phi_n(0) = 1 - \frac{m^n(1 - s_0)}{m^n - s_0}, \, m \neq 1. \quad (4.6.7)$$

If $m < 1$ then as $n \longrightarrow \infty$ all surnames become extinct. If $m > 1$ then the probability of extinction tends to the limit s_0 as $n \longrightarrow \infty$. If $m = 1$ so that $b = (1 - c)^2$, then c is the probability p_0 of extinction in one generation. In this case the probability of extinction at the nth generation is

$$\phi_n(0) = nc/[1 + (n - 1)c]. \quad (4.6.8)$$

Define T as the random variable indicating the first generation at which the surname goes extinct. Then if $m > 1$

$$P[T = n] = m^{n-1}s_0 \frac{(m - 1)(1 - s_0)}{(m^n - s_0)(m^{n-1} - s_0)},$$

$$n = 1, 2, \ldots, \quad (4.6.9)$$

while if $m = 1$

$$P[T = n] = \frac{c(1 - c)}{[1 + (n - 1)c][1 + (n - 2)c]}. \quad (4.6.10)$$

The above formulae are valid if in the original generation there is a single individual with the surname under investigation. If there were m individuals with the surname at the original generation, and each reproduced independently of the others, then the probability of extinction at the nth generation is $[\phi_n(0)]^m$.

Figure 4.6.1, taken from Yasuda et al. (1974, p. 137), shows the probabilities of extinction of surnames represented 300 years before, in the Parma Valley of Italy, by 1, 2, 3, and so on individuals. There is rather good agreement with expectation computed either by assuming $m = 1$ or by taking a value of m estimated from the data, which is slightly higher than 1.

The above-mentioned data from the Parma Valley of Italy are

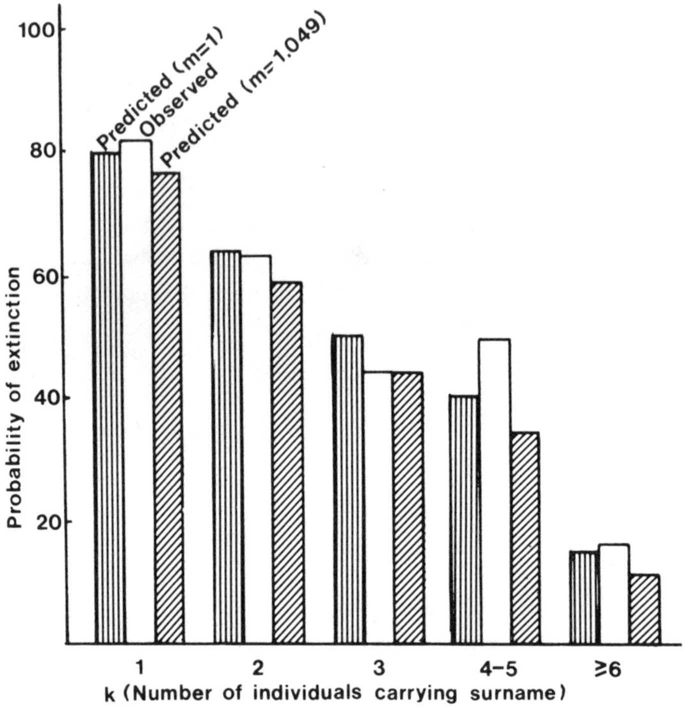

FIGURE 4.6.1. Probability of extinction after ten generations of a surname originally carrried by k individuals (shaded bars) and its comparison with the observed frequency of extinction (white bar, center). The probabilities are computed assuming the population remains constant in size ($m = 1$, bar on left) or increases at the rate observed ($m = 1.049$, bar on right). The analysis is by Yasuda et al. (1974) from data of the upper Parma valley.

from parish records, which provide demographic information for a period of about 350 years. From it the distribution of surnames, that is, the number of individuals carrying a given surname in a population, can be found. The rules of inheritance of surnames are strict and new surnames rarely arise. In particular in the period and place under study a new surname would arise only in illegitimate children (who took the name of the mother or, if this were not known, a name indicating illegitimacy), from a misspelling not corrected in later transcriptions, or from a translation from Latin or a dialect into Italian. Each of these events can be considered as a mutation, but they are all rare. Immigration from other areas is another source of new surnames. Mutation and immigration can usually be pooled in the quantitative treatment of evolution, as we have already seen.

Surnames are a cultural character but their vertical transmission is almost as strict as that of a haploid genetic trait. If females are ignored, transition rules and evolutionary properties can be viewed as those of a haploid genetic system with a single sex. The great number of surnames in existence suggests the analogy between surnames and a single gene with very many alleles. The distribution of the number of alleles at a gene locus may therefore be used to describe the distribution of the number of surnames in a population. Theoretical treatments by Kimura and Crow (1964) and Ewens (1972) of the distribution of the number of alleles use limiting results that are not satisfactory for surnames. A more general treatment was given by Karlin and McGregor (1967) and includes the analyses of Kimura and Crow and Ewens as special cases. This treatment was found by Yasuda et al. to fit the surname data from Parma very well.

The Karlin-McGregor model assumes that there is a population of N individuals (males in the present case) of whom m_i carry the ith surname. There are r different surnames altogether. The vector

$$\mathbf{m} = (m_1, m_2, \ldots, m_i, \ldots, m_r) \text{ with } \sum_{i=1}^{r} m_i = N \quad (4.6.11)$$

represents the state of the population at a given point in time. Most conceivable surnames will not be represented in the population at any time, that is, r is much larger than N; many of the m_i values will be zero. Now a Poisson process of birth and death is assumed such that in a unit time (much smaller than a generation) a single randomly chosen individual dies and is replaced by another with a constant probability per unit time. If he is replaced by an individual of the same surname, **m** will not change. If he is replaced by an individual with a different surname, the frequency of some surname, say m_i, goes down by one and that of another, say m_j, goes up by one:

$$\mathbf{m}' = (m_1, m_2, \ldots, m_i - 1, \ldots, m_j + 1, \ldots m_r). \qquad (4.6.12)$$

Either way, the total population size stays constant at N.

The probability that at the birth-death event there is a change from a specific surname to another is β, and is assumed to be the same for all types. The probability that at the birth-death event there is no change of type is, therefore, $1 - (r - 1)\beta$. Consider for example the transition from **m** to **m**′. With probability m_i/N an individual of type i dies. With probability m_j/N an individual of type j is born, and with probability $1 - (r - 1)\beta$ he does not mutate. With probability $(N - m_j)/N$ a non-j-type is born and with probability β he mutates to j. The probability of the transition $\mathbf{m} \longrightarrow \mathbf{m}'$ is therefore

$$\frac{m_i}{N}\left[\frac{\beta(N - m_j)}{N} + (1 - (r - 1)\beta)\frac{m_j}{N}\right]$$
$$= \frac{m_i}{N^2}(1 - r\beta)\left(m_j + \frac{N\beta}{1 - r\beta}\right). \qquad (4.6.13)$$

The probability that **m** remains unchanged after the birth-death event in unit time, that is, $\mathbf{m}' = \mathbf{m}$ is then

$$1 - \left(\frac{1 - r\beta}{N^2}\right)\sum_i \sum_{\substack{j \\ i \neq j}} m_i\left(m_j + \frac{N\beta}{1 - r\beta}\right) \qquad (4.6.14)$$

Karlin and McGregor derived the equilibrium distribution of the number $n(k)$ of surnames for which $m_i = k$ for the above model as well as with more general fertility distributions. Under the assumption $r \longrightarrow \infty$, $r\beta \longrightarrow \mu$, the expected value of this random variable is

$$E[n(k)] = \frac{1}{k}\frac{N\mu}{1-\mu}\frac{\begin{pmatrix}\dfrac{N}{1-\mu} - k + 1\\ n-k\end{pmatrix}}{\begin{pmatrix}\dfrac{N}{1-\mu} - 1\\ N\end{pmatrix}} \tag{4.6.15}$$

The expected number of different surnames in the population of size N is

$$E[N_s] = \frac{N\mu}{1-\mu}\left\{\frac{\Gamma'\left(\dfrac{N}{1-\mu}\right)}{\Gamma\left(\dfrac{N}{1-\mu}\right)} - \frac{\Gamma'\left(\dfrac{N\mu}{1-\mu}\right)}{\Gamma\left(\dfrac{N\mu}{1-\mu}\right)}\right\}, \tag{4.6.16}$$

where $\Gamma'(\cdot)$ represents the derivative of the gamma function. Yasuda et al. (1974) tabulated $E[N_s]$ for N up to 100 and μ from 0.01 to 0.42.

A simpler though less efficient way of relating N and μ to the number of surnames in the population uses the probability of identity for the trait in question of two individuals taken at random from the population. These arguments also derive from genetics where, for a population of alleles, the ith of which has frequency x_i, the probability that two alleles taken at random are identical is Σx_i^2. This is also called, in genetics, the probability of homozygosity. (If the individuals are diploid, an homozygote is an individual having two identical genes or alleles.) We set $F = \Sigma x_i^2$ and again let μ be the mutation rate to completely novel alleles. At equilibrium the proportion of homozygotes Σx_i^2 has expectation

$$F = E(\Sigma x_i^2) = (N\mu + 1 - \mu)^{-1} \tag{4.6.17}$$

With surnames, (4.6.17) is the probability that two random individuals have identical surnames (i.e., the probability of isonymy). A familiar approximate equilibrium formula in genetics is $F = (1 + 4N\mu)^{-1}$. The factor 4 is due to the diploid population assumed (i.e., two genes per individual) and two sexes; with surnames there is only one gene per individual and only males are counted. The probability of isonymy is easily computed from observed data, and from it μ can be obtained if N is known, or vice versa.

Suppose there are n_s surnames in the population; then, in general

$$F = \Sigma x_i^2 = n_s[M_x^2 + V_x] \tag{4.6.18}$$

where $M_x = \Sigma x_i/N_s$ is the average frequency of a surname and V_x is the variance in the frequency of the surname. Now if all surnames are equally frequent, then $M_x = x = 1/n_s$, $V_x = 0$. Thus, in this case

$$1/F = 1/\Sigma x_i^2 = n_s. \tag{4.6.19}$$

Often the value $1/\Sigma x_i^2$ is written n_e and called "the effective number of alleles" (or surnames) in the population. But if $V_x \neq 0$, then

$$1/F = 1/\Sigma x_i^2 = n_s/[1 + n_s^2 V_x]. \tag{4.6.20}$$

Thus the estimate $n_e = 1/\Sigma x_i^2$ underestimates n_s.

It may seem that estimation of the true number of alleles or surnames that exist in a population should not be difficult because they can be counted. For alleles, however, the problem is a serious one. Considering the alleles at a single gene locus, the investigator is almost always confined to a sample of the population, and a small one at that. The number of alleles in the sample may well be smaller than in the whole population. The problem of inference from the number of alleles in the sample to that in the population has been investigated by Ewens (1972), whose methods can be related to those described above.

For surnames there may sometimes exist a similar problem of inference. But with small populations there may be nothing to prevent investigation of the whole population, in which case the inference problem does not arise. To link n_s with N and μ requires the use of (4.6.16), which is not easily handled numerically. At the cost of some efficiency, the relationship between N and μ can be tested using (4.6.17). This formula can also be employed with data from an adequate population sample when data from the whole population are not available.

Yasuda et al. (1974) applied the Karlin-McGregor model to the surname distributions from the Parma Valley. Two questions bearing on the validity of this model when applied to the present data should be addressed. One concerns the Poisson process for replacement of individuals, which generates an exponential distribution of lifetime length. This is obviously not correct in man, but the deviation may be of little consequence. The second problem concerns the assumption of one child per parent. This may be a reasonable approximation to reality in a traditional rural society in which, together with the surname, property was inherited undivided by the eldest son. Daughters inherited the surname but lost it at marriage. Sons other than the eldest had a smaller chance of establishing a family and continuing the surname, especially in the village of their birth. Thus the population size N basically refers to property holders and remains constant if the property is not divided. In the event that no son is alive to inherit the property, another relative or an unrelated man who marries a daughter takes the property. In terms of surnames, this is then equivalent to a mutation or immigration event.

Using the parish books of marriages from many parishes of the Parma Valley representing the last three centuries, Yasuda et al. estimated population size N by taking the number of marriages celebrated in the parish during a generation (30 years). Only one spouse's surname was counted (the bride, who was more likely to be born in the parish where the marriage is recorded). Parishes were selected which, over the period of generation, had N as close to 30 as possible; 27 such parishes were available. Altogether there were 339 different surnames in the 27 parishes, giving

$339/27 = 12.55$ as an estimate of the number of different surnames n_s for a village of $N = 30$. From (4.6.16), for $N = 30$ and $n_s = 12.55$, a mutation + immigration rate of $\mu = 20.2\%$ is obtained.

For comparison, the immigration rate of males (who are the only important individuals in surname transmission) in the same villages was estimated directly. This was done by calculating the fraction of children born in a parish whose fathers were also born in the same parish, and subtracting from unity. The result was an estimate of $17.9 \pm 1.9\%$ for the rate of immigration. This is reasonably close to the value $\mu = 20.2\%$ estimated from the theory. This value μ, however, includes a fraction due to mutation, the magnitude of which is not precisely known, but must be low. Quite plausibly it could be of the order of the difference between the two immigration rates cited. The fit of the expected surname frequency distribution to the observed one (4.6.15), with n_s as estimated, is shown in Figure 4.6.2 and is quite satisfactory.

One caveat is worth discussion. The theory above is an equilibrium theory and, as such, would require for application that there has been sufficient time for equilibrium to be attained. In situations in which frequent immigration is a recent occurrence, as for example in flourishing industrial areas, it may well be that equilibrium has not been attained. In such communities it might be expected that an excess of relatively rare surnames be found. The frequency of surnames represented by one or few individuals should be correspondingly low in these places. For the Karlin-McGregor theory the rate of approach to the equilibrium state has not been determined.

Another practical problem is that, in some countries, for example the United States, surnames have in recent history been subject to a high mutation rate. On the other hand, in cultures like the Spanish in which the mother's name is also transmitted, quite new theoretical developments for quantitative analysis of surnames would be required.

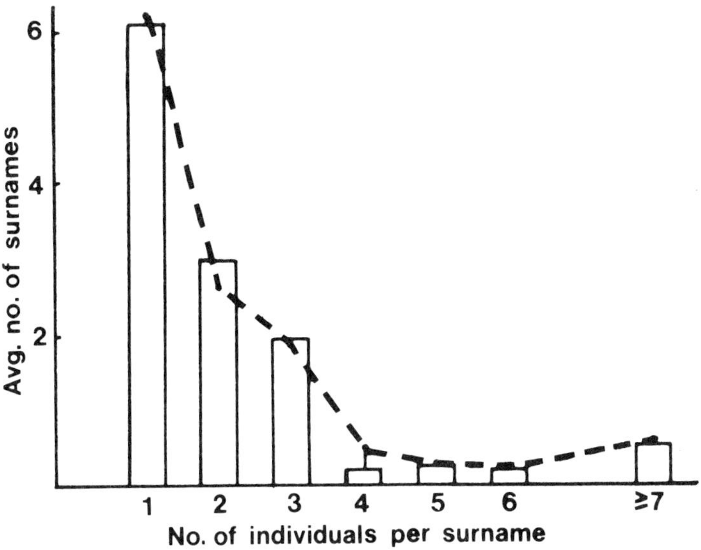

FIGURE 4.6.2. Distribution of the number of surnames by the number of individuals carrying them. The abscissa, k, is the number of individuals carrying the same surname in a parish of $N = 30$ individuals (approximately). (For estimation of the population size N, see text.) The number of surnames with k carriers is averaged over parishes. Bars are observed averages (over 27 parishes). The dotted line refers to the expected frequency (from the Karlin-McGregor theory) for a population of size $N = 30$ and $n_s = 12.55$ surnames per parish (averaged over parishes). The probability of isonymy here is $F = 0.1277$ and the effective number of surnames is 7.38 (analysis by Yasuda et al. (1974) of Parma Valley data).

In summary, in most western cultures, surnames are transmitted vertically only through males. There is a great variety of surnames, though some true recurrence (multiple independent origin) is unavoidable with certain surnames. The similarity of the transmission with that of genes in a haploid asexual organism allows transfer of a large body of theory originally developed for the latter to the study of surnames. The properties of the number of different surnames, and the frequency with which each can be found, depend on the population size N and μ, the rate of mutation or immigration (more generally their sum). From this theory the probability of homozygosity of genes can be identified with the probability of isonymy for surnames. From the theory, and an observed number of surnames, N can be estimated given μ or μ can be estimated given N. Although their application is limited to males, potentially the properties of surnames can help in understanding genetic evolution.

CHAPTER FIVE

Cultural Transmission For a Continuous Trait

5.1 HISTORICAL CONSIDERATIONS ON "BLENDING" INHERITANCE

Many cultural traits can be scored naturally on a continuous scale. The theoretical treatment of their evolution differs from that which we have seen so far for discontinuous traits. Theories about continuous and discontinuous traits are, of course, to some extent interconvertible, but some conceptual and technical differences are inevitable. We will follow the same order of consideration as was used for discontinuous traits, and discuss first vertical, then oblique and horizontal transmission, paying special attention also to the oblique or horizontal transmission of the teacher-leader type. This will be followed by a consideration of natural selection as well as correlations between relatives. Attention will be paid to the consequences of social stratification, such as occurs in class and caste partitioned societies. We will also consider cultural drift and those aspects of the balance between cultural mutation, cultural drift, and cultural selection peculiar to continuous traits. In some respects continuous traits are easier to analyze than discontinuous traits; in others they are more difficult. Frequency distributions must be used, and these often will involve at least two parameters, for example mean and variance. When it is possible to assume that the underlying trait frequency distribution is normal, the treatment simplifies considerably: we will usually assume that the trait under study is normally distributed. A reasonably comprehensive analysis of a specific problem, however, may demand a multivariate treatment—it is not always easy or possible to transform a vector of variables into approximate multivariate normality, but multivariate normality may be a reasonable assumption, at least asymptotically, in some cases.

In the study of biological transmission it is customary to use linear correlation coefficients between relatives as a measure of similarity. In fact, one of the stimuli to the invention of correlation and regression coefficients was the measurement and description of inheritance. It may be appropriate to recall here a few historical facts that may not be widely known outside the field of statistical genetics.

Francis Galton collected measurements and observations of a number of human families, with respect to traits ranging from stature to eye color, artistic tendencies, temper, and disease. He believed that for many traits inheritance was of the "blending" type, and gave skin color as an example ("the children of the white and the negro are of a blended tint; they are neither wholly white or wholly black, neither are they piebald, but of a fairly uniform mulatto brown. The quadroon child of the mulatto and the white has a quarter tint, etc." (F. Galton, *Natural Inheritance,* 1889, pg. 12). Galton realized that the inheritance of some traits could be explained by "particulate" inheritance, but he was unaware of Mendel's particulate theory, published in 1865 Galton had noted that some traits are inherited in a mutually exclusive way. That is, one person inherits from one progenitor and not others, and he cited eye color as an example. Although ignorant of Mendelian laws of inheritance he was thus nevertheless fairly close to understanding, or at least describing, phenomena like segregation, dominance and recessiveness, and even mutation. With reference to the latter he discusses "sports," (that is, new variants, observed and propagated by plant and animal breeders) and their stability in transmission. It is of interest for the purposes of this book that, in explaining why a sport may be found unchanged generation after generation, Galton chose to make recourse to an example of a "cultural" sport: "The hansom cab was originally a marvelous novelty. In the language of breeders, it was a sudden and remarkable 'sport,' yet the suddenness of its appearance has been no bar to its unchanging hold on popular favour." He goes on to cite other examples of innovations, for example, an early electrical battery invented by Sir William Grove that continued to be in use for a long time without alteration. But in spite of several citations of examples of "particulate" inheritance, his interest remained centered on traits that exhibited the inheritance he termed "blending." Skin color, which he cited specifically, is today considered to be due to at least three or four genes, each contributing to the determination of total pigmentation. Thus the "blending" behavior can be explained in terms of Mendelian—that is, particulate—inheritance. Galton was aware of this possibility: "Skin color is therefore a good example of what I call blended inheritance. It may be nonetheless 'particulate' in its origins, but the result may be regarded as a fine mosaic too minute for its elements to be distinguished in a general view." This statement remains valid, and is probably also true for a continuous trait most extensively investigated by Galton, namely stature. For the purposes of studying inheritance, Galton developed what are today called tables of correlation between the average stature of two parents (midparent) and the stature of their child, and showed the "law of regression" towards the population mean. To this end he first eliminated sex differences by showing that the female distribution could be superimposed on the male if individual female stature were multiplied by 1.08. He estimated the regression between midparent and child to be about 2/3 and thought that this was also the theoretically expected value. He showed that the mean and variance of adult stature and other traits in two successive generations remain the same, and partitioned the total variance into two portions, one being the variance between means of sibships (that is, groups formed by brothers and sisters or "fraternities"), the other, the variance within sibships. He estimated the former to be 2/3 of the total variance and used this as one of the estimates of the regression between sibs. His general equation for kinship was written

$$p^2 w^2 + f^2 = p^2$$

where p^2 is the total variance, w is the regression between specific types of relatives, and f^2 the variance within these same relatives. In his 1889 book, Galton did not measure the father-child regression, but hypothesized that it should be 1/2 of the 2/3 estimated for the midparent-child regression, that is, 1/3. He also predicted that the similarity of grandparent to grandchild should be $(1/3)^2$, and that of uncles to nephews (or vice versa) should be $2/3 \times 1/3 = 2/9$, etc. He measured assortative mating and found it negligible; he speculated that much of natural selection would be of the stabilizing kind, destroying those individuals far from some optimum while retaining those close to it, and that this type of selection would reduce the variance of traits. In his analysis Galton evaluated the fraction of the total variance explained by regression, $r^2\sigma^2(w^2p^2$ in his terms) and that not explained by it $(1 - r^2)\sigma^2$. In situations such as those he was investigating, the variances of the two variables x and y (first and second sib for example) were the same. In such cases the regression and correlation coefficients are identical, so that his formulae are indeed valid. The linear correlation coefficient, in modern form:

$$r = \frac{\sum_{i=1}^{N} (x_i - \bar{x})(y_i - \bar{y})}{\left[\sum_{i=1}^{N} (x_i - \bar{x})^2 \sum_{i=1}^{N} (y_i - \bar{y})^2\right]^{1/2}}$$

was introduced by K. Pearson not long afterwards (1896), along with a rigorous treatment of correlation surfaces and their application to inheritance and evolutionary problems.

The blending theory can be set in modern terms by assuming that only parents (not grandparents or earlier ancestors) contribute directly to the makeup of the child. The theory then takes the following form:

The value X_{t+1} of a trait in the offspring is given by the arithmetic mean of the parental values, X_t and X'_t, plus an "error" term. The last is a random variate ϵ with mean zero and constant variance σ^2, which is usually taken to be normally distributed. The notation $\epsilon \sim N(0, \sigma^2)$ is commonly used. The Galton-Pearson blending model would nowadays be expressed as

$$X_{t+1} = \frac{X_t + X'_t}{2} + \epsilon. \qquad (5.1.1)$$

The error term may be viewed as random phenotypic "mutation," and σ^2 as "mutation variance."

With this model (5.1.1), under the usual assumption that the expected phenotypic value remains constant, it is easy to see that under random mating (i.e. X_t and X'_t independent)

$$V_{t+1} = E[(X_{t+1} - EX_{t+1})^2] = (V_t + V'_t)/4 + \sigma^2. \quad (5.1.2)$$

If the variances of paternal and maternal values are the same, $V_t = V'_t$, then

$$V_{t+1} = V_t/2 + \sigma^2, \qquad (5.1.3)$$

and at equilibrium,

$$\hat{V} = 2\sigma^2. \qquad (5.1.4)$$

Again assuming random mating and zero mean, the covariance between parent and offspring W_{PO} is easily obtained:

$$W_{PO} = E(X_t X_{t+1}) = V_t/2, \qquad (5.1.5)$$

and at equilibrium

$$W_{PO} = \hat{V}/2 = \sigma^2.$$

Thus the equilibrium parent-offspring correlation is

$$\hat{r}_{PO} = \hat{W}_{PO}/\hat{V} = 1/2,$$

rather than 1/3 as originally stated by Galton. Under the same assumptions, the midparent offspring correlation is

$$r_{PO} = \frac{E\left(\dfrac{X_t + X'_t}{2} \cdot X_{t+1}\right)}{\left[E\left(\dfrac{X_t + X'_t}{2}\right)^2 E(X_{t+1})^2\right]^{1/2}} = \frac{W_{PO}}{(V_t V_{t+1}/2)^{1/2}}, \quad (5.1.6)$$

and at equilibrium

$$\hat{r}_{PO} = \frac{\hat{W}_{PO}\sqrt{2}}{\hat{V}} = 1/\sqrt{2} \quad (5.1.7)$$

which differs very little from the 2/3 originally suggested by Galton.

The correlation between sibs is, under the same assumptions,

$$r_{SS} = \frac{E(X_{t+1}X'_{t+1})}{V_{t+1}}, \quad (5.1.8)$$

where

$$X_{t+1} = \frac{X_t + X'_t}{2} + \epsilon$$
$$X'_{t+1} = \frac{X_t + X'_t}{2} + \epsilon', \quad (5.1.9)$$

and $E(\epsilon\epsilon') = 0$. Hence

$$r_{ss} = \frac{V_t/2}{V_{t+1}} = 1/2 \quad (5.1.10)$$

at equilibrium, which differs from Galton's result.

Since Galton's original observations and calculations, empirical evaluations of correlation coefficients between parent and offspring, between sibs, and occasionally between other sets of relatives have confirmed these expectations for some anthropometric traits.

Even after the rediscovery of Mendel's laws in 1900, the inheritance of continuous traits was still explained by blending theory, until R. A. Fisher (1918) in his famous paper "On the Correlations between Relatives after Supposition of Mendelian Inheritance" showed that the observed correlations between relatives could be explained easily under a nonblending, simple Mendelian set of assumptions. Fisher also pointed out (1930) that blending inheritance would demand that the mutation variance, that is, the error variance upon transmission, of a trait be, at equilibrium, half as large as the total variance of the trait (see equation 5.1.4). When measurements of mutation rates became available it was clear that mutation of single genes was a very rare phenomenon. Thus, according to Fisher, mutation could hardly be as high as demanded by blending theory. Fisher proposed a model in which the variance of a continuous trait as determined by Mendelian factors is composed, in part, of a "genetic" or transmissible contribution that remains constant over generations. In this formulation of Fisher's, Mendelian segregation of different alleles that contribute to the phenotype produces a "segregation variance" whose presence guarantees the constancy of a large genetic contribution to the variance of the trait over generations. If mutation variance is added, however, as it undoubtedly is, though in small quantities, the variance of the trait in the population under Fisher's formulation would increase indefinitely though very slowly. Empirical evidence (e.g. from fossil records) suggests that this doesn't happen. Thus some mechanism must keep the total variance within bounds. It is commonly assumed that this is accomplished by stabilizing selection.

Blending inheritance would, on the face of it, appear to be an inappropriate model for any trait that is largely influenced by genetic factors. However, it should be noted that Fisher's model for gene action produces the same correlations among relatives as the blending model, so that a choice between blending and

Mendelian models must rest on considerations other than these correlations. The blending model might also provide an adequate description for traits that are under cultural vertical transmission. It will be generalized for our purposes in the next section.

Most "cultural" traits for which theory is available in the literature have to do with economics. Wealth, rather than income, is obviously subject to vertical transmission. The word "inheritance" *tout court* applies in fact primarily to this kind of transmission, and only by analogy to biological inheritance. Indeed, Galton used the term "natural" inheritance to make the distinction from "true" inheritance clear. The vertical component in the transmission of wealth is certainly significant even today. It was obviously more significant in previous times, when transmission of wealth was mostly from father to eldest or favorite son. Primogeniture must have been largely a device for avoiding the dilution of wealth over generations and the introduction of the legitim has probably had the effect of modifying the distribution of wealth. A tendency toward sharing wealth equally among children would certainly decrease the wealth over generations in simply predictable ways. One device that has helped keep the variance (or in economic terminology, the concentration) of wealth very high is assortative mating. Assortative mating for wealth is certainly practiced in part consciously and in part as a consequence of the fact that marriages are most likely to take place within rather than between social strata. Considerations such as these point to wealth as a most interesting cultural object, with special rules of vertical transmission, although some of its characteristics make it difficult to use it as a paradigm. An important limitation is that existing data turn out to be crude. It is somewhat unexpected that individual wealth proves to be difficult to assess, but it is known that, for various reasons, people tend to make inaccurate statements about their wealth. More knowledge is available and much theoretical discussion has centered on the distribution of income (see e.g. Atkinson, 1975). There have recently been open claims that genetic variation is very important in determining income (Taubman, 1976). Unfortunately, the claim is based on the study of twins. These specially related individuals supply the weakest evidence of genetic transmission (see Goldberger, 1977 for a detailed critique of these claims). An example of a continuous trait from a totally different field that has stimulated some of the ideas used in this chapter derives from the measurement of stone tools. This will be discussed in the last sections of the chapter, with reference to cultural drift and cultural selection.

5.2 LINEAR TRANSMISSION

Our consideration of blending inheritance in the previous chapter assumed equal contributions from the two parents, and random mating. A simple generalization allows differential contributions from the two parents, and the existence of correlation between the trait values of mates. Instead of (5.1.1) we write

$$X_{t+1} = b_m X_{m,t} + b_p X_{p,t} + \epsilon \tag{5.2.1}$$

where $X_{m,t}$ and $X_{p,t}$ (which were denoted X_t and X'_t in (5.1.1)) are the female and male parental variates respectively. In the present model these have identical distributions. In (5.1.1) $b_m = b_p = 1/2$. Again it is assumed that $E(\epsilon) = 0$, $\mathrm{Var}(\epsilon) = \sigma^2$. Assortative mating is described by a normal correlation surface with correlation a expressing the similarity between $X_{m,t}$ and $X_{p,t}$. Then the mean μ_t, and variance V_t satisfy

$$\mu_{t+1} = E(X_{t+1}) = (b_m + b_p)E(X_t) \tag{5.2.2}$$

$$\begin{aligned} V_{t+1} &= E(X_{t+1} - \mu_{t+1})^2 \\ &= (b_m^2 + b_p^2 + 2b_m b_p a)V_t + \sigma^2 \\ &= CV_t + \sigma^2, \text{ say.} \end{aligned} \tag{5.2.3}$$

Thus the mean of the trait in the population remains constant if and only if $b_m + b_p = 1$, which is now assumed to hold. The variance approaches the finite limit

$$\lim_{t \to \infty} V_t = \sigma^2/(1 - C) \tag{5.2.4}$$

where $C = (b_m^2 + b_p^2 + 2b_m b_p a)$, provided $C < 1$. Notice that since $a \leq 1$, if the mean converges then so does the variance (except in the degenerate case $b_m = 1$ or $b_p = 1$); the reverse implication is not true. The case $b_m = b_p = 1/2$ is the classical blending model of section 5.1. If $b_m = 1$, or $b_p = 1$, then the mean remains unchanged over time and the variance diverges. As we shall see later, in a finite population the latter conclusion may not be valid, depending on the distribution of the number of offspring per mating.

Oblique transmission can be incorporated into (5.2.1) by assuming that after vertical transmission the parental generation as a whole contributes a fraction g to the offspring phenotype. This can be expressed as

$$X_{t+1} = (1 - g)(b_m X_{m,t} + b_p X_{p,t}) + g\mu_t + \epsilon, \tag{5.2.5}$$

where $\mu_t = E(X_t)$. Then

$$\mu_{t+1} = [(1 - g)(b_m + b_p) + g]\mu_t,$$

so that, as with (5.2.2), if $(b_m + b_p) = 1$ there is no change in the mean. The variance, however, is reduced under this assumption if $g > 0$,

$$V_{t+1} = BV_t + \sigma^2; \lim_{t \to \infty} V_t = \sigma^2/(1 - B) \tag{5.2.6}$$

where $B = (1 - g)^2(b_m^2 + b_p^2 + 2b_m b_p a) = (1 - g)^2 C$. In other words, oblique transmission increases homogeneity.

In these models, which incorporate a group effect, the parameter g represents the contribution of the parental generation. It is as if *every* individual in the parental generation contributes the same fraction g/N to the trait of every individual in the offspring generation (except the parent itself whose contribution is (1 −

$g + g/N$) to its own offspring). This can be generalized to allow that $X_{j,t}$ determines a fraction g_{ij} of $X_{i,t+1}$. Such a model is later treated also by matrix methods (see section 5.10).

Horizontal transmission might be modeled by replacing μ_t by μ_{t+1} in (5.2.5). Then

$$\mu_{t+1} = (b_m + b_p)\mu_t \tag{5.2.7}$$

and

$$V_{t+1} = BV_t + \sigma^2; \quad V_t \rightarrow \sigma^2/(1 - B), \tag{5.2.8}$$

as in (5.2.6).

The effect of generations earlier than parental can be modeled analogously:

$$X_{t+1} = (1 - g^{(t)} - g^{(t-1)})(b_m X_{m,t} + b_p X_{p,t}) + g^{(t)}\mu_t + g^{(t-1)}\mu_{t-1} + \epsilon, \tag{5.2.9}$$

so that

$$V_t \rightarrow \sigma^2/(1 - D), \tag{5.2.10}$$

where $D = (1 - g^{(t)} - g^{(t-1)})^2(b_m^2 + b_p^2 + 2b_m b_p a)$. The variance at equilibrium is therefore smaller than in the case of parental transmission alone. In fact the effect of earlier generations, whether it is manifest through the written record or oral tradition, will generally be a reduction of the overall variance.

Another type of generalization allows not just two but a number n of the individuals in the parental generation to contribute to the trait of an individual in the filial generation. If each contributes an amount b_i with $\Sigma_{i=1}^n b = 1$, then

$$X_{j,t+1} = \sum_{i=1}^{n} b_i X_{i,t} + \epsilon, \tag{5.2.11}$$

where for each j the n parental individuals are chosen at random from a very large population (this allows us to neglect the population size). The resulting variance is expected to be

$$V_{t+1} = V_t \sum_{i=1}^{n} b_i^2 + \sigma^2. \tag{5.2.12}$$

Define

$$\sigma_b^2 = \frac{\Sigma b_i^2}{n} - \frac{1}{n^2}. \tag{5.2.13}$$

Then (5.2.12) becomes

$$V_{t+1} = V_t \left(n\sigma_b^2 + \frac{1}{n} \right) + \sigma^2. \tag{5.2.14}$$

For $n\sigma_b^2 + 1/n < 1$ the equilibrium variance is

$$\hat{V} = \sigma^2 \Bigg/ \left(1 - n\sigma_b^2 - \frac{1}{n} \right). \tag{5.2.15}$$

If each transmitting parent makes the same contribution, $\sigma_b^2 = 0$ and $\hat{V} = n\sigma^2/(n - 1)$, which is close to σ^2 if n is large. In fact the larger the number of parents in this system of n-parental transmission, the smaller the ultimate variance. For $n = 2$ we have biparental transmission and, as before, $\hat{V} = 2\sigma^2$. Uniparental transmission ($n = 1$) does not allow convergence and will be discussed further in section 5.8.

The rate at which the variance converges to its equilibrium value is easy to obtain because the recursions are usually linear:

$$V_{t+1} = CV_t + \sigma^2, \tag{5.2.16}$$

which converges for $C < 1$ to $\hat{V} = \sigma^2/(1 - C)$. The rate of approach to the equilibrium is

$$\lim_{t \to \infty} \frac{V_{t+1} - \hat{V}}{V_t - \hat{V}} = C \tag{5.2.17}$$

per generation; the closer C is to 1, the slower is the approach.

Thus, the greater the number of "parents," n, the more homogeneous is the population, in the sense that the equilibrium variance is smaller and the rate at which this equilibrium is approached is faster.

It is worthwhile pointing out that these univariate analyses can be simply extended to the case where, for example, it is necessary to keep track of trait values separately in the two sexes. To this end denote the values in females and males by Y and Z respectively. Then assume that

$$\mathbf{X}_{t+1} = \mathbf{B}\mathbf{X}_t + \boldsymbol{\epsilon} \tag{5.2.18}$$

where $\mathbf{X}_t^T = (Y_t, Z_t)$, $\boldsymbol{\epsilon}^T = (\epsilon_M, \epsilon_P)$ with $E(\epsilon_M) = E(\epsilon_P) = 0$, $E(\epsilon_M^2) = \sigma_M^2$, $E(\epsilon_P^2) = \sigma_P^2$ and

$$\mathbf{B} = \begin{pmatrix} b_{mm} & b_{mp} \\ b_{pm} & b_{pp} \end{pmatrix}.$$

It is then necessary to follow the bivariate distribution of (Y, Z) over time.

In summary, for the simple univariate model, the contribution of generations earlier than one's own might be said to be homogenizing, in that such a contribution generally reduces population variance below that expected with parental contributions. The general blending rule does not guarantee the convergence of mean and variance although convergence of the former implies convergence of the latter and the greater the number of "parents," the more homogeneous the resulting population. In the limit with a large number of parental individuals contributing to the make up of an offspring (many-to-one transmission), the equilibrium variance is only slightly greater than the mutation variance.

5.3 CORRELATIONS BETWEEN RELATIVES

Under linear transmission correlations between relatives can be computed in the presence of assortative mating of the type introduced in section 5.2. The model (5.2.1) entails divergence of the population mean if $b_m + b_p > 1$ and its convergence to zero if $b_m + b_p < 1$. In some applications it may be legitimate to assume that $b_m + b_p = 1$, so that the mean is stable and not zero. In this way one also assumes that the only transmission is of this cultural type. Even with this assumption, it should be noted that variances and covariances will be subject to change over time. Again the model will be

$$X_{t+1} = b_m X_{m,t} + b_p X_{p,t} + \epsilon, \quad (b_m + b_p = 1), \qquad (5.3.1)$$

with $E(\epsilon) = 0$, $E(\epsilon^2) = \sigma^2$.

Three parent-offspring covariances can be defined:

Father-offspring: $$E(X_{m,t} - \mu_t)X_{t+1} = (b_p + ab_m)V_t = W_{FO} \qquad (5.3.2a)$$

Mother-offspring: $$E(X_{p,t} - \mu_t)X_{t+1} = (b_m + ab_p)V_t = W_{MO} \qquad (5.3.2b)$$

Midparent-offspring: $$E\left(\frac{X_{m,t} + X_{p,t}}{2} - \mu_t\right) X_{t+1} = \frac{W_{FO} + W_{MO}}{2} = W_{\bar{P}O} \qquad (5.3.2c)$$

It is generally true that the *regression* of offspring on parent, computed by dividing the covariance by the appropriate parental variance, is *independent of time*. Using b as the symbol for the regression coefficient we have

$$\begin{aligned} b_{FO} &= W_{FO}/V_t = b_p + ab_m \\ b_{MO} &= W_{MO}/V_t = b_m + ab_p \\ b_{\bar{P}O} &= W_{\bar{P}O}\Big/\left(\frac{1+a}{2}\right)V_t = b_m + b_p \end{aligned} \qquad (5.3.3)$$

The correlations, however, are time dependent. Those of the father or mother with offspring can be obtained by multiplying the regression coefficient by $(V_t/V_{t+1})^{1/2}$, while for the midparent offspring correlation the multiplier is $\{[(1 + a)/2]V_t/V_{t+1}\}^{1/2}$. Thus under conditions allowing equilibrium $(V_t = V_{t+1})$ the correlation coefficients are

$$\hat{r}_{FO} = b_p + ab_m, \tag{5.3.4a}$$

$$\hat{r}_{MO} = b_m + ab_p, \tag{5.3.4b}$$

$$\hat{r}_{\overline{P}O} = (b_p + b_m)\sqrt{(1 + a)/2}, \tag{5.3.4c}$$

and

$$\hat{V} = \sigma^2/2b_m b_p(1 - a). \tag{5.3.5}$$

For the correlations between sibs the phenotype of a second child of the same parents as in (5.3.1) is written

$$X'_{t+1} = b_m X_t + b_p X_t + \epsilon' \tag{5.3.6}$$

with

$$E(\epsilon') = 0, \quad E(\epsilon')^2 = \sigma^2.$$

A problem of potential importance concerns the value of $E(\epsilon\epsilon')$. If there is no correlation in nontransmitted contributions to the phenotypes this will of course be zero. On the other hand the potential for mutual interaction in the development of culturally interesting traits would argue that $E(\epsilon\epsilon') \neq 0$. Further, it may be that $E(\epsilon^2) \neq E(\epsilon')^2$, for example, if there is a birth-order effect on the trait. Intrafamilial transmission has been invoked by Zajonc (1976) to explain the observed variation of IQ with family size and birth order.

The covariance between full sibs at time $t + 1$ is

$$W_{SS,t+1} = E(X_{t+1} - \mu_{t+1})X'_{t+1}$$
$$= (b_m^2 + b_p^2 + 2b_m b_p a)V_t + r\sigma^2, \tag{5.3.7}$$

where r is the correlation between ϵ and ϵ'. Both the corresponding regression and correlation are time dependent:

$$r_{SS,t+1} = b_{SS,t+1} = W_{SS,t+1}/V_{t+1}.$$

At equilibrium, $\hat{r}_{SS} = (b_m + b_p)^2 - 2b_m b_p(1 - a)(1 - r)$ and, if $b_m + b_p = 1$, this reduces to:

$$\hat{r}_{SS} = \hat{b}_{SS} = 1 - 2b_m b_p(1 - a)(1 - r). \tag{5.3.8}$$

If $b_m = b_p = 1/2$, there is the further simplification:

$$\hat{r}_{SS} = [1 + a + r(1 - a)]/2. \tag{5.3.9}$$

Note that if $r > 0$, then (5.3.9) is larger than the conventional (genetic) parent-offspring correlation $(1 + a)/2$ from, for example, (5.3.4a) (with $b_m = b_p = 1/2$).

For half-sib correlations write

$$X''_{t+1} = b_m X'_{m,t} + b_p X_{p,t} + \epsilon'' \tag{5.3.10}$$

as the trait value of a paternal half-sib of X_{t+1} in (5.3.1), where $X'_{m,t}$ and $X_{m,t}$ are unrelated. In this pattern of relationship, correlations among the various mates may be complicated. There may be three mating correlations; the first, a_1, is between the spouses of trait values $X_{m,t}$ and $X_{p,t}$ in (5.3.1), the second, a_2, is between $X'_{m,t}$ and $X_{p,t}$ in (5.3.10), while a_3 might denote the correlation between $X_{m,t}$ and $X'_{m,t}$. With these definitions, the covariances between half-sibs are

$$\begin{aligned} W_{FHS,t+1} &= [b_p^2 + b_m b_p(a_1 + a_2) + b_m^2 a_3]V_t + r'\sigma^2 \\ W_{MHS,t+1} &= [b_m^2 + b_m b_p(a_1 + a_2) + b_p^2 a_3]V_t + r'\sigma^2 \end{aligned} \tag{5.3.11}$$

for the paternal and maternal half-sibs respectively where $r' = E(\epsilon\epsilon'')$. If $r' = 0$ then the quantities in brackets on the right side of (5.3.11) are the appropriate correlation and regression coefficients. Some simplification is possible if it is assumed that the correlation a_3 is due only to correlations between $X_{m,t}$ and $X_{p,t}$ and $X'_{m,t}$ and $X_{p,t}$. In other words, $X'_{m,t}$ and $X_{m,t}$ are conditionally independent (given $X_{p,t}$). In this case the partial correlation coefficient

$$r_{X_{m,t}X'_{m,t} \cdot X_{p,t}} = 0$$

and then

$$a_3 = r_{X_{m,t}X'_{m,t}} = a_1 a_2.$$

It is often assumed in comparisons with observed data that $a_1 = a_2$ and $r' = 0$ in which case (5.3.11) reduces to

$$\hat{r}_{FHS} = (b_p + ab_m)^2 = \hat{r}^2_{FO}$$
$$\hat{r}_{MHS} = (b_m + ab_p)^2 = \hat{r}^2_{MO}. \quad (5.3.12)$$

Correlations between half-sibs thus tend to be relatively low, in fact of the order of those that might be computed for uncle-niece, aunt-nephew or grandparent-grandchild, for examples. More remote relationships such as first cousins will produce even lower correlations under vertical transmission.

The most commonly employed "polygenic" or "multifactorial" model of genetic determination of a trait, originated by R. A. Fisher in 1918, produces exactly the same regressions and correlations as have been given above, with $b_m = b_p = 1/2$, provided that gene action is entirely additive. The random variate ϵ, which we interpret above as a variance augmenting cultural mutation component, in the polygenic model is the sum of two components. One is the non-transmissible contribution of environmental variables and the other is a "segregation variate" with zero mean and variance which, in the case of random mating, is exactly one half of the genetic variance of the previous generation. With assortative mating the analogy breaks down to some extent (see Feldman and Cavalli-Sforza, 1979b). Here it suffices to stress the quantitative similarities between the linear treatment above and the simple genetic model, despite the conceptual differences between the formulations at the level of trait determination.

Various extensions of the above computations can be made to include oblique or horizontal transmission, social stratification and natural selection. Often it is desirable or essential to study more than a single trait at the same time. The linear approach above can be directly generalized to multivariate traits, as is indicated in the next section. To exhibit the role of group transmission, (5.2.5) is used together with the assumption that $X_{m,t}$ and $X_{p,t}$ are correlated to the extent a. As a result the parent (generation t) offspring (generation $t + 1$) covariances are

$$\begin{aligned} W_{FO} &= [(1 - g)(b_p + b_m a)]V_t \\ W_{MO} &= [(1 - g)(b_m + b_p a)]V_t \\ W_{\bar{P}O} &= [(1 - g)(1 + a)/2]V_t. \end{aligned} \tag{5.3.13}$$

The resulting correlations at equilibrium are smaller by the factor $1-g$ than the corresponding values under purely vertical transmission.

The sib-sib covariances using the simple g-model of group transmission is

$$W_{SS,t+1} = (1 - g)^2(b_m^2 + b_p^2 + 2ab_m b_p)V_t, \tag{5.3.14}$$

and for half-sibs the covariances are

$$\begin{aligned} W_{FHS,t+1} &= (1 - g)^2[(b_p^2 + b_p b_m(a_1 + a_2) + b_m^2 a_3)]V_t + r'\sigma^2 \\ W_{MHS,t+1} &= (1 - g)^2[(b_m^2 + b_p b_m(a_1 + a_2) \\ &\qquad + b_p^2 a_3)]V_t + r'\sigma^2. \end{aligned} \tag{5.3.15}$$

In all of the above the random variable ϵ, which generates random deviation from the expected parental effects, is assumed to have constant variance σ^2, independent of the family (i.e. independent of $X_{m,t}$ and $X_{p,t}$). Wagener (1975) made the hypothesis that this variance should depend on the squared difference between the parents in a simple linear way:

$$E(\epsilon^2) = C(X_{m,t} - X_{p,t})^2 + \eta^2 \tag{5.3.16}$$

She showed that this condition has no influence on the parent offspring or sib-sib correlations, or on the trait mean, but does increase the variance (under random mating):

$$V_{t+1} = \eta^2 + (2C + b_p^2 + b_m^2)V_t \rightarrow \eta^2/(1 - 2C - b_p^2 - b_m^2) \quad (5.3.17)$$

if $2C + b_p^2 + b_m^2 < 1$.

In summary, parent offspring regression allows estimation of b_m, b_p and g (using 5.3.13). Sib-sib and half-sib comparisons in principle allow the further estimation of intrafamilial parameters like r and r' in (5.3.7) and (5.3.11). However, the use of half-sibs requires knowledge of three assortative mating coefficients unless certain assumptions are made. The above procedures do not distinguish between genetic and cultural transmission: for this purpose correlations of both biological and foster relatives must be used. This ambitious task was pioneered for IQ by S. Wright (1931b) using path coefficients (see also Wright 1978). There is no indication that much headway has been made since Wright's efforts: only an inconclusive scatter of results has been reached in the past years, using extensions of Wright's early approach. These will not be discussed here, as the theory developed above is not meant to handle joint genetic/cultural transmission. However, when correlations of foster relatives are available, the above formulas can be used for evaluation of cultural transmission. As an application we will consider data of IQ in adopted children.

From the summary of IQ data by Jencks (1973) the following correlations are of interest:

adopted father and adopted child:

$r_{AFAO} = W_{FO}/V_T = 0.21$ (total pairs $n = 536$, mean of 3 studies)

adopted mother and adopted child:

$r_{AMAO} = W_{MO}/V_T = 0.24$ (total pairs $n = 645$, mean of 3 studies)

assortative mating of adoptive parents:

$a = 0.53$ (total pairs $n = 274$, mean of 3 studies)

With $g = 0$, formulas (5.3.13, 5.3.16) produce $b_m = 0.18$ and $b_p = .12$.

We note, however, that the studies on which the average correlation coefficients given above are based show significant heterogeneity. Moreover, there is an effect of selective placement of adopted children that may have increased the observed correlations, and an attenuation effect due to decreased range that has the opposite effect by amounts which are difficult to assess precisely. While the former has been amply reviewed, the latter has been given very little consideration, although it is a well known fact that adoptive parents are highly selected, having IQ higher than the mean of the general population, and smaller variance. One way to estimate this effect will be found in section 5.6. In this case $b_p + b_m < 1$ because the variance used includes sources of variation other than purely cultural. The above formulas should therefore be used with caution.

5.4 MULTIVARIATE LINEAR MODELS

In almost all branches of quantitative science, linear models have great popularity. Even if they are not strictly valid in the sense of conforming to reality, they may provide sufficiently good approximations that useful prediction is possible. Linear relations among normal random variates preserve normality. Simple transformations can often be made that reduce previously skewed distributions to normality and at the same time produce linear relations among the transformed variables to a satisfactory level of approximation. The great advantage of the assumption, of course, is that explicit solutions are possible where non-linearity might have precluded solution.

There is usually no attempt to validate the linearity hypothesis. Indeed, when many variates are involved it is less likely that all relations among them are linear or can be made linear by simple transformation. But non-linear models are in a relatively elementary stage of development insofar as their application to the interpretation of multivariate relationships is concerned. At this time the linear models to be discussed in what follows should be regarded as only one of many ways of formulating the problems, but one which has the obvious advantage in the tractability of the mathematics. Under scaling assumptions of unit variance for all variables, this linear approach can be made equivalent to the well known method of path analysis.

Usually in the study of multivariate linear evolutionary models, one or a few variables constitute the major focus of interest. Such variables will be called *target variables*. Examples in the cultural evolutionary context include educational attainments (which usually measure some sort of skill that may or may not be intellectual), economic or social status, values, beliefs, activities, time allocation. Usually in the study of the target variables other variables come to the notice of the investigator that are not of direct interest to the study but that, there is reason to suspect, influence the target variables. It may, therefore, become important to measure these along with the target variable and to quantify their potential importance to the target. These variables are termed *concomitant*. Clearly, target variables in one study may be concomitant in another, and vice versa.

Finally there is a third type of variable, whose existence is inferred on some grounds usually extrinsic to the study at hand, and which is not directly measurable. A typical example of such a variable, which we call *latent*, is the "genotypic contribution" to continuous variation. It is ordinarily guessed that there may be a biologically inherited contribution to a trait. The reasons for the guess may involve evidence from other organisms, or even political prejudice. (This is not to say that "environmental" contributions to the trait may not be just as intangible.) The classical concept of heritability represents an attempt to measure the proportion of variance of a trait (among individuals in a specific population) due to genotypic differences among individuals. We do not intend to pursue the ambiguities that limit the usefulness of such a quantity. Suffice it to point out that any variable (other than genotype) that is vertically transmitted and not measured will very frequently be indistinguishable from genotype. It is perfectly possible, and indeed in some cases has appeared to be the case, that transmitted variables that are not measured have been interpreted as genetic simply because, in the absence of reliable adoption data, the estimation procedures take as genetic anything which is transmitted and not measured. The investigator should be wary of attributing to any specific latent variable variation that cannot be assigned to chosen target or concomitant variables.

Suppose that the $(k + l + m)$ component column vector $\mathbf{x}$ represents the values of all members of the three sets of variates such that $x_1, x_2, \ldots x_k$ are the k latent variates, $x_{k+1} \ldots x_{k+l}$ are the concomitant variates, and $x_{k+l+1} \ldots x_{k+l+m}$ are the m target variates. At time t the vector is $\mathbf{x}_t = (x_{1,t}, x_{2,t}, \ldots x_{k+l+m})$. The evolution of these variables is described by the linear relationship

$$\mathbf{x}_{t+1} = \mathbf{B}\mathbf{x}_t + \boldsymbol{\epsilon}, \tag{5.4.1}$$

with $\boldsymbol{\epsilon}$ a vector of independent random components, one for each of the $k + l + m$ original variables. It might be noted that the interpretation of the matrix $\mathbf{B}$ in (5.4.1) depends on whether the process of change in trait values occurs within a generation or between generations. In the former case the model would be one of trait development during some part of a generation between two specified points of time, t and $t + 1$. In the latter case t and $t + 1$ are the same census points in successive generations. If transmission is linear and biparental then the simplest case of equal contributions from the parents we can write is

$$\mathbf{x}_{t+1} = \tfrac{1}{2}(\mathbf{B}\mathbf{x}_t) + \tfrac{1}{2}(\mathbf{B}\mathbf{x}'_t) + \boldsymbol{\epsilon}, \tag{5.4.2}$$

where $\mathbf{x}_t$ and $\mathbf{x}'_t$ are the two parental trait vectors. The contributions may well be different from 1/2 and different for every variable, in which case the factors 1/2 multiplying $\mathbf{B}$ in (5.4.2) would be replaced by matrices. Of course, this could be summarized by writing

$$\mathbf{x}_{t+1} = \mathbf{B}_1 \mathbf{x}_t + \mathbf{B}_2 \mathbf{x}'_t + \epsilon. \tag{5.4.2a}$$

In the same way as was done in §5.2, oblique transmission can be introduced by writing

$$\mathbf{x}_{t+1} = \sum_{i=1}^{N} \pi_i \mathbf{B}_i \mathbf{x}_t^{(i)} + \epsilon, \tag{5.4.3}$$

where $i = 1, 2, \ldots N$ varies across the individuals of the parent generation having trait vectors $\mathbf{x}_t^{(i)}$, and π_i is the relative contribution of the ith individual in the parental generation.

The structure of the **B** matrix can be written in nine blocks in the case of vertical transmission. L, C and T are used to denote the latent concomitant and target variables respectively, while the arrow represents linear dependence of the righthand term on the lefthand term.

	One parent to child transmission of latent variables $L_t \rightarrow L_{t+1}$	Influence of parent's variables on latent variables of child $C_t \rightarrow L_{t+1}$	Influence of parent's target variables on latent variables of child $T_t \rightarrow L_{t+1}$
B =	Influence of parent's latent variables on concomitant variables of child $L_t \rightarrow C_{t+1}$	One parent to child transmission of concomitant variables $C_t \rightarrow C_{t+1}$	Influence of parent's target variables on concomitant variables of child $T_t \rightarrow C_{t+1}$
	Influence of parent's latent variables on target variables of child $L_t \rightarrow T_{t+1}$	Influence of parent's concomitant variable on target variables of child $C_t \rightarrow T_{t+1}$	One parent to child transmission of target variables $T_t \rightarrow T_{t+1}$

Latent variables are usually *defined* as transmissible. On the other hand, concomitant and target variables may not necessarily be vertically transmitted. If not, the corresponding diagonal block of **B** would be a zero matrix. In most analyses the assumption is made that the only latent variable is genotypic contribution to the trait value. The validity of this assumption will not be addressed here. Elsewhere we have considered the evolutionary behavior of linear systems in which the latent variable was "genotype" (G), the concomitant variable was called "environment" (E), and the target variable "phenotype" (F) (Cavalli-Sforza and Feldman, 1978b, Feldman and Cavalli-Sforza, 1979b).

In the classical genetic model, genotype and environment determine the phenotype, but only genotypic value is transmitted from parent to child. The relevant transmission matrix is $\mathbf{B}_{\text{gen}}$, acting on the column vector (G, E, F):

$$\mathbf{B}_{\text{gen}} = \begin{pmatrix} 1 & 0 & 0 \\ 0 & 0 & 0 \\ 1 & 0 & 0 \end{pmatrix}$$

or, more generally when there is more than one G, E, or F variable:

$$\mathbf{B}_{\text{gen}} = \begin{pmatrix} \mathbf{I} & 0 & 0 \\ 0 & 0 & 0 \\ \mathbf{N} & 0 & 0 \end{pmatrix}$$

where $\mathbf{I}$ is the identity matrix and $\mathbf{N}$ has some positive elements. Of course, some assumption on $\mathbf{B}$ should be made, if the trait means are to converge to a non-zero value.

We have introduced (Cavalli-Sforza and Feldman, 1973b, 1978b; Feldman and Cavalli-Sforza, 1977b) models in which the phenotype, (or target value) was vertically transmitted in addition to genotype. In the multivariate case the corresponding transmission matrix for such a model, which we call the GF model, is

$$\mathbf{B}_{GF} = \begin{pmatrix} \mathbf{I} & \mathbf{0} & \mathbf{0} \\ \mathbf{0} & \mathbf{0} & \mathbf{0} \\ \mathbf{N}_1 & \mathbf{0} & \mathbf{N}_2 \end{pmatrix},$$

models in which environment is also vertically transmitted we have termed (EFG), for which

$$\mathbf{B}_{EFG} = \begin{pmatrix} \mathbf{I} & 0^* & 0^* \\ 0 & \mathbf{I} & 0 \\ \mathbf{N}_1 & \mathbf{N}_2 & \mathbf{N}_3 \end{pmatrix},$$

and this is described schematically in Figure 5.4.1. Note that starred zeroes will be $\mathbf{0}$ if the latent variables of the child are unaffected by the environment or phenotype, as would be the case if the latent variables were chromosomally transmitted. (In certain cases of maternal or cytoplasmic inheritance, there can be environmental influences on the vertical transmission of what would generally be considered latent variables.)

Natural selection affects the genotype through the environment, but not by altering the genotype itself. It alters the chance that a given genotype will be represented in the offspring generation. Its effect is therefore nonlinear. However, it is still possible

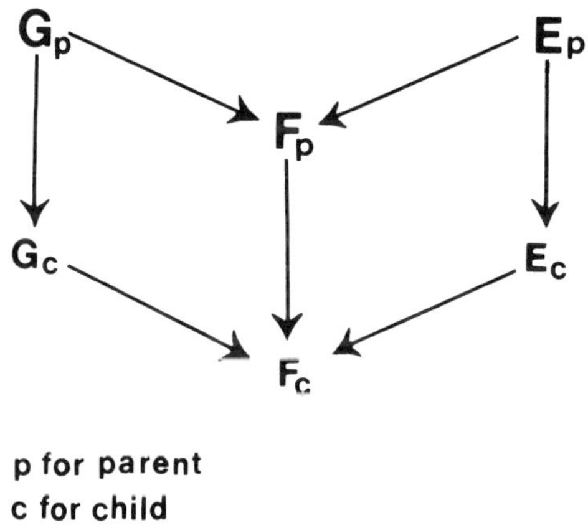

FIGURE 5.4.1. An example of complex transmission. G (for genotype) are the latent variables, E (for environment) the concomitant ones, which jointly determine F, the phenotype. All three variables, or sets of variables, are transmitted directly in the model indicated in the figure.

to use relatively simple algebraic representations of the selection process, as has been done by Slatkin (1970), Lande (1976), Cavalli-Sforza and Feldman (1976, 1978b), and Feldman and Cavalli-Sforza (1977b). A general mathematical theory for the analysis of selection on linearly transmitted continuous traits is given by Karlin (1972b, 1979).

The linear structure can be retained even if developmental effects within generations are superimposed on vertical transmission across generations, provided both are linear. For example, if $\mathbf{x}_{b,t}$ represents the trait vector at some juvenile stage and $\mathbf{x}_{a,t}$ at adulthood, then it may be that

$$\mathbf{x}_{t+1} = \mathbf{x}_{a,t+1} = \mathbf{D}\mathbf{x}_{b,t+1} + \epsilon^*$$

where

$$\mathbf{D} = \begin{pmatrix} I & 0 & 0 \\ 0 & K & K' \\ 0 & \Lambda' & \Lambda \end{pmatrix}.$$

In this context, $\mathbf{x}_{b,t+1} = \mathbf{B}\mathbf{x}_{a,t} + \epsilon$, so the product matrix $\mathbf{DB}$ describes the combined process of transmission and development.

Correlations between relatives in this multivariate context can be carried out using standard matrix manipulations. The use of such treatments in estimation of the transmission parameters has been detailed by Goldberger (1978), Feldman and Cavalli-Sforza (1979b) and Karlin (1979).

The basic classification into target, latent, and concomitant variables may best be regarded as reflecting the interest and observation range of the investigator rather than the intrinsic properties of the variables so classified. The mode of transmission is usually postulated and its adequacy considered by fitting expected to observed variance-covariance matrices. There is a degree of arbitrariness in this procedure in that many models may provide adequate fits. This arbitrariness, in conjunction with the heterogeneity in adequacy of the existing data (even for the most intensely studied traits, such as IQ), forces all conclusions to be statistically rather weak, and perhaps explains the variety of opinions and the relevance of considerations extrinsic to the data that are characteristic of this area.

5.5 SOCIAL STRATIFICATION, CLASS AND CASTE

Most human societies are stratified. The criteria by which subgroups are identified as strata usually depend on such variables as the trait under study, the point in history at which the study is made, and so on. For some traits the strata may be sharply defined at any point in time, yet there may be passage from one to another stratum over time; an example of this type of trait is age-class. For other traits the strata are permanent and there is almost no transition from one to the other; this is a *caste* society. For most traits of interest in cultural evolution, it is likely that classes are not completely closed or may themselves be regarded as a continuum. A few simple models allowing stratification into subgroups will now be considered, in particular the consequences of stratification on two of the most commonly used measures of vertical transmission, the parent-offspring and the sib-sib correlations.

(1) *Selective oblique transmission.* In the models of oblique transmission introduced above, either all members of the parental generation or a random sample of them contributed to an offspring's trait value. But if the sample of oblique transmitters were selected to be close in value to the parental value(s), some degree of stratification might result. Assume then that the trait value of an offspring is determined to the extent $(1 - g)$ by vertical transmission from its parents, while the fraction due to oblique transmission is g. The oblique transmitters, however, are chosen to have the same mean as the parents, but a variance that is a fraction f of the total variance, V_t, in the parental generation.

Then the trait value of a randomly chosen individual in the offspring generation will be

$$X_{t+1} = (1 - g)\left(\frac{X_t + X_t'}{2}\right) + g\xi + \epsilon, \qquad (5.5.1)$$

where ϵ, with zero mean and variance σ^2, is the usual mutation random variate and, given the parents X_t and X_t', ξ has mean $(X_t + X_t')/2$ and variance fV_t. From (5.5.1) the offspring variance V_{t+1} satisfies

$$V_{t+1} = V_t\left(\frac{1 + a}{2} + g^2 f\right) + \sigma^2, \qquad (5.5.2)$$

which converges if $(1 + a)/2 + g^2 f < 1$ to

$$\hat{V} = 2\sigma^2/(1 - a - 2g^2 f), \qquad (5.5.3)$$

where a is the assortative mating correlation. This is greater than $2\sigma^2/(1 - a)$, the value in the absence of oblique transmission, which in turn is greater than $2\sigma^2/[1 - a + (2g - g^2)(1 + a)]$ from (5.2.6), obtained with nonselective oblique transmission. With this model the parent-offspring and sib-sib covariances at equilibrium are the same as in the absence of oblique transmission.

(2) *Discrete social classes with cross migration.* Suppose there are two social classes, for simplicity of equal size, and label them A and B. The variates for the trait values will be X_t and $\tilde{X}_t$ in A and B with expectations μ_t and $\tilde{\mu}_t$ at time t. The overall mean is $(\mu_t + \tilde{\mu}_t)/2 = \bar{\mu}_t$, say. We suppose that cross migration is measured by the parameter m in such a way that a fraction $(1 - m)/2$ of the matings are within population A, $(1 - m)/2$ within B, and m is the fraction of cross-population mating. Now assume that a social force, external to the system, operates through the addition of a constant quantity D to population A and its subtraction from B at every generation. This might represent the imposition of different behavioral rules in A and B which alter the trait values.

Thus we have the following mating arrangement

marriages	frequency	trait value of child
endogamous in A	$\frac{1-m}{2}$	$X_{t+1} = \frac{X_t + X_t'}{2} + \epsilon - D$
exogamous in A	$\frac{m}{2}$	$X_{t+1} = \frac{X_t'' + \tilde{X}_t''}{2} + \epsilon - D$
endogamous in B	$\frac{1-m}{2}$	$\tilde{X}_{t+1} = \frac{\tilde{X}_t + \tilde{X}_t'}{2} + \epsilon + D$
exogamous in B	$\frac{m}{2}$	$\tilde{X}_{t+1} = \frac{\tilde{X}_t''' + \tilde{X}_t'''}{2} + \epsilon + D$ (5.5.4)

where $E(\epsilon) = 0$, $E(\epsilon^2) = \sigma^2$. Now the general mean $\bar{\mu}_t$ remains constant. But the mean difference between classes

$$\Delta_t = \tilde{\mu}_t - \mu_t \tag{5.5.5}$$

changes according to

$$\Delta_{t+1} = (1 - m)\Delta_t + 2D, \tag{5.5.6}$$

and converges to

$$\Delta = 2D/m, \tag{5.5.7}$$

provided $m > 0$. Thus unless there are the differing environmental social effects in the two strata, producing the D contribution, the groups equalize in mean. The total variance in the population at time $t + 1$ is

$$V_{\text{Tot},t+1} = \tfrac{1}{2}E(X_{t+1}^2) + \tfrac{1}{2}E(\tilde{X}_{t+1}^2) - \bar{\mu}_{t+1}^2, \tag{5.5.8}$$

which can be shown to satisfy

$$V_{\text{Tot},t+1} = \tfrac{1}{2}V_{\text{Tot},t} + (1 - m)\Delta_t^2/4 + D^2 + D(1 - m)\Delta_t + \sigma^2. \tag{5.5.9}$$

This converges to

$$\hat{V}_{\text{Tot}} = 2\sigma^2 + \frac{2D^2}{m^2}(1 + m - m^2). \qquad (5.5.10)$$

Note that even if X_t is not correlated with spouse X'_t, nor $\widetilde{X}_t$ with $\widetilde{X}'_t$, and even if the across-population matings are random, the subdivision determines an assortative mating system with covariance between mates given by

$$a_t = \left(\frac{1}{4} - \frac{m}{2}\right)\Delta_t^2, \qquad (5.5.11)$$

which at equilibrium is

$$\hat{a} = \frac{D^2}{m^2}(1 - 2m). \qquad (5.5.12)$$

The parent (t)-offspring $(t+1)$ covariance is

$$W_{PO} = \frac{1}{2}V_{\text{Tot},t} + \left(\frac{1}{8} - \frac{m}{4}\right)\Delta_t^2 + \frac{D\Delta_t(1 - m)}{2}, \qquad (5.5.13)$$

so that at equilibrium

$$\hat{W}_{PO} = \frac{1}{2}\hat{V}_{\text{Tot}} + \frac{D^2}{m^2}\left(\frac{1}{2} - m^2\right)$$
$$= \sigma^2 + \frac{D^2}{m^2}\left(\frac{3}{2} + m - 2m^2\right). \qquad (5.5.14)$$

The sib-sib covariance is

$$W_{SS,t+1} = \frac{1}{2}V_{\text{Tot},t} + \left(\frac{1}{8} - \frac{m}{4}\right)\Delta_t^2 + D\Delta_t(1 - m) + D^2 \qquad (5.5.15)$$
$$= \sigma^2 + \frac{D^2}{m^2}\left(\frac{3}{2} + 2m - 2m^2\right) \qquad (5.5.16)$$

at equilibrium, which is just slightly higher than $\hat{W}_{PO}$.

The regressions (or correlations) at equilibrium are

$$\hat{r}_{PO} = \hat{b}_{PO} = \frac{\hat{W}_{PO}}{\hat{V}_{\text{Tot}}} = \frac{1}{2}\left[1 + \frac{D^2(1/2 - m^2)}{m^2\sigma^2 + D^2(1 + m - m^2)}\right] \qquad (5.5.17)$$

$$\hat{r}_{SS} = \hat{b}_{SS} = \frac{\hat{W}_{SS}}{\hat{V}_{\text{Tot}}} = \frac{1}{2}\left[1 + \frac{D^2(1/2 + m - m^2)}{m^2\sigma^2 + D^2(1 + m - m^2)}\right] \quad (5.5.18)$$

so that both are greater than 1/2. They become equal to 1/2 when $D = 0$ and approach 1 as $m \rightarrow 0$. With $D = 0$ and $m = 0$, from (5.5.6) the expected mean difference Δ_t remains constant. From (5.5.9) the equilibrium variance in this case would be $2\sigma^2 + \Delta^2/4$. If $m = 0$, $D \neq 0$, then the mean difference accumulates without limit and, from (5.5.9), so does V_{Tot}.

(3) *Social class as a continuous concomitant variable.* In this model linear theory is employed with a concomitant variable (in general, variables) expressing social class. The target variable, f, is transmitted directly but is influenced by the social class variable, j, which is also directly inherited. For simplicity, f and j will be considered as univariate with symmetric (in the two parents) biparental transmission. Then we can write

$$j_{t+1} = \frac{j_t + j'_t}{2} + \epsilon$$

$$f_{t+1} = \delta\left(\frac{f_t + f'_t}{2}\right) + \eta\left(\frac{j_t + j'_t}{2}\right) + \xi, \quad (5.5.19)$$

with $E(\epsilon) = E(\xi) = 0$, $E(\epsilon^2) = \sigma^2$, $E(\xi^2) = \tau^2$. Constants δ and η measure the effect of parental variables f and j, respectively, on the target variable f in the offspring.

Three models of assortative mating will be considered:

(i) The primary assortative mating is for social class with a correlation c between the spouses' j values, that is, $c = r_{jj'}$. The correlation between f and f' is secondary, being induced only from that between the j values, that is, the partial correlation coefficient $r_{ff'\cdot jj'} = 0$. This last requirement allows the deductions

$$r_{ff'} = cr_{fj}^2$$
$$r_{fj'} = r_{f'j} = cr_{fj}, \quad (5.5.20)$$

(ii) The primary assortment is by phenotype with correlation $r_{ff'} = \alpha$. Assortment for social class is secondary. Using the same argument as for (i), we have

$$r_{jj'} = \alpha r_{fj}^2; \quad r_{fj'} = r_{f'j} = \alpha r_{fj}. \tag{5.5.21}$$

(iii) Neither type of assortment is primary and the separate correlations are

$$r_{ff'} = b, \quad r_{fj'} = d, \quad r_{jj'} = e.$$

Let J_t, F_t and W_t be var (j_t), var (f_t) and cov (f_t, j_t). Then the covariance matrix of the adult population at time t is

$$\mathbf{\Sigma}_t = \begin{pmatrix} J_t & W_t \\ W_t & F_t \end{pmatrix}. \tag{5.5.22}$$

The covariance matrix between spouses will be represented by $\mathbf{\Theta}_t$. Then the three matrices $\mathbf{\Theta}_t$ in the three assorting models above are, respectively,

$$\text{(i)} \quad \mathbf{\Theta}_t = c\begin{pmatrix} J_t & W_t \\ W_t & W_t^2/J_t \end{pmatrix}. \tag{5.5.23a}$$

$$\text{(ii)} \quad \mathbf{\Theta}_t = \alpha\begin{pmatrix} W_t^2/F_t & W_t \\ W_t & F_t \end{pmatrix}. \tag{5.5.23b}$$

$$\text{(iii)} \quad \mathbf{\Theta}_t = \begin{pmatrix} eJ_t & d\sqrt{J_tF_t} \\ d\sqrt{J_tF_t} & aF_t \end{pmatrix}. \tag{5.5.23c}$$

Now, the recursion giving the offspring variance covariance matrix has been shown (see e.g. Feldman and Cavalli-Sforza, 1979b) to be

$$\mathbf{\Sigma}_{t+1} = \tfrac{1}{2}\mathbf{B}(\mathbf{\Sigma}_t + \mathbf{\Theta}_t)\mathbf{B}^{\mathrm{T}} + \mathbf{\Omega}, \tag{5.5.24}$$

where $\mathbf{B}$ is the transmission matrix

$$\mathbf{B} = \begin{pmatrix} 1 & 0 \\ \eta & \delta \end{pmatrix} \tag{5.5.25}$$

and

$$\boldsymbol{\Omega} = \begin{pmatrix} \sigma^2 & 0 \\ 0 & \tau^2 \end{pmatrix}. \tag{5.5.26}$$

Using the arguments of Feldman and Cavalli-Sforza (1979b), the equilibrium values of the parent-offspring and sib-sib covariances for the concomitant and the target variables can be obtained for any Θ. Thus,

$$\begin{aligned} W_{PO}^{11} &= (\hat{\Sigma}_{11} + \hat{\Theta}_{11})/2 \\ W_{PO}^{22} &= \eta(\hat{\Sigma}_{12} + \hat{\Theta}_{12})/2 + \delta(\hat{\Sigma}_{22} + \hat{\Theta}_{22})/2, \end{aligned} \tag{5.5.27}$$

and

$$\begin{aligned} W_{SS}^{11} &= (\hat{\Sigma}_{11} + \hat{\Theta}_{11})/2 \\ W_{SS}^{22} &= \eta^2(\hat{\Sigma}_{11} + \hat{\Theta}_{11})/2 + \eta\delta(\hat{\Sigma}_{12} + \hat{\Theta}_{12}) \\ &\quad + \delta^2(\hat{\Sigma}_{22} + \hat{\Theta}_{22})/2, \end{aligned} \tag{5.5.28}$$

where the 11 superscript indicates covariance in the concomitant variable (j) and 22 refers to the target variable (f). The values to be inserted in (5.5.27) and (5.5.28) are obtained by solving (5.5.24) with (5.5.23). For example, in model (i)

$$\hat{\Sigma}_{11} = 2\sigma^2/(1 - c) \tag{5.5.29a}$$

$$\hat{\Sigma}_{12} = \eta(1 + c)\hat{\Sigma}_{11}/[1 - \delta(1 + c)/2] \tag{5.5.29b}$$

$$\hat{\Sigma}_{22} = [1 - \delta^2/2]^{-1}\left\{\tau^2 + \frac{\eta^2(1 + c)}{2}\hat{\Sigma}_{11} + \eta\delta(1 + c)\hat{\Sigma}_{12} + \frac{c\delta^2}{2}\hat{\Sigma}_{12}^2/\hat{\Sigma}_{11}\right\}. \tag{5.5.29c}$$

Substitution of (5.5.29) into (5.5.23a) and then both into (5.5.27) and (5.5.28) produces the explicit covariances in terms only of c, σ^2, τ^2, η and δ. For the other two assorting schemes the formulas analogous to (5.5.29) are somewhat more complicated since the equilibrium entries of $\hat{\Sigma}$ involve roots of a quadratic equation. In principle, however, the method is exactly the same as for assorting scheme (i).

The three models suggested in this section have quite different results. Selective oblique transmission eliminates the decrease in W_{PO} and W_{SS} that would be caused by random oblique transmission. The two-population stratification allows a greater range of correlation between sibs and parents up to complete correlation. The third model in which social class is introduced as a concomitant variable in a general linear scheme may be the most flexible of the three. It allows effects on the target variable induced by the concomitant variable, both in assortative mating and at transmission. A larger number of parameters is introduced, which allows for greater flexibility but also makes estimation more complicated.

5.6 NATURAL SELECTION, RANGE ATTENUATION, AND THEIR EFFECTS ON THE CORRELATIONS BETWEEN RELATIVES

Three forms of natural selection on continuous traits have been distinguished by evolutionary biologists. The three—stabilizing, directional, and disruptive—were depicted graphically in Figure 1.8.1. Under stabilizing selection those individuals whose trait values fall toward the extremes of the frequency distribution will be selected against, the more central type being favored. If both ends suffer the same degree of loss, then the mean will remain unchanged and the variance will tend to decrease. Stabilizing selection is also called balancing or optimizing. Under directional selection individuals toward *one* end of the frequency distribution have a greater chance of being represented in the next generation than the others in the population. Clearly, this will result in a shift of the mean of the distribution toward the end favored by selection. Finally, it is conceivable that extremes at both ends of the distribution be favored. This is disruptive selection, which, if it is sufficiently strong or persistent, will produce bimodal distributions. In the long run the result may be completely separated distributions, that is, a fully differentiated trait.

Various monotonic functions have been used to model directional selection (see Figure 5.6.1). If the distribution prior to selection is normal, it may be desirable, especially for tractability of the mathematics, to assume a functional form which retains the normality. One such directional selection function is

$$s_1(x) = e^{kx}. \tag{5.6.1}$$

If the density function of the trait before selection was $\phi(x)$, then after selection it is

$$\phi^*(x) = \frac{\phi(x)s_1(x)}{\int_{-\infty}^{\infty} \phi(x)s_1(x)dx}. \tag{5.6.2}$$

It can easily be shown that if $\phi(x)$ is normal density, with mean m and variance V, then $\phi^*(x)$ is normal with the same variance and mean

$$m^* = m + k/2 \tag{5.6.3}$$

(see Fig. 5.6.1A).

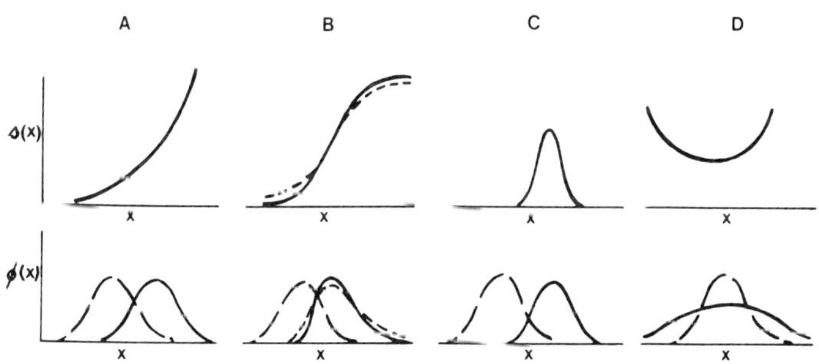

FIGURE 5.6.1. Selection functions (above) and their effect on a normal distribution (below). The distribution prior to selection is dotted in the figures below. *A:* directional selection, exponential function; *B:* directional selection, logistic and probit; *C:* stabilizing selection, Gaussian; *D:* disruptive selection, inverse of Gaussian.

Now the selection function (5.6.1) might be appropriate to describe differential fertility, but would not be applicable to describing differential mortality (or its converse viability), since $s_1(x)$ is not bounded. The function

$$s_2(x) = e^{kx}/(e^{kx} + 1) \tag{5.6.4}$$

is bounded and is used to describe differential mortality in certain cases. It is the logistic integral, well known in ecology (Fig. 5.6.1B). Unfortunately when it is applied to the normal $\phi(x)$ (as was $s_1(x)$ in (5.6.2)) normality is not preserved.

A sigmoidal function similar to (5.6.4) is the normal probability integral (also called the probit function, Fig. 5.6.1B),

$$s_3(x) = \int_{-\infty}^{x} e^{-(y-\mu)^2/2S}\, dy. \tag{5.6.5}$$

With selection function (5.6.5) the distribution after selection is also not normal but the deviation from normality can be evaluated. The proportion of survivors, s, the mean m^*, and the variance V^* of the population after selection have been given by Cavalli-Sforza and Bodmer (1971). Assuming $m = 0$, $V = 1$ with parameters μ and S for the selection as in (5.6.5), they are:

$$\begin{aligned} s &= (\sqrt{2\pi})^{-1} \int_{-\infty}^{-\mu/\sqrt{1+S}} e^{-y^2/2}\, dy \\ m^* &= \{\exp[-\mu^2/2(1+S)]\}/s\sqrt{2\pi(1+S)} \\ V^* &= 1 - m^*\left[m^* - \frac{\mu}{1+S}\right] \end{aligned} \tag{5.6.6}$$

A recursion for higher moments that allows the evaluation of the departure from normality has been derived by K. H. Chen (1980).

The model (5.6.5), with $S = 0$, represents an extreme form of directional selection called *culling selection*, in which only individuals above a certain threshold value of the trait survive to reproduce. This type of selection is sometimes used in artificial selection (See section 5.10).

Stabilizing selection is classically represented by a Gaussian survival-probability function (Fig. 5.6.1D)

$$s_4(x) = ke^{-(x-\mu)^2/2S}. \tag{5.6.7}$$

After selection of this form a previously normal population distribution remains normal. The mean m and variance V before selection are changed to m^* and V^*, respectively, with

$$\begin{aligned} m^* &= (mS + \mu V)/(S + V) \\ V^* &= SV/(S + V) \end{aligned} \tag{5.6.8}$$

with survival probability

$$s = k[S/(S + V)]^{1/2} \exp[-1/2(m - \mu)^2/(S + V)] \tag{5.6.9}$$

This stabilizing selection scheme was introduced independently by several researchers. Galton clearly described the model qualitatively and speculated that it should be the most common type of selection in nature, because most species are already close to their optimum for most traits so that changes away from this value should ordinarily be selected against. K. Pearson (1903) originated most of the mathematical analyses of this type of selection.

Disruptive selection might be modeled by the reciprocal of (5.6.7), namely

$$s_5 = k'e^{(x-\mu)^2/2S}, \tag{5.6.10}$$

if $S > V$, for which the mean and variance after selection are $m^* = (Sm - V\mu)/(S - V)$ and $V^* = SV/(S - V)$. (See Fig. 5.6.1D.)

The Gaussian model of stabilizing selection has also been used to predict results of weak directional selection simply by choosing a selective optimum remote from the mean of the population (Cavalli-Sforza and Feldman, 1978b). This gives results similar to those of a probit directional selection model so long as the population remains remote from the selection optimum while retaining normality of the population distribution.

Natural selection affects the commonly evaluated correlations between relatives. The effect was first evaluated quantitatively for traits determined by an arbitrary number of Gaussian variables by K. Pearson (1903). In modern algebraic notation the variance-covariance matrix for a multidimensional trait after selection, $\mathbf{W}^*$, can be written (Karlin 1972b, 1979, Lande 1976, Feldman and Cavalli-Sforza 1977b)

$$\mathbf{W}^* = (\mathbf{W}^{-1} + \boldsymbol{\Sigma}_S^{-1})^{-1}, \tag{5.6.11}$$

where $\mathbf{W}$ is the variance-covariance matrix of the intercorrelated variables before selection and $\boldsymbol{\Sigma}_S$ is the variance-covariance matrix of the multidimensional selection process, analogous to the stabilizing selection given for one dimension in (5.6.7):

$$s(x_1, s_2 \ldots x_n) = ke^{-(\mathbf{x}-\boldsymbol{\mu}_S)'\boldsymbol{\Sigma}_S^{-1}(\mathbf{x}-\boldsymbol{\mu}_S)/2} \tag{5.6.12}$$

The parent-offspring correlation after selection on the parents, and before selection on (but after transmission to) children, is denoted r_{PO}, while after selection on the children it is r_{PO}^*. Now let V_P, V_O denote variances in parents and children before selection, with V_P^*, V_O^* after selection. W_{PO} and W_{PO}^* are the parent-offspring covariances evaluated in the same way. Then r_{PO}^* can be obtained from (5.6.11) with

$$\mathbf{W} = \begin{pmatrix} V_P^* & W_{PO} \\ W_{PO} & V_O \end{pmatrix}, \quad \boldsymbol{\Sigma}_S^{-1} = \begin{pmatrix} 0 & 0 \\ 0 & 1/S \end{pmatrix},$$

$$\mathbf{W}^* = \begin{pmatrix} V_P^* & W_{PO}^* \\ W_{PO}^* & V_O^* \end{pmatrix}. \tag{5.6.13}$$

It is found that at equilibrium (see for example Feldman and Cavalli-Sforza 1977a),

$$r_{PO}^* = r_{PO}\left[\frac{S}{S + \hat{V}(1 - r_{PO}^2)}\right]^{1/2} \tag{5.6.14}$$

where $r_{PO} = W_{PO}(V_P^* V_O)^{-1/2}$ and $\hat{V}$ is the variance before selection at equilibrium. Clearly $r_{PO}^* < r_{PO}$.

Similarly, the correlation between sibs after selection can be obtained using (5.6.11). Here both sibs have been independently subject to selection so the selection matrix is

$$\mathbf{\Sigma}_S = \begin{pmatrix} S & 0 \\ 0 & S \end{pmatrix}. \tag{5.6.15}$$

The correlation between sibs after selection at equilibrium is then

$$r_{SS}^* = \frac{Sr_{SS}}{S + \hat{V}(1 - r_{SS}^2)}. \tag{5.6.16}$$

At equilibrium, with vertical transmission involving paternal and maternal transmission coefficients b_m and b_p, we have the correlations and regressions

$$\begin{aligned} r_{FO} &= (b_p + b_m a)(\hat{V}^*/\hat{V})^{1/2}; \quad b_{FO} = b_p + b_m a \\ r_{MO} &= (b_m + b_p a)(\hat{V}^*/\hat{V})^{1/2}; \quad b_{MO} = b_m + b_p a \\ r_{SS} &= 2(b_m^2 + b_p^2 + 2b_m b_p a)\hat{V}^*/\hat{V}; \quad b_{SS} = b_m^2 + b_p^2 + 2b_m b_p a \end{aligned} \tag{5.6.17}$$

where a is the assortative mating correlation. The equilibrium variance before selection is

$$\hat{V} = \frac{\sigma^2}{1 - (b_m^2 + b_p^2 + 2b_m b_p a)}. \tag{5.6.18}$$

After selection the variance is

$$\hat{V}^* = \frac{S\hat{V}}{S + \hat{V}} = \frac{\hat{V}}{1 + \hat{V}/S} < V. \tag{5.6.19}$$

In general, natural selection reduces all correlations between relatives, since V^*/V is always less than unity. The correlation between sibs is reduced by a factor which is the square of the reduction factor applicable in the parent-offspring case. Intuitively this can be explained as follows: it is implicitly assumed that parents are measured after reproduction (and therefore after selection if, as is ordinarily the case, natural selection occurs before or during reproduction). When correlations are taken after selection in the parent-offspring case, the effect of selection is manifest in the offspring variance only, whereas in the case of sibs both are subject to the selection.

The formulas derived above to illustrate the effect of selection on correlations between relatives can be used to solve a statistical problem of some importance. We have already remarked that regressions and correlations for IQ between adoptive parents and adopted children may be decreased by the attenuation of range resulting from self-selection of adoptive parents, who tend to have IQ's towards the high part of the range (see section 5.3). In order to evaluate the magnitude of this phenomenon in a real situation we take data concerning approximately 650 adopted Minnesota children studied by Scarr and Weinberg (1979). The relevant information is given below:

	Mean	Variance	Correlation with child's IQ
Adoptive father's IQ	116.53	129	0.15
Adoptive mother's IQ	112.43	104	0.04
Adopted child's IQ	106.19	80	
Father-mother average	114.48	116.5	0.095

IQ's are standardized to give a population mean of 100 and variance of 225. The attenuation of range among adoptive parents is therefore such that the variance is roughly halved. The average adoptive parents' IQ is almost one standard deviation above the population mean. What role has self-selection of adoptive parents played in producing the low observed correlations? If the selection is stabilizing (5.6.11) may be used here. In general it is possible that selection takes place on both parents and children in a correlated manner. But in the present example there is little information given about the adopted children prior to adoption.

Scarr and Weinberg give some evidence supporting their claim that the children were not selected. If we operate under this assumption, then the selection is one-dimensional and the r^*_{PO} in (5.6.14) can be used.

Returning to Scarr and Weinberg's data, the parents are selected from a population with mean 100 and variance 225 and the selection results in a mean of 114.5 and variance 116.5 (ignoring sex differences). From (5.6.8) with $V^* = 116.5$, $V = 225$, we estimate $S = 242$. Then (5.6.14), with observed $r^*_{PO} = 0.095$, $\hat{V} = 225$, can be used to estimate the corrected $r_{PO} = 0.131$, a 40% increase. Of course a more complicated selection matrix would have to be estimated if selection occurred on both parents and children.

If the assumption that the selection is Gaussian is valid, then the distribution after selection should remain normal. Such a check can be made when the full distribution of the original data (or at least the higher moments) is available.

A complete explanation of these data should also deal with the low variance of IQ in adopted children, as well as the fact that their mean is increased by six points over the general population. This demands a treatment in terms of a more complex transmission model involving target, concomitant and latent variables (section 5.4). A preliminary attempt (Cavalli-Sforza and Feldman, 1978a) suggested that the observed increase in the mean is too large to be explained entirely by phenotypic transmission of IQ of the strength estimated on the basis of parent-offspring correlations. Either the correlations are too low or the increase in the mean is inflated for some other reason.

5.7 MUTATION AND CULTURAL DRIFT FOR CONTINUOUS TRAITS

Traits that vary continuously are not necessarily perceived to do so. The enunciation of vowels is an example, as was discussed in section 1.8. In principle every intermediate sound is possible between, say, the vowel "a" (as in "part") and "e" (as in "bet"). Different vowels are generated by different positions of the tongue. But even if they could in principle be generated, not all possible intermediate sounds are actually made. We are trained to produce only a limited number of vowels; and we are trained to perceive a limited number. Every language has its own limited set of distinct vowel sounds. Thus discontinuity is introduced into a potentially continuous variate. Whether phonetic values evolve as continous or discrete traits is perhaps too early to decide. It is possible that both evolutionary mechanisms coexist, the continuous operating at a subperceptual level, and the discrete at the perceptual level.

Differences are found between the same vowel pronounced by different individuals, or even for the same vowel in different words, or in the same word by the same individual but in different circumstances (for instance in formal and in casual speech). Dialects often emphasize these differences as a function of geographic variation. In Figure 1.8.1 we presented a plot of the pronunciation of the word "pot" by 20 English and 20 American people (10 of each sex in each group). This gives some idea of the extent of individual and group differences on what is clearly a continuous scale. At the perceptual level, however, these sounds are not a continuum, and may be perceived as different.

The dimensions of stone tools comprise another example of continuous traits. Clearly stone tools can be cut in all sizes; their use restricts their linear measurements. A hand axe must have a circumference that will make it possible to hold it by the hand. The length must be sufficient to have the tip clearly protruding from the fist. Stone arrow-heads cannot be too big or else they will shorten the trajectory of the arrow. Unmounted small scrapers must be wide enough that they can be held between fingers. In some cases stone tools were manufactured with quite some precision. In Stentinello culture (an early Neolithic culture from

Sicily), obsidian blades probably used for sickles were carefully shaped rectangles of fairly constant size. But there is always some variation between individual tools.

L. Binford (1963) introduced the term *cultural drift* to describe the minor variation found between projectile points of different material in different burials (of Red Ochre cemeteries, late-Archaic period, Great Lakes region). Although there was no difference in the overall style of the points, the metric traits evaluated (length, basal width, maximum width, and thickness) varied significantly when points of different types of stone from the same burial, or points of the same stone from different burials, were compared. It is probable that different types of stones indicate different groups or at least different individuals. The variance between types of stone or burial was from one to nine times greater than that within types, with the ratio of between to within significant in most comparisons.

Binford analyzes previous uses of the concept of linguistic and, by extension, cultural drift by E. Sapir (1921), and by other anthropologists who base their use of this term on the presence of directionality of the evolutionary change. Binford points out that it is more reasonable to restrict the term drift to situations similar to those in which the term is applied in classical population genetics, in which divergence between daughter and parent populations occurs simply as a result of random error, and hence cannot be directional. We will adhere to Binford's proposal and make the following further comments.

Binford carefully pointed out that his material is only suggestive of drift and that the data he presented did not constitute an adequate sample, nor did it provide the controls necessary to test the drift hypothesis. We may add that like any other precise hypothesis, drift can only be disproved. "Proof" of drift is limited in reality to (i) there being no direct evidence for an alternative hypothesis (basically, selection) and (ii) expectations of variation based on knowledge of the factors important in determining drift (in the case of genetic drift, population size, migration and the like) being in reasonable agreement with observed variation. We will examine other similar data, which, like Binford's, also concern stone tools, in the context of the interplay between drift and selection.

The use of the word "drift" for the consequences of random sampling in genetic populations has itself led to some confusion. In common parlance, the term connotes a directional type of variation such as happens for instance with the deviation of a ship from its course under the action of a current or wind. It is true that the vagaries of currents and winds suggest the randomness basic to the term "drift" as used in genetics, but an element of directionality is still often attached to this word. In fact, the term "drift" is employed in applied physics to indicate a directional component in an otherwise random process, and is thus at complete variance with genetic usage. To avoid such confusion, in the previous chapters we have specified "*sampling* drift" to eliminate any implication of directionality and thus choose to align our use of the term with the genetic concept.

Burton and Bick (1972) have suggested the term "tradition drift" for differences between demes (local populations of a species) arising "through behavioral founder effect, local innovation, or the influence of individual personality." Burton and Bick refer mostly to primatological studies and, as usual, employ the word "tradition" as equivalent to "culture" when referred to animals rather than humans. Whether or not the equivalent of cultural phenomena in animals should be called culture or tradition is, however, not the main issue, but rather the definitions of the terms "cultural drift" and "tradition drift." Burton and Bick propose the use of the term tradition drift to indicate *observed* differences between demes; they suggest that these might result from an analogue of genetic (sampling) drift, or from an actual adaptive value of the behavioral shift, or a combination of both. Clearly, as they say, "the adaptive value of the behavioral shift would enhance its acceptance." Among examples of "tradition drift," they cite the well-known acceptance by a colony of Japanese macaques (on Koshima Island) of the innovation of washing potatoes in the sea. This new behavior was invented by an eighteen-month-old female and has spread over the years to almost the whole colony. Coming from a low-rank young female it was slow to spread. The propagation of another discovery, the flotation method for separating wheat grains from sand, has also been observed and documented in the same colony (Kawai, 1965). The fact that potato washing was at all adopted in spite of coming from a juvenile may indicate that it has an intrinsic value through the medium of increased sodium-chloride intake in improving the taste, or in serving some physiological function, although as was emphasized in chapter 1 such speculation may completely miss the mark in assigning adaptive merits. Several other new behaviors have been observed in some colonies of primates and their origin traced to a single innovator who was imitated by the rest. Typically, they are not found among other colonies of the same species. These examples might lend themselves to experimentation; carriers of innovative behavior could be inserted into naive groups. In general, it has been stressed that the personality and position in the social hierarchy of the innovator is very important in determining the adoption of a new behavior.

The term "tradition drift" is an innovation that itself has had some acceptance among the scientific community (see, for example, E. O. Wilson, 1975, p. 13). In our opinion, the use of this term in labeling *observed* situations such as that of the Japanese macaques is to be discouraged. Observed behavior differences between demes do not need to have a specific "scientific" label; they can be called what they are ("behavior differences"). Similarly, diffusion of a behavior through a population could be called "diffusion of behavior." It is appropriate, in our view, to reserve the term drift, and in particular, sampling drift, for an underlying nonadaptive mechanism. In practice, it is very difficult to decide in individual cases—like those of potato washing and wheat flotation in macaques—if an innovation diffuses because it is attractive or because of the status of the innovator or by some other accident. Even if the distinction is empirically difficult to make, just as in biological evolution (see, e.g., Kimura, 1979), it nevertheless remains conceptually fundamental.

As we have seen for vertically transmitted discontinuous traits, the effect of sampling is governed by the size of the population and the distribution of family size. But the laws of transmission may themselves be important; our analysis has so far been limited to very simple cases. The addition of oblique and horizontal transmission can only complicate matters further. For continuous traits there is an additional complication, *transmission error,* measured as mutation variance. Whether we consider speech or tool making, the two examples quoted earlier, there is first a process of learning in which auditory or visual cues, respectively, are used in a process of imitation by the learner. The process may never be absolutely complete in the life of an individual. But there may be an identifiable final stage after which changes are exceptional or, if any, are degenerations due to age, disease, and like factors. For speech, the process of formation takes many years, and the crystallization of a speaking pattern, which is hardly altered thereafter, is largely the consequence of the deterioration of the specific learning mechanisms with puberty (although some later adjustments are possible). Musicians will agree that there are similar early-age periods during which learning of digital manipulation is much easier than later on.

With the end of the learning process, products of the tongue (phones) or of the hands (tools and the like etc.) are basically crystallized. Future learners will learn by imitation of the finished products, which are proposed to them as examples by their teachers. The learner is exposed to a variety of models, perhaps all made by one person but possibly also from different sources. The learner will probably choose among them, and perhaps imitate only one, or reject them all and go on to innovate. Generally, however, change of model will result in a changed product. Even ignoring this complication, the units of a given product made by a single individual will still vary one from the other and these varying traits may be interesting and important personality traits of the individuals. Here, as before, we shall ignore this type of variation in the life of an individual and will be concerned with just the average product of the individual once his learning period is over. This can be regarded as the equivalent of the mental image, the engram of this product in the brain of the individual. We will be interested in differences between the engrams of individuals as revealed by differences in (average) actual products. For simplicity it will usually be assumed that variability between the units of product of a given individual is negligible, so that one unit of product tells us as much about the maker as we want to learn.

The deviation between the engram of an individual and that of the model he has used is an error of copy. Insofar as the child's engram rather than his parents' will be copied by the children of that child, the error accumulates over generations. It will be considered as a "random" error, that is, its magnitude has some probability distribution $\phi(\cdot)$. For simplicity we usually assume that this distribution is normal with zero mean and variance ϵ^2. The error distribution can be viewed as that of a continuous analogue to mutation.

In summary, the term "cultural drift" (or "tradition drift," which some scientists prefer for animals other than humans) should be reserved for the mechanism of variation due to sampling accidents. In small populations the size of the population is an important determinant of random variation. For a continuous trait the difference between a copy and the model (be it a stone tool, a sound, or anything the production of which has to be learnt by an individual) is an error of copy that has the property of a random variate and is the equivalent of mutation in a discrete trait. Sample size and random copy error generate random variation responsible for cultural drift. Although this may contribute to the differentiation between individuals and between groups, it is better to restrict the terms cultural drift and tradition drift to the nonadaptive, random mechanism rather than to observed differentiation among groups. The latter could be due to a variety of mechanisms, adaptive and nonadaptive, between which it is usually difficult to distinguish. We will discuss later the interplay of drift and selection for continuous traits. First, we will consider the general properties of drift for a continuous trait in finite groups under cultural transmission.

5.8 UPPER LIMITS TO INDIVIDUAL VARIATION UNDER CULTURAL DRIFT

Consider a group of N individuals, each of which reproduces by giving rise to one descendent that learns a trait from its single parent. Let $X_{i,t}$ be the trait value of the ith individual at generation t. Then we postulate

$$X_{i,t+1} = X_{i,t} + \epsilon_i, \tag{5.8.1}$$

where ϵ_i is a random variable with mean 0 and variance σ^2. The latter takes the role of mutation in this framework of continuously varying traits. It is also assumed that ϵ_i is independent of ϵ_j $(i \neq j)$. The expected population average at generation $t+1$, given our knowledge of the previous generation, is

$$E[\Sigma X_{i,t+1}/N \mid t] = E[\overline{X}_{t+1}/t] = \overline{X}_t \tag{5.8.2}$$

so that

$$E(\overline{X}_{t+1}) = E(\overline{X}_t) = \ldots = \overline{X}_0. \tag{5.8.3}$$

Denote by V_{t+1} the expected value of the within population variance at generation $t+1$, that is with respect to the mean X_{t+1}:

$$V_{t+1} = E\left[E\left\{\sum_{i=1}^{N} (X_{i,t+1} - \overline{X}_{t+1})^2/(N-1)\,|\,t\right\}\right]$$

$$= V_t + \sigma^2 \tag{5.8.4}$$

$$= V_0 + t\sigma^2. \tag{5.8.5}$$

Thus, with one child per parent, the variance increases without bound. In the next section it will be seen that the squared deviation of the observed mean from the original mean also increases indefinitely.

The model (5.8.1) is very special, and under more realistic conditions the within-group variance would be expected to remain finite. For example, within any population group there would be no reciprocal understanding unless most aspects of languages had limited variation, at least on a local level. Social relations would be chaotic if a certain number of rules were not accepted by everybody with but minor variation. One of the central questions of cultural evolution revolves around exactly how stabilization of the variance is achieved in the face of accumulated mutation or error variance, σ^2. *Although some kind of selection might be invoked, it is within the process of cultural transmission itself that an interesting mechanism for restriction of variance among the individuals of a population resides.*

There are two modifications of (5.8.1), each of which is sufficient to prevent the variance from diverging. One of them invokes contributions from more than one individual of the parental group to each individual of the offspring generation. The other is that the number of children per parent is not constant, but has some appropriate probability distribution. One assumption we have made is that every one of the N individuals of the $(t + 1)^{\text{st}}$ generation has a probability $1/N$ of having had a specific individual of generation t as its parent. Thus the population size remains constant, N, while a given parent produces a random number of offspring with mean unity. The procedure used is that of Cavalli-Sforza and Feldman (1973a), except that for mathematical simplicity the order of events in that paper, mutation followed by transmission, is reversed. This results in the present substitution of σ^2 for $(1 - g)^2\sigma^2$ in the earlier treatment. The model can be written

$$X_{i,t+1} = (1 - g) \sum_{j=1}^{N} \delta_{ij} X_{j,t} + \epsilon_i + g\overline{X}_t, \qquad (5.8.6)$$

with

$$\delta_{ij} = 1 \text{ if } j \text{ is the parent of } i$$
$$= 0 \text{ otherwise,}$$

and the chance that $X_{i,t+1}$ is the offspring of $X_{j,t}$ is $1/N$ for every j. Using this device, the distribution of progeny number per parent is binomial. Then

$$V_{t+1} = \frac{1}{(N-1)} E \sum_{t} [(X_{i,t+1} - \overline{X}_t)^2/t] \qquad (5.8.7)$$

$$= \left(1 - \frac{1}{N}\right)(1 - g)^2 V_t + \sigma^2$$

$$\rightarrow \hat{V} = \sigma^2 \Bigg/ \left[1 - \left(1 - \frac{1}{N}\right)(1 - g)^2\right]. \qquad (5.8.8)$$

When $N \rightarrow \infty$ we obtain the result

$$\hat{V} = \sigma^2/[1 - (1 - g)^2]. \qquad (5.8.9)$$

Thus if $g = 1$, so that every individual is expected to be equal to the mean of the parental generation, the variance is exactly σ^2. Even in the absence of a group effect ($g = 0$) from (5.8.8), the limiting variance is bounded, $\hat{V} = N\sigma^2$. This upper bound comes about because, implicit in the model, every individual at time t is equally likely to be the descendant of any of the N members of the parental generation. Thus there is a chance that any individual exerts transmission by having more than one descendant later on.

Cavalli-Sforza and Feldman (1973a) have also extended this model of stochastic relationship between generations to the biparental case with b_m and $b_p = 1 - b_m$ the paternal and maternal contributions respectively. Under random mating it was shown that

$$\hat{V} = \sigma^2/[1 - (b_m^2 + b_p^2)(1 - 1/N)(1 - g)^2]. \quad (5.8.10)$$

Again, as $N \longrightarrow \infty$, (5.8.10) corresponds to the result of section 5.3. It reduces to (5.8.8) for $b_m = 0$ or $b_p = 0$. The smallest value for $\hat{V}$ is obtained when $b_m = 1/2$.

Thus, even with uniparental transmission, there is no divergence of the variance under this model of variation in the number of offspring per parent. But if the only process that controls the evolution is binomial sampling in uniparental transmission in a finite population of size N, then the equilibrium variance is large, N times the mutation variance. This should be contrasted with the case of biparental or n-parental or group transmission (section 5.2) where the limiting within group variance is ordinarily higher than the mutation variance by a factor of 2, of $n/(n-1)$, or of $1/(1-g)^2$, respectively.

5.9 VARIATION BETWEEN GROUPS

As long as there is some degree of communication across generations, the *within group variance* reaches an upper bound expressible as a constant multiple of the mutation variance. The communication may be de facto induced by the binomial transmission model of the previous section. We now turn attention to the *group mean*.

If there is no selection, just sampling drift, the expected value of the mean in the group does not change (see (5.8.3)). But the *observed* mean of a finite group will change over time. There are two ways in which the evolution of the group mean might be simply quantified:

(1) The variation of the group mean at time t can be studied relative to the original group mean $\overline{X}_0$. The quantity of interest is

$$E(\overline{X}_t - \overline{X}_0)^2. \tag{5.9.1}$$

(2) Variation among the means of groups that started with a common value and evolved independently. The quantity of interest is

$$E(\overline{X}_t - \overline{\overline{X}}_t)^2, \tag{5.9.2}$$

where $\overline{\overline{X}}_t$ is the average of the group means at time t.

The difference between (5.9.1) and (5.9.2) is expected to become negligible as the number of groups becomes very large. Here we focus on (5.9.1). Now

$$E(\overline{X}_t - \overline{X}_0)^2 = E\sum_{i=1}^{t}(\overline{X}_i - \overline{X}_{i-1})^2 = tE(\bar{\epsilon}^2), \tag{5.9.3}$$

since the contribution from mutation in any generation is independent of that in any other generation. In the simplest case of uniparental transmission (with or without group effect),

$$E(\overline{X}_t - \overline{X}_0)^2 = t\sigma^2/N. \tag{5.9.4}$$

With biparental transmission in terms of b_m and b_p, the lower bound to this squared deviation from the original mean is

$$\frac{t\sigma^2}{N}\left\{1 - (b_m^2 + b_p^2)\left(1 - \frac{1}{N}\right)(1 - g)^2\right\}, \tag{5.9.5}$$

(Cavalli-Sforza and Feldman, 1973a), while (5.9.4) is an upper bound.

In general, the variation of the group mean about the original group mean increases linearly in time with a constant of proportionality expressible in terms of σ^2/N, modified by the transmission coefficients in the different models discussed previously.

Cases other than biparental transmission are considered in the cited paper. The main result is that *the variance between group means increases indefinitely,* unless some kind of stabilizing selection or communication between the groups limits their divergence.

5.10 CULTURAL SELECTION VERSUS CULTURAL DRIFT

In section 5.7 cultural drift was discussed in terms of Binford's example of projectile points that came from various archaeological caches and are made with different materials. The data revealed significant differences between average linear measurements of arrowheads made of different stones but found within the same cache, and between those from different caches but made from the same stone. One interpretation is that the different stone types were contributed by different population subgroups or by different individuals. In Table 5.10.1 the variances within and between stone types are recorded for the cache having the greatest variety of stones (Huron Beach). There is a significant fraction of the variance between stone types in two of the four measurements. The possibility that the material of which the points are made may have an effect should be considered. However, the fact that points made of the same material but from different caches also

TABLE 5.10.1. Analysis of variance of four linear traits in projectile points from Red Ochre caches at Huron Beach (Binford's data). The mean squares between stone types (V_B) have 4 degrees of freedom and the mean squares within (V_W) 54 degrees of freedom. F is the ratio between them. Asterisks indicate significance (* for $P < 5\%$, ** for $P < 1\%$, ***$P <$ 0.1%). r is the intraclass correlation coefficient.

	V_B	V_W	F	r
Length	0.8512	0.3030	2.81*	0.159
Basal width	0.8258	0.1929	4.88**	0.256
Max. width	0.6227	0.1592	3.91**	0.234
Thickness	0.0119	0.0099	1.20	0.021

differed significantly, on average, suggests that this may be of less importance than other effects.

These data speak to the question of restriction of variation among group means, which as we will see in section 5.12, can grow without limit unless restrained by some form of selection. Such selection would be stabilizing in the sense defined in section 5.6. It would have to act to eliminate extreme types while retaining those whose characteristics are near some optimum.

To model this conflict between drift and cultural selection it is first assumed that the trait in question is normally distributed, that the transmission is uniparental with group effect, and that the usual mutation contribution ϵ_i is normal with mean zero and variance σ^2. After transmission and mutation the trait value $X_{i,t}$ of the ith individual in a population of size N would satisfy

$$X_{i,t+1} = (1 - g)X^*_{i,t} + g\overline{X}^*_t + \epsilon_i. \tag{5.10.1}$$

Here the asterisk indicates that the measurement from the earlier generation being copied has been taken after cultural selection. That is, such tools have been evaluated, considered acceptable and might be used as models by somebody who is learning to make them. Otherwise they would have been discarded.

We will consider two specific hypotheses: (i) stones from the same cache and of the same type are made by a single individual; (ii) they are made by a single group of individuals. In the absence of selection, assuming (i), the variance within stone types would be equal to σ^2, the mutation (copy error) variance, and that between stone types (i.e. individuals) would grow indefinitely, but only if $g = 0$ and if there were no variation in progeny size. Otherwise (§5.8) it would stabilize to a level depending on g and N. In case (ii), under drift alone the variance within the cache would stabilize to a quantity determined by σ^2, g, and N, while that between caches would grow indefinitely with time. It should be noted that the variance "between," as computed in the standard analysis of variance of Table 5.10.1, is equal to n times the variance between as given in section 5.9, where n is the average number of stones per stone type found in the cache. The effect of the observed number of stones per stone type is eliminated in the intraclass correlation coefficient given in the table, which expresses the similarity between stones of the same type independently of n. The variances "between" of Table 5.10.1, divided by n, are always smaller than the variances within. That the variation between is relatively small lends weight to the intuition that some type of stabilizing selection played a role in the composition of the caches.

Of the many possible mathematical representations of stabilizing selection (see Figure 5.10.1), we assume here that the probability of acceptance is a normal function of the trait of the cultural object under discussion, i.e. (see also section 5.6)

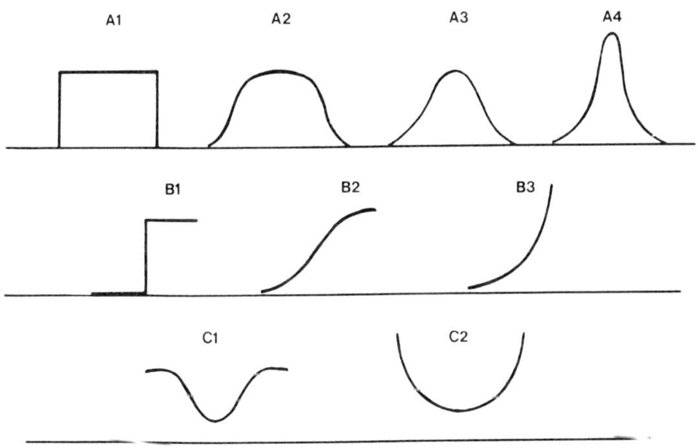

FIGURE 5.10.1. Cultural selection for a continuous trait. Examples of curves of probability of acceptance (ordinate) of a cultural trait as a function of the metric value of the trait. *A:* stabilizing selection—*A*1: step function; *A*2: platykurtic curve; *A*3: normal Gaussian curve; *A*4: leptokurtic curve. *B:* directional selection—*B*1: step function; *B*2: probability integral, or logistic; *B*3: exponential. *C* disruptive selection—*C*1: Gaussian turned upside down; *C*2: inverse of Gaussian.

$$\phi(x) = Ke^{-(x-\mu)^2/2S}, \tag{5.10.2}$$

where μ is the optimal trait value, S a variance that indicates the strength of selection and K a constant normalizing $\phi(\cdot)$. Weak selection occurs if S is large, strong selection entails that S be small. If the mean and variance of the trait before selection are m and V, respectively, then after selection through (5.10.2), the distribution of the trait is normal with mean m^* and variance V^*:

$$m^* = \frac{\mu V + mS}{V + S}, \quad V^* = SV/(S + V). \tag{5.10.3}$$

Clearly the effect of selection has been to reduce the variance,

$$SV/(S + V) < V, \tag{5.10.4}$$

and to bring the mean closer to μ;

$$(\mu V + mS)/(S + V) - \mu < m - \mu. \tag{5.10.5}$$

Suppose that the population trait value in adults at generation t is represented by X_t^*, which is normally distributed with mean μ_t and variance V_t, and that after transmission,

$$V_{t+1} = (1 - g)^2 V_t^* + \sigma^2, \tag{5.10.6a}$$

$$\mu_{t+1} = \mu_t^*. \tag{5.10.6b}$$

Then after selection, according to (5.10.3), these become

$$V_{t+1}^* = SV_{t+1}/(S + V_{t+1}), \tag{5.10.7a}$$

$$\mu_{t+1}^* = \frac{(\mu V_{t+1} + \mu_t S)}{S + V_{t+1}}, \tag{5.10.7b}$$

On substitution of (5.10.6a) into (5.10.7a), a recursion of V_t^* is obtained which converges to the equilibrium

$$\hat{V}^* = \frac{S(2g - g^2) - \sigma^2 + \{[S(2g - g^2) + \sigma^2]^2 + 4\sigma^2 S(1 - g)^2\}^{1/2}}{2(1 - g)^2} \tag{5.10.8}$$

with $\hat{V} = \hat{V}^* + \sigma^2$. For $g = 0$, this simplifies to

$$\hat{V}^* = \sigma^2[(1 + 4S/\sigma^2)^{1/2} - 1]/2. \qquad (5.10.9)$$

Under hypothesis (i) above, the variance within (of stones of the same type made by an individual) after selection, is given by (5.10.7a) for $V = \sigma^2$ as

$$V_W = S\sigma^2/(S + \sigma^2).$$

In order to predict the variance between, we assume that the stones made by an individual are a realization of an image which that individual has in mind at the end of a learning process during which he has seen a probably large number, m, of stones produced by his teacher. This observation and experimentation are assumed to have reduced σ^2 to a stable low value. If we call $X^*_{i,t+1}$ the image characteristic of the ith individual of the $(t+1)^{\text{th}}$ generation, formed by some process of averaging on the realizations of the same artifact by the teacher $X^*_{i,t}$, then the error variance of the images is some multiple C of (5.10.7a), that is,

$$\text{Var}\,(X^*_{i,t+1}\,|\,t) = \frac{C\sigma^2 S}{S + \sigma^2}\,. \qquad (5.10.10\text{a})$$

If the learner made perfect measurements and took an arithmetic average of the m realizations of the teacher he observed, $C = 1/m$. The process of learning could hardly be so perfect and we might expect $C > 1/m$. From (5.10.7b) we have, in the same way,

$$E(X^*_{i,t+1}\,|\,t) = \frac{\mu\sigma^2 + SX^*_{i,t}}{S + \sigma^2}\,. \qquad (5.10.10\text{b})$$

Now the variance among, say, k individuals conditioned on the earlier generation is expected to be

$$V_{\tilde{B},t+1|t} = E\left[\sum_{i=1}^{k} (X^*_{i,t+1} - \overline{X}^*_{t+1}\,|\,t)^2\right]\Big/(k - 1), \quad (5.10.11)$$

where the tilde on $\hat{B}$ indicates the conditioning. Expanding

(5.10.11), we have

$$V_{\tilde{B},t+1|t} = \frac{1}{k-1}\left\{\sum_{i=1}^{k} [\mathrm{Var}\,(X^*_{i,t+1}|t) + [E(X^*_{i,t+1}|t)]^2] - k\,\mathrm{Var}\,(\overline{X}^*_{t+1}|t) - k[E(\overline{X}^*_{t+1}|t)]^2\right\}$$

$$= \frac{C\sigma^2 S}{S+\sigma^2} + \frac{S^2}{(S+\sigma^2)^2}\sum_{i=1}^{k}(X^*_{i,t} - \overline{X}^*_t)^2 \qquad (5.10.12)$$

by (5.10.10a) and (5.10.10b). Removing the conditioning in (5.10.11), we have the recursion

$$V_{B,t+1} = \frac{C\sigma^2 S}{S+\sigma^2} + \frac{S^2}{(S+\sigma^2)^2} V_{B,t}, \qquad (5.10.13)$$

which converges to

$$\hat{V}_B = C\frac{S(S+\sigma^2)}{2S+\sigma^2}. \qquad (5.10.14)$$

There are three parameters of this model to be estimated, σ^2, C, and S, and only two quantities (V_B and V_W) from which to make these estimates. Thus even in this simple case a complete estimation is impossible. However, it is highly likely that σ^2 is small at the end of the learning process compared with S. If not, the proportion of arrowheads discarded would be high. The proportion p of arrowheads made that are considered acceptable is given by (5.6.9) as

$$p = k[S/(S+\sigma^2)]^{1/2} \exp\,[-\tfrac{1}{2}(X^*_i - \mu)^2],$$

and for p to be high with $k = 1$, X^*_i (the image) must be close to μ and σ^2 small with respect to S. If σ^2 and S were the only important factors here, then for the proportion of arrowheads discarded to be as low as 5% the variance within must be $V_W = 0.9\,\sigma^2$. For a discard fraction of 10%, $V_W = 0.81\,\sigma^2$ and for 20% discarded (which seems very high) $V_W = 0.64\,\sigma^2$.

The intraclass correlation coefficient is

$$r = (\hat{V}_B - V_W/n)/(\hat{V}_B - V_W/n + V_W) = \frac{1}{1 + \dfrac{\sigma^2}{SC}\left(2 + \dfrac{\sigma^2}{S}\right)}. \tag{5.10.15}$$

If $\sigma^2 \ll S$,

$$r \simeq \frac{1}{1 + 2\sigma^2/SC}, \tag{5.10.16}$$

so that $r = 0.3$ corresponds to $CS = 0.85\,\sigma^2$, $r = 0.1$ to $CS = 0.22\,\sigma^2$, and $r = 0.01$ to $CS = 0.02\,\sigma^2$. High values of r are observed (Table 5.10.1) for linear measurements other than thickness, and a low value for thickness. The suggestion is that thickness of the projectile (which is probably an important factor in determination of weight) is subject to especially strong selection. When selection is weak (as appears to have been the case for shape and other linear measurements) and $S \gg \sigma^2$, CS is of the same order of magnitude as σ^2, suggesting that C can be small, a natural expectation since in the ideal case $C = 1/m$, where m is the number of projectile points used to produce the image.

The details of the analysis under hypothesis (ii) are not pursued here; even more parameters require estimation in this case. Suffice to say that a shift of variance from the between category to the within would result and then, all other things being equal, the ratio between them should be lower than that expected under hypothesis (i).

Quantitative data on early artifacts have also been obtained by G. L. Isaac (1977), and are represented in Figure 5.10.2. These tools were taken from a variety of sediments of different ages at Olorgesailie near Nairobi. The highest and lowest levels are separated by a substantial depth of sediment. There are no accurate dates assigned to the levels, but the time separation between the highest and lowest layers may be large. Of the sites in the middle stratigraphic set only three could be placed in proper chronological sequence. The data in Figure 5.10.2 refer to various measurements (length/breadth index, weight, thickness/breadth index, and the like) on all large bifacial tools found. The horizontal bars represent 95% confidence intervals around the means (under a normality assumption). The means vary greatly as shown by an analysis of variance (and as confirmed visually by the behavior of confidence intervals in the figure). However, as Isaac noted, none of the traits shows a clear-cut time trend. The absence of clear time trends extends to much longer periods and covers wide areas. The variation of means of stone tools from other parts of Africa is depicted at the bottom of Figure 5.10.2. The maximum range might indeed be fixed biologically (by hand size for example), in which case one might again infer a balance between cultural drift and some mode of stabilizing selection.

One important and interesting complication in the Olorgesailie data is that the material is not homogeneous, in the sense that the "bifaces" were shaped in somewhat different ways. Presumably this was done to accommodate the different tasks for which the tools were to be used. This raises the possibility that each collection of bifacial tools might contain different tools, in which cases the observed fluctuations in the means might merely represent differences in the proportions of different tools. The problem turns out to be more complex than even this difficulty would suggest. A more detailed analysis of the material produces a suggestion of multimodality in the assembly of artifacts. Only 39% of the bifaces correspond to the description of a handaxe. Others are cleavers, scrapers, and like tools. However, if the handaxes above are tested for heterogeneity between layers (Fig. 5.10.2B) the result is very similar to that already observed for all bifaces (Fig. 5.10.2A). The heterogeneity between layers is not (or not entirely) due to heterogeneity of frequencies of different tools. It is also important that the separation between categories is never a sharp one in these Acheulean tools, although it becomes sharper in tools made at times closer to modern man.

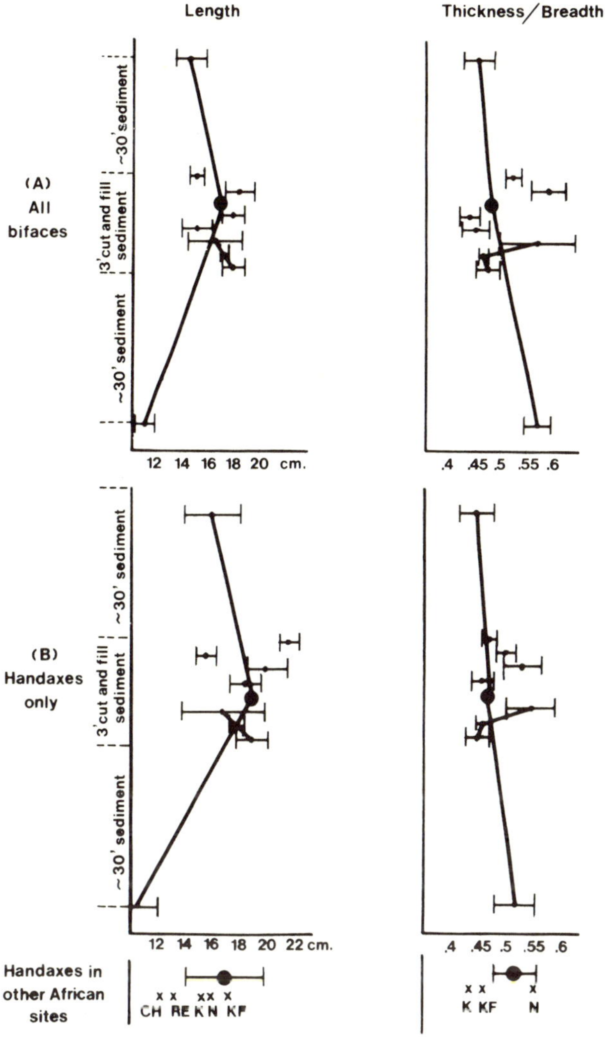

FIGURE 5.10.2. Length and thickness/breadth measurements of all bifaces (above) and handaxes only (below) found at Olorgesailie, near Nairobi, after Isaac (1972, 1977). The ordinate shows stratigraphic information. The horizontal segments show the mean measurements and the 95% confidence limits. Note significant variation of means without clear time trends. Mean measurements of handaxes from other African sites (CH, RE, etc.); their mean and confidence interval shown at bottom.

Some process of specialization of tasks must have occurred, but gradually, so that continuity of tool types was preserved. Perhaps there was weak disruptive selection producing different classes. In Table 5.10.2 the variances between and within layers and the intraclass correlation coefficients are recorded. The latter (the ratios of variance between the total variance, after elimination of the random-sampling component in the variance between) are comparable to those observed for Binford's projectile points.

The large numbers of tools characteristic of the Olorgesailie data suggest that several individuals contributed to the tools found in any layer. If this were the case, then the variance within layers contains not only variance within the production of an individual but also between individuals. This would result in a reduced intraclass correlation in Isaac's data compared with Binford's. On the other hand, variation in individual optima, due for instance to the overall size of the individual, would tend to increase the correlation but only under hypothesis (i). The very different nature of the tools in the two cases suggest caution in comparing the statistical properties of the two data sets.

TABLE 5.10.2. Analysis of variance of measurements on bifacial tools between nine levels at Olorgesailie (data by Isaac, 1977). Between variances have 8 degrees of freedom, and within variances have a variable but large number (on average 510). One site (DE 189 B) was excluded because of excessive uncertainty in the number of observations per layer from the published data. See Table 5.10.1 for other information.

	V_B	V_W	F	r
Length	42,252	1,773	23.8***	.29
Breadth/width	1,077	83.3	12.9***	.17
Thickness/breadth	1,819	119.6	15.2***	.20
Ovoidal shape index	772	93.7	8.2**	.11
	5,421	510	10.6***	.15
	2,750	470	5.8**	.08
Scar No.	399	989	4.0*	.005
Trimming	4,113	291	14.1***	.19

In light of the variance properties of our models, a reasonable conclusion from the statistical significance of the variance ratios in the two sets of data is that drift, perhaps between individuals in Binford's data and between groups of individuals in Isaac's, has been important. However, the fact that the between variance remains small might well have been the result of forces such as stabilizing selection and/or cultural transmission.

The specialization of stone tools by contrast is a problem that can be put in terms of disruptive cultural selection. Figure 5.10.3 is an idealized scheme presented by Isaac of the evolution of tools over 3–4 million years. Four discrete steps are represented. The

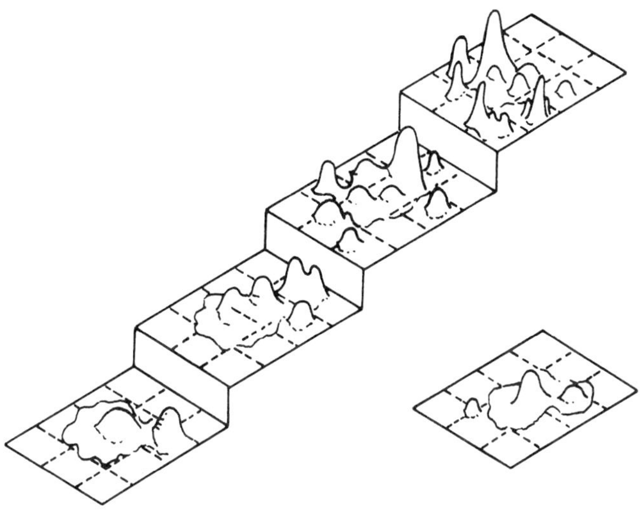

FIGURE 5.10.3. A diagrammatic representation of the variation of stone tools, and the increase of differentiation over time. The two dimensions for each step represent two different measurements. The four steps represent six different times, in order from bottom left: earliest paleolithic (Oldowan); Acheulean; Musterian; latest paleolithic. The peaks indicate what are potentially at least different tools. With time there is an increase in the number of tools, and also in the sharpness of differentiation. The bottom right diagram shows tools by a modern Australian aboriginal tribe. From G. L. Isaac 1976.

first corresponds to Oldowan tools, which are found in the oldest layers all over the world and in Australia in recent layers. Among these only two types of tools can be distinguished, core-choppers, used primarily as choppers but also to generate flakes, and the flakes themselves. Step two is the Acheulian period, between approximately 500,000 and 200,000 years ago, to which the Olorgesailie material belongs. Here there is some multimodality, indicating some separation in progress. The separation continues into step three (late Acheulian and Musterian) between 200,000 and 40,000 years ago, when specialization becomes quite marked, though not as pronounced as among tools of the last 40,000 to 50,000 years (covering the appearance and spread of modern man). Naturally, archaeologists working in layers corresponding to step four and, to some extent, step three tend to use elaborate typologies to describe their artifacts; these are less and less justified the further back studies are made.

When compared to some biological changes, for example in brain size, that occurred during the same period, the extreme slowness of cultural change is certainly puzzling. It has been suggested that lack of biological change in neural mechanisms may have been responsible for the slowness of cultural change. But the fact that in respect of brain size and the potential for specific intellectual capabilities, including language, today's Australian aborigines are perfectly comparable to the other representatives of modern man, while their artifacts are closer to those of our ancestors of 3–4 million years ago, indicates that the stimulus toward cultural specialization was missing or weak. The tool requirements of hunters and gatherers are easily met, and in the absence of a specific stimulus we might suspect that very little cultural change would happen. In such a situation biological divergence may even occur more quickly than would cultural. Only when specific new tasks were invented, or became necessary, as occurred for instance with the development of agriculture, was the stage set for rapid technological and cultural evolution.

We have discussed stabilizing and disruptive selection in reference to stone tools. Can we document a cultural example for which the analogue of directional selection seems appropriate? The example we suggest refers to a biological object, maize, but the mechanism of selection is artificial and therefore cultural. At the beginning of domestication, wild corn and the first domesticates had very small cobs, less than one centimeter long. Most probably because the corn set aside as seed for next year was not a

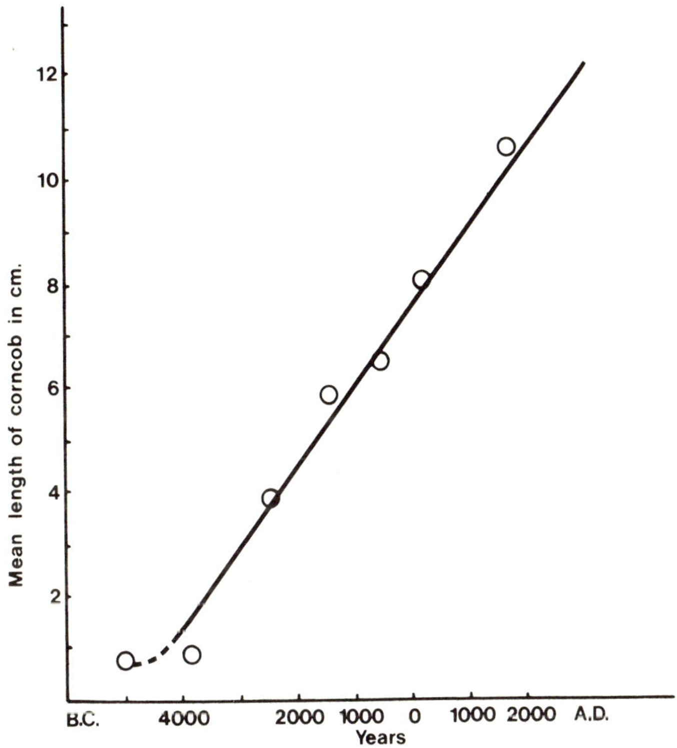

FIGURE 5.10.4. Evolution of mean corncob length in the last 7000 years based on data from Tehuacan presented by Mangelsdorf, MacNeish, and Galinat, and as graphed by Kirkby (1973).

random sample, but was selected because it was of better quality than the corn used for consumption (for instance by setting aside larger cobs), the average size of cobs has steadily increased. The increase has been slow, but regular, presumably limited by the available genetic variation for cob size. The result has been a very regular increase of about 1 cm every 500 years (see Figure 5.10.4), as determined from archeological data from ancient Mexico dating from 7000 years ago to the present. Although the evolutionary change is a genetic one, the pressure applied has been *cultural.* The data suggest that the cultural selection pressure, at least in gross terms, remained remarkably constant over this long period of time, so that the form of the cultural pressure responsible for maize evolution has been resistant to change during this period.

5.11 SIMPLE SOCIAL HIERARCHIES AND COMPARTMENTS

As has already been suggested, the parental mode of cultural transmission is probably the most ancient. Even so it may have been supplemented from very ancient times by transmission from other members of the nuclear family (perhaps even through the evolution of primates). The model in which the whole group participates equally in the transmission is just one of the many conceivable models. Other special individuals may play even more important roles than parents in cultural transmission, with results described above. Leaders constitute one important class of such special individuals, whose influence may have become especially significant with the development of agricultural societies.

It is of interest to explore other patterns of societal structure, namely through the formation of hierarchies. Obviously in the limit of such compartmentalization there would be no communication and the group variance would increase without limit. Some simple hierarchical relations in terms of transmission matrices were suggested in an earlier study (Feldman and Cavalli-Sforza, 1975). For the purpose of establishing some simple paradigms, we limit our study to populations of three individuals: real populations will obviously be much more complex. Individuals are numbered from 1 to 3, and in Figure 5.11.1 the hierarchical relations are indicated by the graph and the corresponding stochastic matrix. The directed paths represented by the arrows in Fig. 5.11.1 and the entries of the matrices show the relative influence of each individual over the others. This is the simplest "linear" heirarchy, to be compared with those of Fig. 1.11.2. The process of transmission may occur repeatedly within a generation or across generations. For larger groups, extensions and combinations of these elements will be required. As an example of the analysis we pursue in more detail the relationship in Figure 5.11.1.

Consider the case of a population consisting of three individuals with trait values $X_{1,t}$, $X_{2,t}$, $X_{3,t}$ in the tth generation. Individual number 1 contributes to its offspring and to that of number 2; number 2 contributes to its offspring and to that of number 3, and number 3 contributes only to its own offspring. The arrangement is linear. The representation of the offspring trait values is then

$$\begin{pmatrix} X_{1,t+1} \\ X_{2,t+1} \\ X_{3,t+1} \end{pmatrix} = \begin{pmatrix} 1 & 0 & 0 \\ p_1 & 1-p_1 & 0 \\ 0 & p_2 & 1-p_2 \end{pmatrix} \begin{pmatrix} X_{1,t} \\ X_{2,t} \\ X_{3,t} \end{pmatrix} + \begin{pmatrix} \epsilon_1 \\ \epsilon_2 \\ \epsilon_3 \end{pmatrix}, \tag{5.11.1}$$

	1	1	0	0
$1 \rightarrow 2 \rightarrow 3$	2	p_1	$1 - p_1$	0
	3	0	p_2	$1 - p_2$

FIGURE 5.11.1. A hierarchy in a population of three individuals. Tree graph and transmission matrix.

with $0 < p_1, p_2 < 1$ and ϵ_i of zero mean and variance. The analysis of the dynamics of the expected variance among the three individuals is detailed in Feldman and Cavalli-Sforza (1975) in terms of the spectral properties of the transmission matrix in (5.11.1). The equilibrium is

$$E(V_\infty) = \frac{\sigma^2}{3(p_2 - p_1)^2}\left[(p_1^2 + p_2^2 - p_1 p_2)\left(\frac{2}{2p_1 - p_1^2} + \frac{2}{2p_2 - p_2^2}\right) - \frac{(p_1 + p_2)^2}{p_1 + p_2 - p_1 p_2}\right], \quad (5.11.2)$$

and it is approached at a rate equal to the square of the larger of $(1 - p_1)$ and $(1 - p_2)$.

If transmission were governed not by the matrix (5.11.1) but by the "teacher-leader" type matrix as in Fig. 1.11.2a,

$$\begin{pmatrix} 1 & 0 & 0 \\ p_1 & 1 - p_1 & 0 \\ p_2 & 0 & 1 - p_2 \end{pmatrix}, \quad (5.11.3)$$

the equilibrium variance would be

$$E(V_\infty) = \frac{\sigma^2}{3}\left[\frac{2}{2p_1 - p_1^2} + \frac{2}{2p_2 - p_2^2} - \frac{1}{p_1 + p_2 - p_1 p_2}\right]. \quad (5.11.4)$$

It is a matter of algebra to show that, if $p_1 \neq p_2$, this last variance is less than that in (5.11.2). Again, in line with the result of the earlier section, the teacher mode of transmission reduces the equilibrium variance, although the rate of approach to the limit (5.11.4) is the same as for (5.11.2).

5.12 TRANSMISSION MATRICES AS MODELS OF VERTICAL AND OBLIQUE TRANSMISSION: TEACHER VS. PARENTS

We have already seen in Chapter 3 that for some purposes it is convenient to use transmission matrices to describe vertical and oblique transmission from one generation to the next. In intergenerational transmission, the elements of the ith row refer to the contributions by all the individuals of the parental generation to the ith offspring. Each diagonal element of the matrix represents the uniparental contribution to the respective offspring. One matrix we have proposed for vertical plus oblique group transmission (Section 3.11) of a *discrete trait* is of the type $\mathbf{P} = \|p_{ij}\|$, with $p_{ij} = (1 - \tau)/(N - 1)$ if $i \neq j$ and $P_{ii} = \tau$:

$$\mathbf{P} = \left\{\begin{matrix} \tau, & \frac{1-\tau}{N-1} & \cdots\cdots & \frac{1-\tau}{N-1} \\ \frac{1-\tau}{N-1}, & \tau & \cdots\cdots & \frac{1-\tau}{N-1} \\ & \cdot & & \\ & & \cdot & \\ & & & \cdot \\ \frac{1-\tau}{N-1}, & \frac{1-\tau}{N-1} & \cdots\cdots & \tau \end{matrix}\right\} \quad (5.12.1)$$

Vertical uniparental transmission is measured by τ and oblique transmission from nonparental members of the tth generation to the $(t + 1)^{\text{st}}$ is measured by $(1 - \tau)/(N - 1)$. The same matrix can be used for a continuous trait in the following way. $\mathbf{x}_t$ is the vector of trait values of the N individuals at time t, and $\mathbf{x}_{t+1}$ the vector after one generation. Then we write

$$\mathbf{x}_{t+1} = \mathbf{P}\mathbf{x}_t + \boldsymbol{\epsilon}_t. \quad (5.12.2)$$

The recurrence (5.12.2) is analogous to the group treatment given earlier in this chapter with $\tau = 1 - g + g/N$. For this special matrix $\mathbf{P}$ it is easy to see that if $\bar{x}_t = \Sigma_i x_{it}/N$, then

$$E[\bar{\mathbf{x}}_{t+1} \mid t] = \bar{\mathbf{x}}_t. \quad (5.12.3)$$

The variance between individuals (within the population) can be computed as follows:

$$V_{W,t+1} = \frac{1}{N-1} E\left\{\sum_i (x_{i,t+1} - \bar{x}_{t+1})^2\right\}. \tag{5.12.4}$$

Substituting $P_{ii} = \tau$, $P_{ij} = (1 - \tau)/(N - 1)$ in (5.12.4) produces

$$V_{W,t+1} = \left(\frac{N\tau - 1}{N - 1}\right)^2 V_{W,t} + \sigma^2, \tag{5.12.5}$$

so that, in the limit,

$$V_{W,t+1} \longrightarrow V_W^\infty = \sigma^2/[1 - (N\tau - 1)^2/(N - 1)^2], \tag{5.12.6}$$

which is identical to $\sigma^2[1 - (1 - g)^2]$. Thus $g > 0$ or $\tau < 1$ is sufficient for the convergence of the within-group variance. With **P** of this form it is easy to see that

$$\bar{x}_{t+1} = \bar{x}_t + \bar{\epsilon}_t, \tag{5.12.7}$$

and hence that the between-group variance is

$$E(\bar{x}_t - \bar{x}_0)^2 = t\sigma^2/N \tag{5.12.8}$$

as before.

The transmission matrix analogous to **P**, which describes the situation when there is a teacher or social leader, is written as **T**:

$$\mathbf{T} = \left\{\begin{matrix} \tau, & \frac{(1-\tau)}{N-1} & \cdots\cdots & \frac{(1-\tau)}{(N-1)} \\ \tau & \frac{(1-\tau)}{(N-1)} & \cdots\cdots & \frac{(1-\tau)}{(N-1)} \\ \cdot & & & \\ \cdot & & & \\ \cdot & & & \\ \cdot & & & \\ \tau, & \frac{(1-\tau)}{(N-1)} & \cdots\cdots & \frac{(1-\tau)}{(N-1)} \end{matrix}\right\} \tag{5.12.9}$$

If $\tau > 1/N$, and therefore $\tau > (1 - \tau)/(N - 1)$, the transmission from the first individual, the teacher is most important. Note that whereas **P** above is doubly stochastic, **T** is in general not doubly stochastic. This difference entails that when

$$\mathbf{x}_{t+1} = \mathbf{T}\mathbf{x}_t + \boldsymbol{\epsilon}_t, \tag{5.12.10}$$

we do *not* have $E(\bar{x}_{t+1}) = \bar{x}_t$. Instead

$$E(\bar{x}_{t+1}) = \bar{x}_t + \sum_j x_{j,t}(T_j - 1/N), \tag{5.12.11}$$

where

$$T_j = \sum_i T_{ij}/N.$$

The within-population variance corresponding to (5.12.10) and (5.12.11) is

$$V_{W,t+1} = \sigma^2 \tag{5.12.12}$$

for all t. This equilibrium value is reached immediately, and is smaller than the corresponding result for the parental and oblique transmission system P by the factor $1/[1 - (N\tau - 1)^2/(N - 1)^2]$.

The variance between groups in the teacher-type transmission system is obtained by first noting that

$$\bar{x}_{t+1} = x_{T,t}\left[\frac{N\tau - 1}{N - 1}\right] + \bar{x}_t\frac{N(1 - \tau)}{N - 1} + \bar{\epsilon}_t, \tag{5.12.13}$$

where the subscript T refers to the individual in the paternal generation who is the teacher. But also

$$x_{T,t+1} = x_{T,t}\left[\frac{N\tau - 1}{N - 1}\right] + \bar{x}_t\frac{N(1 - \tau)}{N - 1} + \epsilon_{T,t}. \tag{5.12.14}$$

Hence $E(\bar{x}_t) = E(x_{T,t})$ in every generation and

$$E(\bar{x}_t - x_{T,t})^2 = \sigma^2(1 - 1/N). \tag{5.12.15}$$

Now

$$x_{T,t+1} - x_{T,t} = \frac{N(1-\tau)}{N-1}[\bar{x}_t - x_{T,t}] + \epsilon_{T,t} \tag{5.12.16}$$

$$= \frac{N(1-\tau)}{N-1}[\bar{\epsilon}_{t-1} - \epsilon_{T,t-1}] + \epsilon_{T,t}, \tag{5.12.17}$$

unless $t = 0$, when the first term of (5.12.16) is irrelevant. Thus

$$x_{T,t+1} - x_{T,0} = \sum_{j=0}^{t} \epsilon_{T,j} + \frac{N(1-\tau)}{N-1} \sum_{j=1}^{t} (\bar{\epsilon}_j - \epsilon_{T,j}). \tag{5.12.18}$$

From (5.12.18), with repeated use of (5.12.16), we have

$$E[x_{T,t} - x_{T,0}]^2 = \sigma^2\left[2 - 2\tau + \tau^2 + \frac{(1-\tau)^2}{N-1}\right] + t\sigma^2\left[\frac{(1-\tau)^2}{N-1} + \tau^2\right]. \tag{5.12.19}$$

Thus, the squared deviation from the original teacher increases with time linearly with a constant $\sigma^2[(1-\tau)^2/(N-1) + \tau^2]$. This should be compared with (5.12.8), from which it can be seen that the present model of teacher-type transmission involves a faster rate of divergence between populations. This is a simple consequence of the fact that, for $\tau > 1/N$, $(1-\tau)^2/(N-1) + \tau^2 > 1/N$. From (5.12.15) the result is similar if, instead of the deviation $\mathbf{x}_{T,t} - x_{T,0}$, we use $\bar{x}_t - x_{T,0}$.

A treatment of divergence within and between populations, in terms of more general transmission matrices, is given by Feldman and Cavalli-Sforza (1975)*. In that paper the role of the order of the action of the two types of transmission is also shown. If parents act first, the subsequent homogenizing effect of teachers is diluted, but if teachers act first the parents' effect washes out. This is due to the nature of the **T** matrix, which is a projection matrix, whereas **P** is a stochastic matrix and **PT** = **T**.

The qualitatively important consequence of these results is that transmission through a teacher or social leader results in greater homogeneity in a population, which is also reached much faster (in one generation) than with the vertical and oblique transmission model perhaps more characteristic of traditional societies. Teacher transmission also results in a much faster divergence between noncommunicating populations.

*Note the important misprint in formula (13) of that paper, where the between group variance is printed as $\sigma^2 K$ instead of $\sigma^2 Kt$.

CHAPTER SIX

Epilogue

6.1 GENERAL CONSIDERATIONS

We are all the agents and witnesses, as well as the beneficiaries and victims, of cultural change. From all indications the rate of cultural change has continuously increased since the beginning of agriculture. In the very wide range of phenomena subsumed by the term cultural, we have included all sorts of activities, from technology to entertainment, and all sorts of beliefs, values, and behaviors. Among these there is wide variation in the rate of change within populations, and for a given trait there is variation from group to group and from place to place. Rates of change are difficult to measure, so that except for a few phenomena it is difficult to have reliable data with any real time depth. The possibilities for diachronic study are therefore limited and almost entirely restricted to "modern" or "western" cultures. As one might expect, there are very few diachronic data available concerning traditional cultures, although there is an abundance of synchronic data about them, and an examination of the differences among surviving, traditional cultures illustrates the tremendous amount of cultural variation in existence. Unfortunately, data from different societies are not easily compared. Moreover, data on intragroup differences are not readily available. There is a genuine dearth of good data and it is to good data that one should continually return in assessing such a theory as we have tried to construct. As any biologist who has an interest in evolution knows, knowledge of individual variation is important for the interpretation of differences between groups. With the recent beginning of evolutionary studies, for example in linguistics, interest in individual variation is beginning to emerge in disciplines outside biology.

The center of our theoretical models is the concept of transmission: for biologists, and in particular geneticists like ourselves, the centrality of the concept of transmission to evolutionary thinking is patently obvious. In biological material, however complex, there exists an almost perfect mechanism of genetic transmission. Evolution is the result of mutational exceptions to perfect transmission and the differential success of the products of this imperfect transmission.

We have used the terminology "second-order organisms" for such cultural objects as technologies, languages, and customs that are entirely dependent on properties of the "first-order organisms," namely humans or other cultural animals. Concepts usually reserved for biological evolution can then be transferred to the second-order organism which will also be affected indirectly by the evolutionary processes acting on the first-order organism. The result is the two orders of selection, natural and cultural.

Cultural selection is made possible by the great flexibility in individual choices available to the first-order organisms. The flexibility might itself be under the direct control of natural selection in the sense that the bounds on choices may to some extent be under such control. The choice of red, white, or black pigments for ceremonial body painting is ordinarily unimportant for natural selection and would be determined entirely by custom or availability of pigments. On the other hand, choices regarding sexual behavior or the satisfaction of hunger are important for reproduction and survival. Yet even these can be dominated by choices that lead to chastity or to fasting for political, social or religious ideals.

Thus cultural selection can act counter to natural selection, although harmony between the two is expected on average. This is a relatively trivial expectation based on the assumption that the neural structures and mechanisms that permit choice by the first order organisms evolve under the control of natural selection, which will therefore indirectly control the scope of choices made. But in spite of the control exercised by natural selection over the neural mechanisms involved in cultural choices, the variety of situations with which people are confronted makes it extremely difficult to reach a "rational" decision every time.

It is worth discussing further the meaning of "rational," which in its baldest terms might be equivalent to maximizing Darwinian fitness. In economics, behavior is called rational if it is designed to maximize economic returns (in terms of allocation of material goods to satisfy material wants, see Becker, 1976). Such a pecuniary view may be more fitting than most others, except that *material* wants measurable in monetary terms, even if important, do not represent the totality of human motivation. The allocation not only of money but of time and energy must be taken into account and we know well that the energy of which we are capable, and the time spent working, depend to a large extent on specific tastes, expectations and motivations. These are determined in a complex way by individual personality, which is the result of a unique genotype developing in a unique environment: both of these are subject to great interindividual variation. There are people determined to risk their life to reach the top of Mt. Everest, and others that spend their life accumulating money, or attempting artistic or scientific creations, or simply trying to do as little as possible. It is difficult to subsume all of these choices under a common schedule admitting no individual variation.

6.2 HARMONY AND CONFLICT OF CULTURAL AND NATURAL SELECTION

Despite enormous individual variation in motivations and the complexity of interaction between individuals with different motivations, it might be argued that natural selection has been rather successful in insuring harmony between cultural and Darwinian selection. This still leaves room for a great deal of disagreement and even direct conflict between the two, in the sense that individual and even collective behavior may be directed by cultural choices towards solutions that have negative Darwinian fitness. It is of interest to analyze situations in which this can happen.

One of the most striking conflicts between a cultural choice and Darwinian fitness is found in the case of drug abuse. Here there might be two important mechanisms: error in perception of the danger associated with the drug habit, because of insufficient or wrong information, and the existence of a true addiction mechanism, which creates a need that overwhelms any perceived danger. The breaking of a habit becomes then extremely difficult. Of course, the maintenance of drug habits must also depend on the availability and promotion of the drug.

Many other behaviors established by custom are difficult to modify even when disadvantages are perceived. Dietetic customs leading to high fat and sugar consumption, common in North America, may contribute to mortality at ages that affect Darwinian fitness. Customs that probably have high acceptability in terms of taste ("marbled" meat, large consumption of milk, ice cream, candy, and the like) may also have been helped in this progress by advertising and are now so well entrenched that epidemiologists trying to demonstrate that "the American way of life need not be hazardous to your health" (Farquhar, 1978) have an uphill battle. Similarly in spite of the findings of danger to life associated with smoking, and the campaign by the Surgeon General of the United States, cigarette smoking in this country has certainly not ceased. Here again there is the difficulty of breaking a habit; but the continued heavy advertising, the hope that the toxicity of the product has been decreased, the high contagiousness of the habit, and its (however superficial) tranquilizing activity in a world of increasing tension preserve the custom or promote its spread. It is unclear if this conflict will simply determine a delay in the final eradication of the habit which spread throughout the world in only a few hundred years, or if smoking tobacco is here to stay. The use of alcohol certainly seems to be, judging from its persistence over millennia and its diffusion throughout a great number of cultures.

The perception of the dangers associated with specific customs may be quite difficult or even impossible until an appropriate scientific analysis is made. Such ignorance can promote the conflict between natural and cultural selection. Necrophagic customs may have been repeatedly responsible, as in the case of Kuru, for the spread of infectious fatal diseases. The habit of eating the dead may persist in spite of the danger, partly because the danger is not perceived at all. A maladaptive trait which has become common can be eliminated by cultural selection only if it is perceived as maladaptive. The anthropological literature is full of examples of mutilations, of which perhaps only one, male circumcision, serves some useful purpose and might emerge as acceptable from a "cost-benefit" analysis made in modern terms. One component of cost, the danger of infection associated with almost any mutilation, was not easy to recognize until the advent of modern bacteriology. Others, like the consequences at childbirth of scars on the external female genitalia provoked by female circumcision, should have been more easily perceived. Yet even today the custom of female circumcision remains extremely common in north and east Africa and other parts of the world under Arabic influence, to which this supposedly pre-Islamic custom has spread. In some north African tribes, up to 95% of the women are circumcised. The custom still finds supporters even among the medical profession, according to an analysis by the World Health Organization. However, these are exceptions, and, as might be expected, the major opposition to female circumcision is from African doctors and midwives who are familiar with the damage to health that the practice entails. Because of the poor dissemination of health care information, or misunderstanding, or disbelief of the information communicated by health care workers, or fear of ostracism upon nonconformity, or simply the belief prevalent in some Islamic cultures that the custom will prevent women from indulging in "sinful behavior," or yet other rationalizations, female circumcision may continue for a long time.

Customs connected with reproductive behavior have a direct influence on Darwinian fitness. Some are clearly aimed at increasing fitness even if they superficially decrease it. The killing of a twin—a custom found in many cultures—may actually increase the average chance of survival of progeny from a twin birth in underdeveloped areas. Late marriage, prolonged lactation, or long post-partum sex taboos result in birth regulation even if they arose for other reasons, lost in antiquity. A perceived, claimed, or rationalized reason for a custom need not be the *original* or the *only* motivation, nor the *main* subconscious motivation, nor the reason why it is adaptive (if it really is adaptive, see section 6.5).

New customs reducing fertility have spread through populations, although they clearly have negative Darwinian fitness. They must spread under the force of cultural fitness, in opposition to Darwinian fitness. The main selective resistance to the spread is not Darwinian fitness, but rather competition from older customs—that is, transmission of traditional values, and lack of recognition that change in death rates, not followed by change of birth rates, may have disastrous consequences. In fact, the motivation behind the changed fertility, although not clearly known, is unlikely to be rationally perceived (decrease in death rate and the need to avoid overcrowding) and may instead be largely economic. However, the economic and technologic changes which have decreased death rates have also increased job availability, resulting in the geographic and social mobility of people, with corresponding decreases in traditional social values. Living away from the family and old social ties may give an individual new incentives and new possibilities, so that more choices become available in the acquisition of values and motivations. Many of these choices are incompatible with high fertility, because of the cost of marriage and of rearing children. The traditional motivations for high fertility also lose their grip because of lack of contact with the older generations, and the substitution of old values by new ones. Thus, eventually, the same causes that have determined a higher standard of life and lower mortality will cause a lower birth rate. However, this is not usually a direct adaptive response, and will occur independently of the fall in death rates, ordinarily later: the delay between the two phenomena may be responsible for the serious demographic imbalance that has resulted in population explosions in many countries.

It is perfectly possible that a trait, cultural or biological, has no effect at all on Darwinian fitness, so that s is exactly zero. Biological transmission is usually perfect; that is, meiotic drive and other disturbances are rare, so that any biological equivalent of the transmission component τ of D would also be zero. But in cultural transmission it is unlikely that τ would be exactly zero, since the mechanism of transmission would not generally be as precise as its biological counterpart. Thus we can expect that cultural transmission is either so strong ($\tau > 0$) or so weak ($\tau < 0$) that D is likely to be different from zero, with the result that, on average, the rate of cultural evolution should be faster than that of biological evolution.

6.3 CULTURAL TRANSMISSION, COMMUNICATION, AND CULTURAL SELECTION

The model for the spread of infectious disease: susceptible → infectious → resistant or dead (removed); and that for the spread of innovation: naive → aware → adopted; are both highly oversimplified. More complex and probably more realistic schemes can be developed that have more intermediate stages. There still remain the problems of individual variation in susceptibility and of the spread of the trait over geographic space, both of which have been modeled in only very simple terms (see Chapter 1).

The effects of signaling (i.e., the step naive → aware) and that of cultural selection (i.e., the step awareness → adoption) are clearly separable during the spread of an innovation. In principle they can be measured independently.

In Chapters 2 and 3 we have shown how cultural transmission can counteract negative natural selection and maintain traits with negative Darwinian fitness in populations. In fact, even traits with fitnesses as low as 50% of a competing trait may remain in the population under vertical transmission pressure (see section 2.6). It may be worth noting that the treatment of natural selection we have used throughout is especially simple for differential mortality. It may be extended to differential fertility only under certain special assumptions (see, for example, Bodmer, 1965). General analysis of fertility entails a much more complicated treatment. Horizontal and oblique transmission counteract natural selection in the same way that vertical transmission has been shown to do, with some differences in the analysis (see sections 2.6 and 3.2 to 3.4). One approximate way to summarize this quantitatively uses the expression D in the diffusion approximation for drift in formula

$$\begin{aligned} D/N &= s + \alpha_1 + \alpha_2 + (1 - m)(b_1 + b_2 - 1) \\ &= s + \tau, \end{aligned} \tag{6.3.1}$$

where s is the natural selection coefficient in favor of trait H, α_1 and α_2 the equivalents of the parameters of oblique and horizontal transmission, b_1 and b_2 the chance that H offspring are produced from the heterotypic matings $H \times h$, $h \times H$, and m is the assortative mating parameter. N is the population size. Cultural selection is incorporated into the transmission parameters α_1, α_2, b_1 and b_2. If mutation is negligible the trait will not be expected to change its frequency with time if $D = 0$, that is, if natural selection and the sum of the forces of cultural transmission—oblique, horizontal, and vertical—balance each other *exactly*. Then only drift will affect the frequency of the trait and fixation, or extinction will occur as for a neutral gene. However, fixation of H will tend to occur if $D > 0$, and extinction if $D < 0$. These processes will be faster the larger the absolute value of D. The quantity $\tau = \alpha_1 + \alpha_2 + (1 - m)(b_1 + b_2 - 1)$ summarizes the forces of cultural transmission (in the simplified model here considered): if $\tau = 0$, then natural selection is the only force determining D. If $\tau \neq 0$, we have an "imbalance" in cultural transmission (including, as usual, cultural selection) similar to meiotic drive in biological transmission; depending on signs, this imbalance can counteract or reinforce the effect of natural selection measured by s. The examples of addictive drugs, and of birth-control discussed earlier, illustrate the potentially antagonistic effects of cultural selection (and hence of transmission) and of natural selection. As discussed earlier, these effects more often are expected to be synergistic.

In the context of epidemics similar analogies to those discussed in connection with (6.3.1) can be made. Thus, if α is the rate of communication and β the rate of cultural selection, the epidemic will not start unless there is an initial number, S_0, of aware individuals greater than $\rho = \beta/\alpha$. This threshold value appears again in the final total number of adoptions, which is approximately $2(S_0 - \rho)$.

An innovation is a specific cultural event that must be perceived as such and adopted by some form of choice. But most of cultural life is not innovative. Rather, it consists of routines that are widely spread. Nonadopters if any, are the exception rather than the rule, and they may contribute the innovation. If a trait exists in more than one form, then conscious choice may still play an important role in adoption. For many customs, however, the chance that conscious choice be exercised is likely to be small. Children who are circumcised usually have no chance to refuse.

Even in the rare societies where circumcision occurs at adulthood, so that there is a better chance of electing not to be circumcised, the social penalty for refusal may be so high that presumably few elect to refuse.

Since much of the teaching of social behavior occurs at a very young age, conscious choice may not be very important in determining which among existing possibilities is adopted. For some traits true imprinting might occur, in which case there is clearly *no* role of cultural selection: the first phase, signaling, determines *ipso facto* adoption.

Cultural transmission is the complete process of passage of a behavior from a "donor" or transmitter to a naive individual. It therefore involves *both* stages, communication and cultural selection. Lest we are accused of "transmissionism," we hasten to add that the use of summary transmission coefficients in the theory is not intended to indicate that the process cannot be subdivided. These coefficients are technical conveniences in terms of which correlations between certain types of individuals are simply expressed and vice versa. In the same way that correlation is not causation in general, the existence of parent-child correlation, in particular, does not imply direct transmission. On the other hand, in the absence of correlation, transmission has either not occurred or has been nullified by a counteractive event. It is much more likely that transmission as a whole can be measured than the components of which it is made.

The limitations of this simplification are worth pursuing further. Psychologists have expended much effort in breaking down the transmission process but in quantitative terms it might be useful to distinguish (i) the contribution of the teacher, that of the learner and interaction between these, and (ii) the cultural fitness, which is in principle a property of the trait itself (but could of course be subject to individual differences). Thus a transmission coefficient from individual i to individual j in a population of size N might be written

$$t_{ij} = e_t e_l(1 + I_{ij})w,$$

where w is the cultural fitness of the trait, e_t and e_l are measures of the efficiency of the teacher and learner respectively, and I_{ij} is a term that expresses the interaction between these. Such breakdowns of the simple coefficients we have introduced earlier would be highly desirable, but there is little hope that data will enable separation of them until we learn more about human motivation. In addition they will most likely be highly trait specific.

If repeated contacts with an innovation are necessary before awareness, the two-stage models do not apply. Nor are they valid if contact with a certain fraction of the population is more influential than with others. These possibilities would further complicate any attempt to measure cultural selection.

In truth, among our catalogue of behaviors the full spectrum of cultural transmission phenomena is sure to be represented—from those due to imprinting where no choice is made but signaling is decisive, to those where full freedom of choice can be exercised.

Both in modeling and estimation a final complication must be faced—that is, the instability through time of a trait *within* an individual. Traits may be acquired, lost, regained, and so on, or may fluctuate in intensity over a lifetime. Thus the point in time at which the frequency of a behavior is censused becomes critical. Instability must be considered in parent-child analysis or even sib-sib analysis. Carefully designed longitudinal studies might contribute to a definition of the pattern of change over life, and corrections made before estimation of transmission itself.

Transformations of individuals under cultural influences have been represented by us in a highly simplified way. Paraphrasing Maimonides'*A Guide for the Perplexed,* "in the beginning we must simplify the subject, thus unavoidably falsifying it, and later we must sophisticate the falsely simple beginning way." We have only considered one initial, naive state of the learning process, and one final state. In reality, the transformations an individual undergoes in the cultural process are probably very complex, even for relatively simple traits. They may not be unidirectional, but may involve considerable teacher-pupil feedback, resulting in effects on the trait value of the teacher. Our transmission coefficients, therefore, summarize a very complex process in a very simple occasionally, even if, naive way.

6.4 MODES OF TRANSMISSION AND THEIR CONSEQUENCES FOR RATES AND EQUILIBRIA UNDER CULTURAL EVOLUTION

The cultural differences between human groups can, in terms of our model, be regarded as the outcome of the balance of several evolutionary forces: (1) *mutation,* which is both purposive (innovation) and random (copy error); (2) *transmission,* which is not as inert as in biology; (3) *cultural drift* (sampling fluctuations); (4) *cultural selection* (decisions by individuals); and (5) *natural selection* (the consequences at the level of Darwinian fitness), simple manifestations of which might involve adaptations to differences in local physical environments. For example, the use of straw or palm leaves or other roofing material (such as tiles or shingles in more advanced societies) depends in part at least on the availability of the raw materials and is under fairly stringent environmental control. The plan of the house—for instance, round, oval, square or rectangular—may be less subject to environmental constraints, so that in different population groups the original choices may be determined by tastes, fashion, or customs, which are then transmitted within each group, but largely under the influence of cultural drift. The concept of "local environment" may also include the cultural environment, for example, customs, values, and the like, which may make it difficult to accept a trait in some cultures and easy in others.

Certain of the qualitative conclusions from the simple models of cultural transmission are worth recalling, using for simplicity the paradigm of a continuous trait.

(1) Under cultural drift *alone,* the mean value of a population does not change. If there is selection for an optimal value of the trait, then the group mean will converge towards it at a rate determined by the intensity of cultural selection. Variance about this mean approaches a limit determined by a balance between drift and the strength of the stabilizing selection.

(2) The variation between individuals of one population (i.e. the within population variance) approaches a finite value that is usually a small multiple of the mutation variance. This is true even in the absence of stabilizing selection. With real data, of course, separation of cultural selection from transmission will prove to be most difficult.

(3) The rate at which the population variance approaches the limit is faster with greater degrees of group transmission.

(4) Non-communicating groups diverge under cultural drift linearly with time. But if there is stabilizing selection the range of variation between groups will be limited.

(5) In the models studied here,

(a) the limiting degree of uniformity within a population,

(b) the rate at which this limit is approached,

(c) the rate of increase of heterogeneity between independent groups, and

(d) the rate of approach to homogeneity for discrete traits

all increase with increasing contribution of teacher-type transmission (or, in general, one-to-many transmissions) relative to parental transmission. This can be regarded as a general property of "cultural inbreeding" in the sense that a single individual may have as his or her "cultural offspring" a large fraction of the population.

(6) Using uniparental transmission as a baseline, biparental transmission ($b_m + b_p = 1$) decreases the variance within while increasing the variance between groups. The same is true for

increasing the strength of oblique transmission (as measured by g). In general, the greater the number of teachers from the earlier generation (many-to-one transmission), the smaller are the quantities 5 a, b, c, and d.

(7) The protection of a trait from extinction is easier with oblique transmission than with horizontal when vertical transmission is weak, with the reverse being true when vertical transmission is strong.

(8) With multiple-state traits the equilibrium configuration of the population can be sensitive to small changes in the transmission scheme. Thus estimates of the transmission scheme from observed correlations between relatives should be handled with circumspection.

(9) The greater the contribution from earlier generations the slower is the rate of cultural evolution, as measured by amount of and rate of approach to the limiting within-population variance.

In view of these comparisons it is interesting to note that hunting-gathering societies have less tendency than most others to recognize hierarchies or leadership. The available anthropological data suggest that uniparental, oblique, and grandparental transmission is important in such societies and teacher/leader transmission rare. In theory, and apparently in fact, this generates considerable conservativism, with very slow change under cultural drift. With the arrival of agriculture and of village chiefs, other faster-evolving transmission modes probably became available. But if the social subdivision had been confined only to the village level, with tribal chiefs having only nominal authority, then substantial between-village differentiation could have resulted. Presumably it took strong centralization of authority to produce larger hierarchically stratified communities. In the absence of the power to communicate with every group member (and ignoring military considerations), hierarchy may be the only way that a single leader can govern a large group within the same culture.

The conservatism with respect to cultural change characteristic of hunting-gathering societies is in striking contrast to the fast changes that characterize western societies today. This is especially so for traits related to or dependent on technology, and for social customs. On the other hand, rates of change are trait and society specific. It might be difficult, for example, to decide which of these two groups of societies is changing faster linguistically. It may in fact be the case that linguistic change is faster in present-day hunting and gathering societies, with their smaller size and often intense exposure to the languages of economically more advanced neighboring tribes. Perhaps the lack of a written record may also contribute to speeding the rate of linguistic evolution of these groups. By contrast, for hunters and gatherers the transmission of technology related to food "production" is differentiated by sex and is probably uniparental, with some influence of elders. These all have conservative influences on transmission. Oblique (group) and grandparental transmission must also exert conservative effects on the evolution of social customs.

With the advent of telecommunications it seems reasonable to suggest that, for many socio-behavioral traits, the teacher-leader type of transmission has come to play a dominant role. Today the audience of a single social-cultural leader can be more than continental in size. For traits under this sort of influence, evolution, as we have seen, becomes extremely fast. Since the awareness increases so quickly via the mass media, it is the rate of acceptance that limits the rate of evolution of these traits.

Prior to telecommunications, rates of communication were slower, but in monarchies the rulers were probably very influential in terms of acceptance of change in customs. As Shakespeare's Henry V answers his future French wife, who refuses to be kissed before being married because "Les dames et demoiselles pour être baisées devant leur noces, il n'est pas la coutume de France": "O Kate, nice customs curts'y to great kings. Dear Kate you and I cannot be confined within the weak list of a country's fashion: we are the makers of manners, Kate; and the liberty that follows our places stops the mouth of all faults,—as I will do yours for upholding the nice fashion of your country in denying me a kiss." Cultural changes determined by the whims and fancies of kings and other social leaders were hardly limited to starting the custom of kissing the bride before the wedding, and have undoubtedly been numerous and important, including, for example, the religious conversions of masses of people.

What was said before about continuous traits can be extended (or perhaps we should say reduced) to discontinuous ones. For instance, reduction in variance of the continuous trait can be viewed as an increased probability of fixation of a discrete character at either pure state. Roughly, the relative infrequency of internal equilibria ("polymorphisms") of discrete traits could be compared with a small variance for a continuous trait. Just as the variance of a population for a continuous trait is affected by the mode of transmission, so also is the chance of finding stable polymorphisms for a discrete one. Internal polymorphisms are not infrequently observed in vertical transmission. By contrast, at the teacher end of the transmission scale given above, no stable heterogeneity is ever likely to persist, except among those groups that communicate poorly or not at all. Signals that are transmitted by a teacher/leader will diffuse very rapidly, to be replaced by new ones that may have arisen only shortly afterwards. In moderately conservative situations, stable heterogeneity is possible, but in the evolutionarily more rapid types of transmission, there is only a steady state of flux of continuously produced innovations that are rapidly accepted and rapidly replaced. A summary of some of these considerations is presented in Fig. 6.4.1.

The mechanisms of transmission can explain the contrast between the patterns of geographical distributions of biological and cultural traits. For the latter, populations are often relatively homogeneous for the trait, but are very different from each other, some showing frequencies close to 0 or 100%, with relatively few in between. Differences in D, due either to s or τ or both, which make $D > 0$ in some populations and $D < 0$ in others, can quickly produce the gross, irregularly checkered, geographic variation typical of many cultural traits. For biological traits, on the other hand, the variance between populations is very often low compared to that within populations of the same species.

The detailed differences between horizontal and oblique transmission have not been stressed here. Ideally this would require the treatment of age effects: these can be included using age-specific transmission rates but do complicate the analysis considerably. One interesting example of the role of age-specific transmission, which contrasts the different roles of horizontal transmission against vertical or oblique, concerns the acquisition of language by children of foreign parents. Until school age it might be hypothesized that the dominant effect on the child is that of the parents, and occasionally older sibs. But in language acquisition, age peers, namely schoolmates and playmates, become important. If we assume that slightly older playmates are more influential on the subjects, then in our idealized modeling context, the trait to be copied comes from a near-age peer so that the transition time between states at t and $t + 1$ is very small, perhaps ten or twenty times smaller than a biological generation. If continued among other children, this process of change will therefore be much faster than a process involving learning only from parents. One does not have to invoke the learning of a new language, but only the acquisition of accents in the native tongue, as well as the rise and demise among children of slang words and expressions, many of which are completely foreign to the parents.

In summary, we postulate that the mode of transmission is of great importance in determining the rate of change of trait frequencies in populations. It will differ with the trait, and for a given trait, with the population. A scale of transmission mechanisms, graduated by the ratio of number of teachers to number of people taught, is useful in predicting qualitatively the rate of evolution and the chance of polymorphism. The differences between biology and culture of rate of evolution and pattern of geographic variation are also predicted by the mechanisms of transmission.

6.5 CHANCE AND PURPOSE IN CULTURAL VARIATION

Galton used the analogue of invention to develop the notion of biological mutation. Theoretically the reverse analogy is also valid (see Gerard et al., 1956; Cavalli-Sforza, 1971). Cultural "mutations" are changes for which no fixed paradigm for detection exists, although major changes may of course be revealed retrospectively by archeology. But the most important distinction between biological and cultural mutation is that the former is assumed, according to the present state of knowledge, to be random. Cultural mutations, on the other hand, may have a purpose. They may be generated as a response to specific stimuli or to solve a problem posed by the environment. The intention may be adaptive but the acceptance of the innovation does not necessarily mean that noticeable or measurable improvement can be perceived or has occurred. Indeed the reason for the existence of a culturally transmitted rule may be completely unfathomable, as we have discussed in connection with ritual mutilation.

A rigorous proof that a cultural innovation is adaptive would be as difficult to give as for any biological mutation—the cost-benefit analysis in these situations is usually extremely complicated. The Latin, Russian and German languages use elaborate declension of nouns and conjugation of verbs, while other languages may use syntactic devices, including affixes or prepositions. The adaptive value of one system over the other is unknown. Similarly it might be asked whether the sentence structure subject-verb-object (as in English) is preferable to that used, for instance, in Japanese: subject-object-verb. If more languages used the former, one might argue for its greater "adaptive value." But this indirect argument would then prove that it is better in general to use herbal medicine than western drugs, given that more people probably still use the former.

That cost-benefit analyses of human cultural differences promise to be difficult is witnessed by the enormous variation that exists among human societies. One of the great achievements of cultural anthropology and ethnography has been the recording of this variation before its almost inevitable disappearance beneath the onslaught of Western-derived customs and technology.

There are three explanations for this variation.

(1) All differences between local customs represent adaptations to local physical environments; hats are worn when and where the sun is especially hot or precipitation heavy. Generally, it is not a problem to produce an adaptive explanation; the problem is that there are often too many equally plausible scenarios.

(2) Adaptiveness or maladaptiveness may not be intrinsic to the trait but be influenced by the cultural background in which the trait exists. Hats (and other clothes) usually follow aesthetic criteria and fashion rather than biological need. The clothes might be perceived to increase attractiveness, but only if they do not fall outside certain limits dictated by the cultural background. Again, in a society that derives most of its livelihood from fishing, a high frequency of words, customs, and skills pertaining to the sea, or boats, or fish is to be expected. Thus a battery of customs might be adaptive by virtue of their linkage to some feature of that group's daily life.

(3) The third explanation is that the differences for a trait between cultures are entirely of random origin and have no adaptive value. Of course, it is also possible that the original stimulus for the differences may have disappeared, leaving no

present adaptive value where, in the past, it might have existed. The wearing of hats by orthodox Jews has an original rationale in antiquity. European Jews have had no real biological reasons for this custom, and indeed an argument might be made that as a symbol of identification it has been maladaptive in the non-Jewish cultural background.

No matter which of these three explanations best describes the variations of a cultural trait, the power of transmission in preserving differences between groups in their manifestation of the trait is obvious. The experience with biological traits suggests that the chance of distinguishing to which of the above three categories many traits belong is small. A major controversy of the past ten years of population genetics research has addressed the question of whether subtle biological differences, such as mutations giving rise to amino acid substitutions in proteins, are adaptive (category 1 or 2) or "neutral" (category 3 above). For many biologists the failure of our knowledge and imagination to provide a satisfactory verifiable *adaptive* explanation for all observed variability has been distinctly displeasing. The fact remains that, among genetic mutations, it now seems very likely that some are selectively neutral (Kimura, 1979). The problem might then be rephrased in terms of the relative frequency of adaptive and random substitutions (Bodmer and Cavalli-Sforza, 1972). Not much progress has been made toward the resolution of the controversy in these terms. However, evidence has accumulated that the fractions of neutral and adaptive substitutions are both non-trivial (Lewontin, 1974; Singh et al., 1976). As we have mentioned above, the parallel assessment for cultural traits promises to be at least as difficult as it has been in genetics.

6.6 OVERLAPS WITH OTHER AREAS OF STUDY

For simple biological traits, Mendel's rules are sufficient to completely describe the transmission process. But even for simple two-valued cultural traits, our theory, which has attempted to provide an analogous framework, must be regarded with considerable circumspection. Our models should be viewed as providing evolutionary speculations by using the answers to the "triple" question "who transmits what to whom?" But, as we have stressed, the combination of teaching and acceptance in a given instance will be very hard to quantify, especially in terms of a simple set of transmission parameters. It may be possible to amplify the elementary theory by using *age-specific* transmission rates. These promise to be very difficult to calculate, but the experimental findings of cognitive, developmental, and social psychology might well suggest a natural age subdivision for a given trait whose qualitative properties, at least, might be modeled.

The decision process is implicitly important in cultural evolution. The results of many studies of this area in psychology and related social sciences might eventually be incorporated into the transmission models. Clearly many decisions also have long-term biological evolutionary consequences. The decision to start smoking, or the level of addiction to it, may influence not only the disease process in the smoker but also in the descendents of the smoker by accumulation of deleterious mutations (Ames, 1979). The popularity of a decision to restrict family size has profound effects on the society but also is obviously a perturbation of the biological evolutionary process. These interactions deserve further study.

It is natural, given a theory with relatively few parameters, to try to estimate them. But it must be kept in mind that correlations between individuals may be generated in many different ways. One is by bona fide transmission between them. A second is by a true assortment, in which like (or unlike) individuals seek each other out. A third is the result of stratification of society, wherein a random choice may be made within a limited social environment. For instance, when a specific trait is measured on married couples well after marriage, all three sources of correlation may be present.

The objectives of our theory are actually often qualitative, despite the quantitative formulation of the models. Economy in the number of parameters and mathematical tractability are clearly considerations in trying to meet these objectives. Our limiting the complexity of the models and of the traits is one way in which we are guilty of reductionism. It is not that one mechanism is favored over another in describing transmission; it is rather that, whatever the mechanisms, the processes are grossly oversimplified in the models.

6.7 INDIVIDUAL SELECTION AND GROUP SELECTION

Throughout this book, natural selection has been studied at the *individual* level, that is, in the classical Darwinian way. There have been repeated controversies concerning the potential importance to biological evolution of *group selection* (Wynne-Edwards, 1962). It may be of value to consider a similar concept of group selection in the context of cultural evolution. Certain innovations might give relative selective advantages, not necessarily to the individuals accepting them, but to whole groups adopting them, relative to nonadopting groups. The advantage might be measured in terms of differences in the relative rates of increase in size of the groups. In this cultural context, an example could be the practice of agriculture. The breeding of plants and animals greatly raised the "carrying capacity" of the land, that is, the maximum possible density of a population. Groups that adopted agriculture earlier increased in size sooner, and the further improvement of agricultural technologies has been accompanied by a continuous increase in population numbers. Even so, some 10,000 years after the beginning of agriculture, there still are in all continents populations (many of them decreasing in numbers) or parts of populations that have not yet converted to the production, rather than the collection, of food. The world human population has increased a thousand-fold since the beginning of agriculture, and the increase differs in the various parts of the earth, the northern temperate zones having seen the greatest demographic development. This early demographic transition could be regarded as an example of group selection determined by cultural changes in food technology. The more recent demographic transitions are other examples, determined by changes in the technologies of hygiene and of birth control.

Another important aspect of group selection under cultural transmission is the potential increase in group efficiency deriving from the subdivision of labor. In other words, efficient social organization could make the sum of individual Darwinian and cultural fitnesses larger than they would be in the absence of such organizations. This approach is theoretically promising, and may answer some of the questions suggested by a consideration of the history of increasing complexity of human societies. Some previous discussions (White, 1959, Sahlins and Service, 1960, and Steward, 1955) tend to reserve the term "cultural evolution" for typologies of historical changes in social and political structure. In view of the theory we have developed, it seems more appropriate to keep the term "cultural evolution" as it is here, for general use, and not limit it to the evolution of group structure.

6.8 CULTURAL ACTIVITY AS AN EXTENSION OF DARWINIAN FITNESS

The dramatic increase in numbers of man with the adoption of agriculture is the most conspicuous testimonial of how a cultural adaptation, and hence culture itself, can promote Darwinian fitness of a species. There is no doubt that cultural mechanisms can be powerful adaptive mechanisms that favor survival and ultimate reproduction. Cultural adaptation, very much like physiological adaptation, is the transformation of the behavior of an individual to meet some potential demands or challenges. In the case of culture, this occurs through learning by imitation, education, or related mechanisms that involve use of the prior experience of other individuals (not necessarily of the same species: we may learn of the presence of another animal from the behavior of our dog). Physiological adaptation is the transformation of a physiological activity in an adaptive direction as a response to a specific external stimulus experienced by the individual: the tanning of skin under exposure to the sun, the formation of callouses on hands of workers, or the production of a specific enzyme capable of transforming a substrate only when that substrate is present, are examples of physiological adaptations that can take place in each individual. On the other hand, *genetic adaptation* is the adaptation of a population (not of an individual) by natural selection—that is, by preferential survival and reproduction of the fitter individuals.

Physiological adaptation is narrow in scope and always very specific, implying a certain response by an individual to a certain, usually narrow, range of stimuli. But cultural adaptation can provide the same individual with adaptive responses to a great variety of situations, from wearing clothes if it is cold to using wood, stoves, and tools of all sorts. The capacity to learn to wear some kind of protective clothing when it is cold is obviously a rather general property, probably similar to that which enables us to make a variety of tools. Presumably the capacity for culture can cover a wide range of challenges by employing similar neural mechanisms and generate potentially great flexibility for adaptation to a variety of situations. The potential for cultural adaptation is one of the most efficient solutions generated by the mechanism of natural selection for extending the potential of living organisms (see also Pulliam and Dunford, 1980).

With the growing importance of culture in a species, more and more important decisions concerning life and reproduction of the individuals have ceased to be under innate control and have had to be surrendered to education through cultural channels. The mechanism of natural selection retains ultimate control, but only in the long run, and for cultural adaptation to be possible, at least some delegation of control is necessary. Control is delegated to a system of poorly understood internal drives and rewards that direct the activity and choices of the individual. This system is, presumably, the same as that in command of standard behaviors (e.g. through hunger, fear, and the like), which do not necessarily involve learning. It may also be entrusted with the control of factors that ultimately produce decisions over activities and choices that are culturally influenced. Our very inadequate knowledge of this steering system prevents us from making finer statements, but it is probably true that the system's overall activity is directed towards maximizing self satisfaction of the individual. Important complications arise because we can satisfy ourselves in many different, competing ways, many of which demand careful advance planning. The wise distribution of effort and access to reward in the process of maximization of self satisfaction, as well as the setting of priorities among competing drives, must themselves be learnt. There clearly are differences between individuals in the kinds of actions that affect this maximization.

The theory of natural selection predicts that organs, as well as activities, have functions under selective control and as such have adaptive value. However, the discovery of adaptive functions may be difficult, and today there are still organs or activities without a clear adaptive value. In some cases their usefulness may have disappeared, or even have become a liability, because of environmental change (e.g., eyes are unnecessary and maybe even a liability for life in caves; also, vision is not so important for animals with nocturnal life). In other cases the precise function may be difficult to evaluate because the cost-benefit situation is not clear-cut, and some organs may look more like liabilities than assets, (such as tonsils, the appendix, or maybe even cranial sinuses). Of course, prior to Harvey's work we did not know the real function of the heart, and the same may be true today for a number of other organs and activities. The approach of considering them adaptive and looking for one or more specific functions has obviously been useful, but especially with respect to behavior, is fraught with experimental difficulties.

In applying this same adaptive approach to predictions about cultural evolution, we suggest that Darwinian selection provides only a second echelon of control, where what is under primary selection is some measure of the degree of self satisfaction, or the chance of attaining it in the future. This could explain why some cultural activities may appear to be counteradaptive. There are of course excellent examples of customs which have apparently adaptive value: laws attempting to punish or prevent murder, theft, and a variety of other crimes, the existence of social hierarchies, or even ceremonies to celebrate marriage, death, initiation and so on. The actual details of these customs are of course often more difficult to accept as entirely adaptive. Cruel and counterproductive punishment is not unusual in almost all societies. Social hierarchies often generate impositions beyond reason, and most ceremonies do not entirely avoid boredom, ridicule, and useless expense. Much of this can be attributed to imperfect adaptation, to historical changes that have undermined the adaptive value of a custom, to changed perceptions, or to lack of understanding because of cross-cultural differences in values. Even so, to call adaptive some customs like the breast-feeding of pigs by Melanesian women, female circumcision, some of the most complex marriage rules of Australian aborigines, or head-hunting, one would have to stretch the meaning of the word "adaptive" farther than most scientists are prepared to go. The explanation here is that the control by natural selection is too indirect to be entirely effective (for an extensive discussion of this topic, see Gould and Lewontin, 1979). The mechanisms that are in direct control of cultural choices must serve as guides in a great variety of situations and can be regarded as adaptive only in a roughly average sensc. Interests in esthetic, religious, and other activities are drives whose importance in terms of Darwinian fitness may escape immediate understanding. Their role is not as strong as that of more elementary ones, like hunger, fear or, in general, drives connected directly with survival and reproduction. However, when these more elementary needs are satiated, they temporarily lose importance and other values and activities, for example, entertainment, take a leading role (*"panem et circenses"*). Again, transmission mechanisms are important because some of these activities demand considerable education, and this education inevitably forms tastes and tendencies that are not easily reversible. The variety of tastes for music in different places and times may give an intuitive feeling for the importance of transmitted custom in forming local taste. The prediction of transmission coefficients is—like that of the coefficients of natural selection—a desirable but technically difficult goal. The main limitations are the boundaries of our present knowledge on the functioning of steering mechanisms in our central nervous system. Although reasonably efficient, such mechanisms are undoubtedly subject to imperfect regulation, with negative feedbacks and inappropriate responses. In spite of this, there would be consensus that positive cultural evolution of human society is possible, and, at least for technology has certainly occurred.

Bibliography

Abramowitz, M., and Stegun, I. A., 1964. *Handbook of Mathematical Functions with Formulas, Graphs and Mathematical Tables*. U.S. Government Printing Office, Washington.

Ames, B. N. 1979. Identifying environmental chemicals causing mutations and cancer. *Science, 204* : 587–593.

Ammerman, A., and Cavalli-Sforza, L., 1971. Measuring the rate of spread of early farming in Europe. *Man, 6* : 674–688.

Ammerman, A., and Cavalli-Sforza, L., 1977. The wave of advance model for the spread of agriculture in Europe. In: *Transformations: Mathematical Approaches to Culture Change*, eds. K. Cooke and C. Renfrew. Academic Press, pp. 275–293.

Ammerman, A., and Feldman, M., 1974. On the "making" of an assemblage of stone tools. *American Antiquity, 39* (4) : 610–616.

Ammerman, A., Matessi, C., and Cavalli-Sforza, L., 1978. Some new approaches to the study of the obsidian trade in the Mediterranean and adjacent areas. In: *The Spatial Organization of Culture*, ed. I. Hodder. Duckworth, London, pp. 179–196.

Atkinson, A. B., 1975. *The Economics of Inequality*. Clarendon Press, Oxford.

Bailey, C. J. N., 1973. *Variation and Linguistic Theory*. Center for Applied Linguistics, 1611 North Kent Street, Arlington, VA 22209.

Bailey, N. T. J., 1975. *The Mathematical Theory of Infectious Diseases*. Hafner, New York.

Bartholomew, D. J., 1973. *Stochastic Models for Social Processes*, 2nd edition. J. Wiley & Sons, New York.

Beal, G. M., and Rogers, E. M., 1960. The adoption of two farm practices in a central Iowa community. *Iowa Agri. and Home Eco. Exp. Sta.*, Spec. Rept. 26, Arnes, Iowa.

Becker, G. S., 1976. *The Economic Approach to Human Behavior.* The University of Chicago Press, Chicago and London.

Bennett, J. H., Rhodes, F. A., and Robson, H. N., 1959. A possible genetic basis for Kuru. *Amer. J. Hum. Genet.* *11* : 169–187.

Bernoulli, D., 1760. Essai d'une nouvelle analyse de la mortalité causeé par la petite virole, et des avantages de l'inoculation pour la prevenir. *Mem. Math. Phys. Acad. Roy. Sci., Paris,* 1–45.

Binford, L., 1963. "Red Ochre" caches from the Michigan area: a possible case of cultural drift. *Southwestern Journal of Anthropology 19* : 89–108.

Blumberg, B. S., 1977. Australia antigen and the biology of hepatitis B. *Science 197* : 17–25.

Blumberg, B. S., Melartin, L., Guinto, R. A., and Werner, B., 1966. Family studies of a human serum isoantigen system (Australia antigen). *Amer. J. Hum. Genet. 18* : 594–613.

Bodmer, W. F., 1965. Differential fertility in population genetics models. *Genetics 51* : 411–424.

Bodmer, W. F., and Cavalli-Sforza, L. L., 1972. Variation in fitness and molecular evolution, Vol. 5: Darwinian, Neo Darwinian and Non Darwinian Evolution. *Proceedings of the VI Berkeley Symposium on Mathematical Statistics and Probability*, Univ. of California Press, Berkeley.

Bonner, J. T., 1980. *The Evolution of Culture in Animals.* Princeton Univ. Press, Princeton, New Jersey.

Bourdon, R., 1973. *Mathematical Structures of Social Mobility.* Jossey-Bass, San Francisco.

Breiman, L., 1968. *Probability.* Addison Wesley, Menlo Park, California.

Burton, F. D., and Bick, M. J. A., 1972. A drift in time can define a deme: the implications of tradition drift in primate societies for hominid evolution. *J. Human Evol. 1* : 53–59.

Campbell, D. T., 1965. Variation and selective retention in sociocultural evolution. In: *Social Change in Developing Areas, a Reinterpretation of Evolutionary Theories,* eds., H. R. Barringer, A. I. Blanksten, and R. W. Mach. Schenkman, Cambridge, Mass.

Campbell, D. T., 1975. On the conflicts between biological and social evolution and between psychology and moral tradition. *American Psychologist,* pp. 1103–1126.

Cavalli-Sforza, L. L., 1958. Some data on the genetic structure of human populations. *Proc. X Int. Cong. Genetics I.* Univ. of Toronto Press, Toronto. pp. 389–407.

Cavalli-Sforza, L. L., 1971. Similarities and dissimilarities of sociocultural and biological evolution. In: *Mathematics in the Archaeological and Historical Sciences,* eds., F. R. Hodson, D. G. Kendall, and P. Tautu. Edinburgh Univ. Press, pp. 535–541.

Cavalli-Sforza, L. L., 1974. The role of plasticity in biological and cultural evolution. *Annals of New York Academy of Sciences 231* : 43–59.

Cavalli-Sforza, L. L., 1975. Cultural and biological evolution: a theoretical inquiry. *Advances in Applied Probability 7* : 90–99.

Cavalli-Sforza, L. L. and Bodmer, W. F., 1971. *The Genetics of Human Populations.* W. H. Freeman and Co., San Francisco.

Cavalli-Sforza, L. L. and Feldman, M., 1973a. Models for cultural inheritance, I: group mean and within-group variation. *Theor. Pop. Biol. 4* : 42–55.

Cavalli-Sforza, L. L. and Feldman, M., 1973b. Cultural versus biological inheritance: Phenotypic transmission from parent to children (a theory of the effect of parental phenotypes on children's phenotype). *Am. J. Hum. Genetics 25* : 618–637.

Cavalli-Sforza, L. L. and Feldman, M., 1976. Evolution of continuous variation: direct approach through joint distribution of genotypes and phenotypes. *P.N.A.S. 73* (5) : 1689–1692.

Cavalli-Sforza, L. L. and Feldman, M., 1978a. Dynamics and statistics of traits under the influence of cultural transmission. In: *Genetic Epidemiology,* eds., N. E. Morton and C. S. Chung. Academic Press, New York, pp. 133–143.

Cavalli-Sforza, L. L. and Feldman, M., 1978b. The evolution of continuous variation, III: joint transmission of genotype phenotype and environment. *Genetics 90* : 391–425.

Cavalli-Sforza, L. L. and Feldman, M., 1978c. Toward a theory of cultural evolution. *Interdisciplinary Science Reviews 3* (2) : 99–107.

Cavalli-Sforza, L. L. and Feldman, M., 1978d. Darwinian selection and altruism. *Theor. Pop. Biol. 14* : 268–280.

Cavalli-Sforza, L. L., Feldman, M. W., Dornbush, S. M., and Chen, K. H. 1981. Theory and observation in cultural transmission (in preparation).

Chen, K. H., 1980. Moment analysis for the probit model of directional selection. *Biometrical Journal* (in press).

Christiansen, F. B., and Fenchel, T. M., 1977. *Theories of Populations in Biological Communities.* Springer-Verlag, Berlin.

Christiansen, F. B., and Frydenberg, O., 1973. Selection component analysis of natural polymorphisms using population samples including mother-offspring combinations. *Theor. Pop. Biol. 4* : 425–445.

Coale, A. J., 1974. The history of the human population. *Scientific American,* Sept. 1974, 41–51.

Coleman, J. S., 1964. *Introduction to Mathematical Sociology.*

The Free Press of Glencoe. Collier-Macmillan Limited, London.

Coleman, J. S., Katz, E., and Menzel, H., 1966. *Medical Innovation. A Diffusion Study.* The Bobbs-Merrill Company, Inc., Indianapolis, New York, Kansas City.

Cooke, K. L., 1979. Mathematical approaches to culture change. In: *Transformations,* eds., C. Renfrew and K. L. Cooke. Academic Press, New York, San Francisco, London, pp. 45–81.

Crow, J., and Kimura, J., 1970. *An Introduction to Population Genetics Theory.* Harper and Row, New York, Evanston, London.

Crumpacker, D. W., Zei, G., Moroni, A., Cavalli-Sforza, L. L., 1976. Air distance versus road distance as a geographical measure for studies on human populations structure. *Geographical Analysis VIII* (2) : 215–223.

Darwin, C., 1859. *On the Origin of Species.* Watts and Co., London.

Davis, J. M., 1973. *Imitation: A Review and Critique.* In: *Perspectives in Ethology,* eds., Bateson, P. and Klopfer, P., Plenam, New York.

Dawkins, R., 1976. *The Selfish Gene.* Oxford University Press.

Dietz, K., 1972. The role of mathematical models in the epidemiology and control of communicable diseases. Presentation to the European meeting of statistics 31 Aug.–5 Sept., 1972, Budapest (manuscript).

Dietz, K., 1976. The incidence of infectious diseases under the influence of statistical fluctuations. In: *Mathematical Models in Medicine,* eds., J. Berger, W. Bühler, R. Repges and P. Tautu, Springer-Verlag, New York, pp. 1–15.

Dobson, A. J., 1978. Evolution times of languages. *J. Amer. Stat. Assoc.* 73 : 58–64.

DuPont, R. L., and Greene, M. H., 1973. The dynamics of a heroin addiction epidemic. *Science 181* : 716–722.

Durham, W. H., 1976. The adoptive significance of cultural behavior. *Human Ecology 4* : 89–117.

Elston, R. C., and Stewart, J., 1971. A general model for the genetic analysis of pedigree data. *Human Heredity 21* : 523–542.

Ewens, W. J., 1979. *Mathematical Population Genetics.* Springer-Verlag, Berlin.

Ewens, W. J., 1972. The sampling theory of selectively neutral alleles. *Theor. Pop. Biol. 3* : 82–112.

Falconer, D. S., 1960. *Introduction to Quantitative Genetics.* The Ronald Press Company, New York.

Farquhar, J. W., 1978. *The American way of life need not be hazardous to your health.* Stanford Alumni Assoc., Stanford, California, p. 370.

Feldman, M., 1966. On the offspring number distribution in a genetic population. *J. Appl. Prob. 3* : 129–141.

Feldman, M., and Cavalli-Sforza, L., 1975. Models for cultural inheritance: A general linear model. *Annals of Human Biol. 2* : 215–226.

Feldman, M., and Cavalli-Sforza, L., 1976. Cultural and biological processes, selection for a trait under complex transmission. *Theor. Pop. Biol. 9* : 238–259.

Feldman, M., and Cavalli-Sforza, L., 1977a. Quantitative inheritance, stabilizing selection and cultural evolution. *Proceedings of the International Conference on Quantitative Genetics, Aug. 16–21, 1976,* Iowa State Univ. Press, pp. 761–777.

Feldman, M., and Cavalli-Sforza, L., 1977b. The evolution of continuous variation, II; complex transmissions and assortative mating. *Theor. Pop. Biol. 11* : 161–181.

Feldman, M., and Cavalli-Sforza, L., 1977c. Random sampling drift under nonMendelian transmission. Proceedings of the 41st Session of the International Statistical Institute, New Delhi, 1977. *Bull. Int. Stat. Inst. XLVII* (2) : 151–164.

Feldman, M., and Cavalli-Sforza, L. L., 1979a. On hereditary transmission in diseases of complex etiology. In: *Genetic Analysis of Common Diseases: Applications to Predictive Factors in Coronary Disease*, eds. C. F. Ging and M. Skolnick, Alan R. Liss, Inc., N.Y., pp. 203–227.

Feldman, M., and Cavalli-Sforza, L., 1979b. Aspects of variance and covariance analysis with cultural inheritance. *Theor. Pop. Biology 15* : 276–307.

Fine, P. E. M., 1977. Analysis of family history data for evidence of non-Mendelian inheritance resulting from vertical transmission. *J. Med. Genet. 14* : 399–407.

Fisher, R. A., 1918. The correlation between relatives on the supposition of Mendelian inheritance. *Trans. Roy. Soc. Edinburgh 52* : 399–433.

Fisher, R. A., 1922. On the dominance ratio. *Proc. R. Soc. Edinburgh 42* : 321–341.

Fisher, R. A., 1930. *The Genetical Theory of Natural Selection*, 2nd. ed. 1958. Dover, New York.

Fisher, R. A., 1937. The wave of advance of advantageous genes. *Ann. Eugen. 7* : 355–360.

Friedlaender, J. S., 1971. Isolation by distance in Bougainville. *Proc. N.A.S. 68* : 704–707.

Friedlaender, J. S., 1975. *Patterns of Human Variation*. Harvard Univ. Press, Cambridge, Mass.

Gajdusek, D. C., 1977. Unconventional viruses and the origin and disappearance of Kuru. *Science 197* : 943–960.

Galef, B. G., Jr., 1976. Social transmission of acquired behavior: a discussion of tradition and social learning in vertebrates. *Adv. in the Study of Behavior 6* : 77–99.

Galton, F., 1889. *Natural Inheritance*. Macmillan, London and New York.

Gardner, R. A., and Gardner, B. T., 1969. Teaching sign language to a chimpanzee. *Science 165* : 664–672.

Gerard, R. W., Kluckhohn, C., and Rapoport, A., 1956. Biological and cultural evolution. Some analogies and explorations. *Behavioral Sci. 1* : 6–34.

Glass, D. V., and Hall, J. R., 1954. A description of a sample inquiry into social mobility in Great Britain. In: *Social Mobility in Britain,* D. V. Glass, ed. The Free Press, Glencoe, Ill., pp. 79–97.

Goldberger, A. S., 1977. *The Genetic Determination of Income.* Social Systems Research Institute, No. 7707, Madison, Wisconsin.

Goldberger, A., 1978. Models and Methods in the IQ Debate: Part I revised. Social Systems Research Institute, University of Wisconsin.

Gomila, J., 1971. *Les Bedik (Sénégal oriental).* Les Presses de L'Université de Montréal, Montréal.

Gould, S. J., and Lewontin, R. C., 1979. The spandrels of San Marco and the panglossian paradigm: a critique of the adaptationist programme. *Proc. Roy. Soc. Lond. B. 205* : 581–598.

Greenberg, J., ed., 1965. *Universals of Language,* 2nd ed. M.I.T. Press, Cambridge, Mass.

Guglielmino-Matessi, C. R., Gluckman, P., and Cavalli-Sforza, L. L., 1979. Climate and the evolution of skull metrics in man. *Amer. J. Phys. Anthrop. 50* : 549–564.

Guglielmino-Matessi, C. R., Viganotti, C., Gluckman, P., and Cavalli-Sforza, L., 1981. Geographic clustering of cultural traits. (Manuscript in preparation)

Hägerstrand, T., 1967. *Innovation diffusion as a Spatial Process.* The University of Chicago Press, Chicago and London.

Haldane, J. B. S., 1924. A mathematical theory of natural and artificial selection. *Trans. Camb. Phil. Soc. 23* : 19–41.

Isaac, G. L., 1972. Some experiments in quantitative methods for characterizing Acheulian assemblages. *Congres Panafricain de Prehistoire, Dakar, 1967: Actes de 6[e] Session,*. ed. H. J. Hogot, Paris, Chambery.

Isaac, G. L., 1976. Stages of cultural elaboration in the Peistocene: possible archeological indicators of the development of language capabilities. In: *Origins and Evolution of Language and Speech,* ed. by S. R. Harmad, H. D. Stentis, and J. Lancaster, Ann. N.Y. Acad. Sci. *280* : 275–288.

Isaac, G. L., 1977. *Olorgesailie–Archeological Studies of a Middle Pleistocene in Kenya.* University of Chicago Press, Chicago and London.

Jencks, C., 1973. *Inequality. Harper Colophon Books, Harper and Row, New York, Evanston, San Francisco, and London.*

Jones, N. B., and Reynolds, V., eds., 1978. Symposia of the Society for the Study of Human Biology, Vol. XVIII, *Human Behaviour and Adaptation.* Halsted Press, New York-Toronto.

Karlin, S., 1966. *A First Course in Stochastic Processes.* Academic Press, New York.

Karlin, S., 1972a. Some mathematical models of population genetics. *Amer. Math. Monthly 79* : 699–739.

Karlin, S., 1972b. Unpublished lecture notes, Weitzman Institute.

Karlin, S., 1979. Models of multifactional inheritance: I. Multivariate formulations and basic convergence results. *Theor. Pop. Biol. 15* : 308–355.

Karlin, S., and McGregor, J., 1964. Direct product branching processes and related Markoff chains. *P.N.A.S. 51* : 598–602.

Karlin, S., and McGregor, J., 1967. The number of mutant forms maintained in a population. *Proc. Fifth Berkeley Symp. Math. Stat. Prob. 4* : 415–438.

Karlin, S., and Taylor, H., 1980. *A Second Course in Stochastic Processes.* Academic Press, New York, San Francisco, and London.

Kawai, M., 1965. Newly acquired pre-cultural behavior of the natural troop of Japanese monkeys on Koshima Islet. *Primates 6* : 1–30.

Kendall, D. G., 1956. Deterministic and stochastic epidemics in closed populations. *Proc. 3rd. Berkeley Symp. Math. Statist and Prob. 4* : 149–165.

Kendall, D. G., 1965. Mathematical models of the spread of infection. In: *Mathematics and Computer Science in Biology and Medicine.* Medical Research Council, London.

Kermack, W. D., and McKendrick, A. G., 1927. Contributions to the mathematical theory of epidenics. *Proc. Roy. Soc., A 115* : 700–721.

Keyfitz, N., and Flieger, W., 1971. *Population–Facts and Methods of Demography.* W. H. Freeman & Co., San Francisco.

Kidd, K. K., and Cavalli-Sforza, L. L., 1973. An analysis of the genetics of schizophrenia. IV Int. Cong. Human Genetics, Paris, Proceedings. *Social Biology 20* : 254–265.

Kimura, M., 1979. The neutral theory of molecular evolution. *Scientific American, 241* (5) :

Kimura, M., and Crow, J., 1964. The number of alleles that can be maintained in a finite population. *Genetics 49* : 725–738.

Kimura, M., and Weiss, G. H., 1964. The stepping stone model of population structure and the decrease of genetic correlation with distance. *Genetics 49* : 561–576.

Kimura, M., and Ohta, T., 1969. The average number of generations until fixation of a mutant gene in a finite population. *Genetics 61* : 763–771.

Kirkby, A. V. T., 1973. The use of land and water resources in the past and present valley of Oaxaca. *Memoirs No. 5. Museum of Anthropology,* University of Michigan, Ann Arbor.

Kroeber, A. L., 1963. *Anthropology:* revised edition, 1948. Harcourt Brace and World, Inc., New York and Burlingame.

Kruskal, J. B., Dyer, I., and Black, P., 1971. The vocabulary method of reconstructing family trees: innovations and large scale applications. In: *Mathematics in the Archeological and Historical Sciences*, eds. F. R. Hodson, D. G. Kendall and P. Tautu, Edinburgh University Press, pp. 361–380.

Labov, W., 1972. *Sociolinguistic Patterns*. Philadelphia, Univ. of Pennsylvania Press.

Labov, W., 1974. The relative influence of family and peers on the learning of language. (Manuscript).

Lande, R., 1976. The maintenance of genetic variability by mutation in a polygenic character with linked loci. *Genet. Res. Camb. 26* : 221–235.

Lewontin, R. C., 1972. The apportionment of human diversity. In: *Evolutionary Biology*, Vol. 6, eds. T. H. Dobzhansky, M. K. Hecht, and W. C. Steere, Appleton-Century-Crofts, New York, pp. 381–398.

Lewontin, R. C. 1974. *The Genetic Basis of Evolutionary Change*. Columbia University Press, New York.

Li, C. C., 1975. *First Course in Population Genetics*. The Boxwood Press, Pacific Grove, California.

Loehlin, J. C., and Nichols, R. C., 1976. *Heredity Environment and Personality*. University of Texas Press, Austin, Texas.

Lotka, A. J., 1931. The extinction of families. *J. Wash. Acad. Sci. 21* : 377.

Mainardi, D., 1980. Tradition and the Social Transmission of Behavior in Animals. (Manuscript)

Malécot, G., 1966. *Probabilités et Hérédité*. Presses Universitaires de France.

Maruyama, T., 1974. The age of an allele in a finite population. *Genetic Res. of Cambr. Vol 23* : 137–143.

May, R., 1973. *Model Ecosystems*. Princeton University Press, New Jersey.

Mendel, G., 1865. Versuche über Pflanzen-Hybriden. *Verh. Naturforsch Ver. Brünn 4* : 3–47.

Menozzi, P., Piazza, A., and Cavalli-Sforza, L., 1978. Synthetic human gene frequency maps of Europe. *Science 201* : 786–792.

Migliorini, B., 1978. *Storia della Lingua Italiana,* 5th ed. Sansoni, Firemze.

Moran, P. A. P., 1962. *The Statistical Processes of Evolutionary Theory.* The Clarendon Press, Oxford.

Moran, P. A. P., and Watterson, G. A., 1959. The genetic effect of family structure in natural populations. *Aust. J. Bio. Sci. 12* : 1–15.

Murdock, G. P., 1967. *Ethnographic Atlas.* Univ. of Pittsburgh Press, Pittsburgh.

Neel, J. V., 1973. Diversity within and between South American indian tribes. *Israel Journal of Med. Sciences 9* : 1216–1224.

Nei, M., and Roychoudhury, A. K., 1974. Genic variation within and between the three major races of Man, Caucasoids, Negroids, and Mongoloids. *Am. J. Hum. Genet. 26* : 421–443.

Pearson, K., 1896. Mathematical contributions to the theory of evolution. III. Regression, Heredity and Panmixia. *Phil. Trans. Roy. Soc. London A 187* : 253–318.

Pearson, K., 1903. Mathematical contributions to the theory of evolution. XI on the influence of natural selection on the variability and correlation of organs. *Phil. Trans. Roy. Soc. London 200* : 1–66.

Prais, S. J., 1955. Measuring social mobility. *J. Roy. Stat. Soc. A118* : 56–66.

Prout, T., 1965. The estimation of fitnesses from genotypic frequencies. Evolution 19 : 546–551.

Pulliam, H. R., and Dunford, C., 1980. *Programmed to Learn.* Columbia Univ. Press, New York.

Quackenbush, E. M., 1968. *From Sonsorol to Truk: A Dialect Chain.* Ph.D. Thesis, The University of Michigan. University Microfilms, Inc., Ann Arbor, MI.

Roberts, Fred S., 1976. *Discrete Mathematical Models.* Prentice Hall, New Jersey.

Rogers, E. M., 1962. *Diffusion of Innovations.* The Free Press of Glencoe, Collier-Macmillan Canada, Ltd., Galt, Ontario.

Rogers, E. M., and Schoemaker, F. F., 1971. *Communication of Innovations: a Crosscultural Approach.* The Free Press, a division of Macmillan Publishing Co., Inc., New York.

Ross, A. S. C., Mitford, N., Waugh, E., "Strix," Sykes, C., and Betjeman, J. 1956. *Noblesse oblige.* Hamish Hamilton, London.

Roughgarden, J., 1979. *Theory of Population Genetics and Evolutionary Ecology: An Introduction.* Macmillan, New York.

Sahlins, M. D., and Service, E. R., 1960. *Evolution and Culture.* University of Michigan Press, Ann Arbor.

Sankoff, D., 1971. Stochastic models for glottochronology. In: *Mathematics in the Archeological and Historical Sciences,* eds. F. G. Hodson, D. Y. Kendall and P. Tautu, University Press, Edinburgh, pp. 381–386.

Sapir, E., 1921. *Language.* Harcourt, Brace and World, Inc., New York.

Scarr, S., and Weinberg, R. A., 1978. The influence of "family background" on intellectual attainment. *American Sociological Review 43* : 674–692.

Scott, J. P., ed., 1978. Critical Periods. In: *Benchmark Papers in Animal Behavior.* Dowden, Hutchinson & Ross Inc., Stroudsburg, Penna.

Sgaramella-Zonta, L., and Cavalli-Sforza, L. L., 1973. A method for the detection of a demic cline. *Proceedings of the Workshop on Population Structure,* Hawaii Univ. Press, pp. 128–135.

Singer, B., and Spilerman, S., 1974. *Social Mobility Models for Heterogeneous Populations.* In: *Sociological Methodology*

1973–1974. Edited by H. Costner. Jossey Bass, San Francisco.

Singh, R. S., Lewontin, R. C., and Felton, A. A., 1976. Genetic heterogeneity within electrophoretic "alleles" of *xanthine dehydrogenase* in *Drosophila pseudo-obscura. Genetics* 84 : 609–629.

Skellam, J. G., 1951. Random dispersal in theoretical populations. *Biometrika 38* : 196–218.

Slatkin, M., 1970. Selection and polygenic characters. *Proc. Nat. Acad. Sci. USA 66* : 87–93.

Sorokin, P. A., 1959. *Social and Cultural Mobility*. The Free Press, New York.

Spielman, R. S., Migliazza, E. C., and Neel, J. V., 1974. Regional linguistic and genetic differences among Yanomama Indians. *Science. 184* : 637–644.

Spuhler, J. N., 1972. Genetic, linguistic and geographic distances in native North America. In: *The Assessment of Population Affinities in Man*, eds. J. S. Weiner and J. Huizinga, Clarendon Press, Oxford. pp. 72–

Steward, J. H., 1955. *Theory of Culture Change*. University of Illinois Press, Urbana, Ill.

Sutter, J., ed., 1962. *Human Displacements*. Entretiens de Monaco en Sciences Humaines. Editions "Sciences humaines" Diffusion, Hachette.

Swadesh, M., 1952. Lexicostatistic dating of prehistoric ethnic contacts. *Proc. Am. Phil. Soc. 96* : 452–463.

Taubman, P., 1976. The determinants of earnings; genetics, family and other environments: a study of white male twins. *Amer. Econ. Review 66* : 858–870.

Tolman, E. C. 1932. *Responsive Behavior in Animals and Men*. Appleton, New York.

Trudgill, P. 1974. Linguistic change and diffusion. Description and explanation in sociolinguistic dialect geography. *Language in Soc. 2* : 215–246.

Uyenoyama, M., Feldman, M., and Cavalli-Sforza, L., 1979. Evolutionary effects of contagious and familial transmission. *P.N.A.S. 76*: 420–424.

Wade, M. J., 1977. An experimental study of group selection. *Evolution 31*: 134–153.

Wagener, D., 1975. Contributions to the Study of the Evolution of Continuous Phenotypes. Ph.D. thesis. Stanford University, Stanford, California.

Wallace, Bruce, 1968. *Topics in Population Genetics.* W. W. Norton & Co., New York.

Waltman, P., 1974. *Deterministic Threshold Models in the Theory of Epidemics.* Springer-Verlag, Berlin, Heidelberg, New York.

Wang, W. S.-Y., 1976. Language change. *Ann. N.Y. Acad. Sci. 280*: 61–72.

Wang, W. S.-Y., ed., 1977. *The Lexicon in Phonological Change.* Mouton, The Hague.

Wang, W. S.-Y., 1978. The three scales of diachrony. In: *Linguistics in The Seventies,* ed. B. B. Kachru, Univ. of Illinois Dept. of Linguistics, Urbana, Ill., pp. 63–75.

Wang, W. S.-Y., 1979. Language change—a lexical perspective. *Ann. Rev. Anthropol. 8*: 353–371.

Wang, W. S.-Y., 1980. The evolution of language. *American Scientist* (in press).

Watson, H. W. and Galton, F., 1874. On the probability of the extinction of families. *J. Anthropol. Inst. Great Britain and Ireland 4*: 138–144.

Weinreich, U., Labov, W., and Herzog, M. I., 1968. Empirical foundations for a theory of language change. In: *Directions for Historical Linguistics,* edited by W. Lehmann and Y Malkiel. University of Texas Press, Austin. pp. 97–175.

Weltman, N., and Remmers, H., 1940. *Pupils', parents' and teachers' attitudes—similarities and differences.* Purdue

University, the Division of Educational Reference. Studies in higher education, lvi, Lafayette, Indiana.

White, L. A., 1959. *The Evolution of Culture.* McGraw-Hill Book Co., Inc., New York.

White, M. J. D., 1978. *Modes of Speciation.* W. H. Freeman & Co., San Francisco, California.

Wilson, E. O., 1975. *Sociobiology, the New Synthesis.* Balknap Press, Cambridge, Mass.

Wilson, S. R., 1974. Simulation of Mendelism by a non-genetic Markov chain model. *Ann. Hum. Genet. Lond. 38* : 225–229.

Wright, S., 1921. Systems of mating. *Genetics* 6 : 111–178.

Wright, S., 1931a. Statistical methods in biology. *J. Am. Stat. Ass. 26* : 155–163.

Wright, S., 1931b. Evolution in Mendelian populations. *Genetics 16* : 97–159.

Wright, S., 1940. Breeding structure of populations in reaction to speciation. *Amer. Nat. 74* : 232–248.

Wright, S., 1978. *Evolution and the Genetics of Populations, Vol. IV.* The Univ. of Chicago Press, Chicago.

Wynne-Edwards, V. C., 1962. *Animal Dispersion in Relation to Social Behavior.* Oliver and Boyd, Edinburgh.

Yamazaki, T., 1971. Measurement of fitness at the esterase-5 locus in Drosphila pseudoobscura. *Genetics 67* : 579–603.

Yasuda, N., Cavalli-Sforza, L., Skolnick, M., and Moroni, A., 1974. The evolution of surnames. An analysis of their distribution and extinction. *Theor. Pop. Biol. 5* : 123–142.

Zadoks, J. C., and Kampmeijer, P., 1977. The role of crop populations and their deployment, illustrated by means of a simulator, EPIMUL76. *Ann. N.Y. Acad. Sci. 287* : 164–190.

Zajonc, R. B., 1976. Family configuration and intelligence. Science *192* : 227–236.

Zei, G., and Cavalli-Sforza, L., 1977. Education and birth control. *Genus 33* : 15–42.

Index

Library of Congress Cataloging in Publication Data

Cavalli-Sforza, Luigi Luca, 1922–
Cultural transmission and evolution.

(Monographs in population biology; 16)
Bibliography: p.
Includes index.
1. Social evolution. 2. Social evolution–Mathematical models. I. Feldman, Marucs W., joint author. II. Title. III. Series.
GN360.C38 306 80–8539
ISBN 0-691-08280-4
ISBN 0-691-08283-9 (pbk.)